BSAVA Manual of Canine and Feline Reproduction and Neonatology

Second edition

Editors:

Gary C.W. England

BVetMed PhD DVetMed CertVA DVR DipVRep DipACT DipECAR FHEA FRCVS

School of Veterinary Medicine and Science, University of Nottingham, Sutton Bonington Campus, Loughborough, Leicestershire LE12 5RD

and

Angelika von Heimendahl

MSc BVM DipECAR MRCVS

Veterinary Reproduction Service Surgery, 27 High Street, Longstanton, Cambridge CB24 3BP

Published by:

British Small Animal Veterinary Association
Woodrow House, 1 Telford Way, Waterwells
Business Park, Quedgeley, Gloucester GL2 2AB

A Company Limited by Guarantee in England.
Registered Company No. 2837793.
Registered as a Charity.

WORLD LAND TRUST™

www.carbonbalancedpaper.com
CBP006075

Carbon Balancing is delivered by World Land Trust, an international conservation charity, who protects the world's most biologically important and threatened habitats acre by acre. Their Carbon Balanced Programme offsets emissions through the purchase and preservation of high conservation value forests.

17388PUBS22

Other titles in the BSAVA Manuals series:

Manual of Avian Practice: A Foundation Manual
Manual of Backyard Poultry Medicine and Surgery
Manual of Canine & Feline Abdominal Imaging
Manual of Canine & Feline Abdominal Surgery
Manual of Canine & Feline Advanced Veterinary Nursing
Manual of Canine & Feline Anaesthesia and Analgesia
Manual of Canine & Feline Behavioural Medicine
Manual of Canine & Feline Cardiorespiratory Medicine
Manual of Canine & Feline Clinical Pathology
Manual of Canine & Feline Dentistry and Oral Surgery
Manual of Canine & Feline Dermatology
Manual of Canine & Feline Emergency and Critical Care
Manual of Canine & Feline Endocrinology
Manual of Canine & Feline Endoscopy and Endosurgery
Manual of Canine & Feline Fracture Repair and Management
Manual of Canine & Feline Gastroenterology
Manual of Canine & Feline Haematology and Transfusion Medicine
Manual of Canine & Feline Head, Neck and Thoracic Surgery
Manual of Canine & Feline Musculoskeletal Disorders
Manual of Canine & Feline Musculoskeletal Imaging
Manual of Canine & Feline Nephrology and Urology
Manual of Canine & Feline Neurology
Manual of Canine & Feline Oncology
Manual of Canine & Feline Ophthalmology
Manual of Canine & Feline Radiography and Radiology: A Foundation Manual
Manual of Canine & Feline Rehabilitation, Supportive and Palliative Care: Case Studies in Patient Management
Manual of Canine & Feline Reproduction and Neonatology
Manual of Canine & Feline Shelter Medicine: Principles of Health and Welfare in a Multi-animal Environment
Manual of Canine & Feline Surgical Principles: A Foundation Manual
Manual of Canine & Feline Thoracic Imaging
Manual of Canine & Feline Ultrasonography
Manual of Canine & Feline Wound Management and Reconstruction
Manual of Canine Practice: A Foundation Manual
Manual of Exotic Pet and Wildlife Nursing
Manual of Exotic Pets: A Foundation Manual
Manual of Feline Practice: A Foundation Manual
Manual of Ornamental Fish
Manual of Practical Animal Care
Manual of Practical Veterinary Nursing
Manual of Psittacine Birds
Manual of Rabbit Medicine
Manual of Rabbit Surgery, Dentistry and Imaging
Manual of Raptors, Pigeons and Passerine Birds
Manual of Reptiles
Manual of Rodents and Ferrets
Manual of Small Animal Practice Management and Development
Manual of Wildlife Casualties

For further information on these and all BSAVA publications, please visit our website: **www.bsava.com**

Contents

Contributors

Eva Axnér DipECAR
Division of Reproduction, Department of Clinical Sciences, Faculty of Veterinary Medicine and Animal Science, PO Box 7054, Swedish University of Agricultural Sciences, SE-750 07 Uppsala, Sweden

Margret Casal Drmedvet PhD DipECAR
Veterinary Hospital, University of Pennsylvania, 3900 Delancey Street, Philadelphia, PA 19104-6010, USA

Autumn Davidson DVM MS DipACVIM
Department of Medicine and Epidemiology, School of Veterinary Medicine, 1 Shields Avenue, University of California Davis, CA 95616, USA

Gary C.W. England BVetMed PhD DVetMed CertVA DVR DipVRep DipACT DipECAR FHEA FRCVS
School of Veterinary Medicine and Science, University of Nottingham, Sutton Bonington Campus, Loughborough, Leicestershire LE12 5RD

Wenche K. Farstad DVM PhD DipECAR
Norwegian School of Veterinary Science, PO Box 8146, Dep. N 0033, Oslo, Norway

Sonia Fernandez DVM
Facultat de Veterinaria de Bellaterra, Centre Veterinari Doc's, Premià de Mar, Barcelona, Spain

Alain Fontbonne DVM PhD DipECAR
Alfort National Veterinary College, 7 avenue du Général de Gaulle, 94700 Maisons-Alfort, Paris, France

Angelika von Heimendahl MSc BVM DipECAR MRCVS
Veterinary Reproduction Service Surgery, 27 High Street, Longstanton, Cambridge CB24 3BP

Michelle Kutzler DVM PhD DipACT
Department of Animal Sciences, Oregon State University, 312 Withycombe Hall, Corvallis, OR 97331, USA

Xavier Levy DVM DipECAR
Centre de Reproduction des Carnivores du Sud-Ouest, 58 Bd des Poumadères, 32600 Isle-Jourdain, France

Catharina Linde Forsberg DVM PhD DipECAR
Division of Reproduction, Department of Clinical Sciences, Faculty of Veterinary Medicine and Animal Science, PO Box 7054, Swedish University of Agricultural Sciences, SE-750 07 Uppsala, Sweden and CaniRep HB, Fjällbo 110, 755 97 Uppsala, Sweden

Cheryl Lopate MS DVM DipACT
Reproductive Revolutions Inc., 1000 Wilsonville Rd #55, Newberg, OR 97132, USA

Gaia Cecilia Luvoni DVM PhD DipECAR
Department of Veterinary Clinical Sciences, Obstetrics and Gynaecology, Università degli Studi di Milano, Via Celoria, 10-20133 Milan, Italy

Josep Arus Marti DVM DipECAR
Arvivet Veterinaris SL, Veterinary Reproduction Department, Avenida Textil 42, Nave E 08224, Terrassa, Spain

Stefano Romagnoli DVM MS PhD DipECAR
Department of Veterinary Clinical Sciences, Faculty of Veterinary Medicine, University of Padova, Agripolis, Legnaro, 35020, Padova, Italy

Hasan Sontas DVM PhD
Department of Obstetrics and Gynaecology, Faculty of Veterinary Medicine, Istanbul University, Avcilar Campus, 34320, Istanbul, Turkey

Daniele Zambelli DVM PhD DipECAR
Veterinary Clinical Department, Animal Reproduction Unit, Faculty of Veterinary Medicine, Alma Mater Studiorum, University of Bologna, Italy

Foreword

When the first edition of the *BSAVA Manual of Small Animal Reproduction and Neonatology* was published in 1998, it was observed that there had been an 'explosion of knowledge' in its subject matter during the previous two decades. Since then there has been a further welcome increase in knowledge across the whole spectrum of small animal reproduction and neonatology. In this edition there are additional chapters on conditions of the neutered bitch and queen, and also those of the male. Separate chapters for the bitch and queen are now devoted to pregnancy diagnosis, normal pregnancy and parturition and also to the clinical approach to the infertile female. All this has been achieved with a modest increase in page numbers.

I attended my first whelping case many years ago and I remember being only too well aware of the many gaps in our knowledge then. Was a birth really as overdue as the owner claimed? Were the unborn puppies or kittens at risk, or was it safe to give the mother a little more time before embarking on a Caesarean section? What were the normal physiological parameters of the neonate? At that time a puppy mortality rate of up to 25% might be anticipated between the commencement of birth and weaning.

Dog and cat breeders rightly expect and demand higher and higher standards of knowledge and care from their veterinarians. We must advise breeders against attempting to breed from subfertile animals or from those with congenital defects. We can guide them in such matters as the correct timing of service in the bitch, or the need for a number of services to induce reflex ovulation in the queen.

Antenatal care in small animals is desirable not only to advise owners on breeding matters but also to ensure that management, preventive medicine and care before, during and after birth are of the highest standard. We must ensure that parturition is accompanied by minimal morbidity and minimal mortality in mother and offspring.

The Editors have assembled an international team of authors who have provided readers with state-of-the-art knowledge in the theory and practice of canine and feline reproduction and neonatology. Those working in these fields will read the text eagerly and apply the information in their clinical work to the benefit of patients and owners.

Peter G.G. Jackson BVM&S MA DVM&S FRCVS
Emeritus Fellow, St Edmund's College, Cambridge
August 2010

Preface

Clinical veterinary reproduction and neonatology are important aspects of general veterinary practice, with a large number of consultations being undertaken by veterinarians for the prevention of breeding as well as for optimising reproductive performance and neonatal survival. There has been substantial development in the field of small animal reproduction since the last *BSAVA Manual of Small Animal Reproduction and Neonatology* was published in 1998, and therefore it is timely that we have produced the second edition of the Manual.

In compiling this new edition, we have retained the essence of the original with initial chapters covering the basic clinically relevant physiology, preceding specific sections on the male and female dog and cat. We have also enlarged the chapters on the prevention of breeding and the determination of breeding status. The Manual then follows a systematic approach to common clinical situations presented in the adult male and female cat and dog. We are pleased that in this new edition we have expanded the sections on the neonate and reproductive biotechnology, reflecting the exponential increase in knowledge in these areas since the last edition was published.

As with the previous edition, the *BSAVA Manual of Canine and Feline Reproduction 2nd edition* is aimed primarily at veterinary surgeons, veterinary nurses and veterinary students, but it is also likely that technicians and experienced dog and cat breeders will find a wealth of practical information within many of the chapters. We hope that those studying for examinations will find the book to be particularly useful, and by inclusion of a wide spectrum of international authors we hope that the Manual will be valuable across many continents.

We are extremely grateful to all of our contributors, a truly international team of experts in their fields, for finding the time to share their expertise and knowledge. We are also indebted to Nicola Lloyd and Ben Dales for expert help during the editing process, and of course to the unwavering support of Marion Jowett.

Gary England
Angelika von Heimendahl
July 2010

Physiology and endocrinology of the female

Gary C.W. England

Introduction

In the majority of domestic species the interval between one ovulation and another is approximately 21 days. For these species the phases of the oestrous cycle comprise: the period of sexual receptiveness when mating is allowed (termed *oestrus*); the period of preparation for pregnancy (termed the *luteal phase* and often divided into *metoestrus* and *dioestrus*); and the period of preparation for return to oestrus (termed *pro-oestrus*). When the animal becomes pregnant the luteal phase is extended, often for the duration of the pregnancy. In some species there are periods of sexual inactivity (termed *anoestrus*), which are normally controlled by photoperiod and therefore related to season of the year.

The cycles of the domestic bitch and queen differ from this general scheme. For the bitch, the periods of pro-oestrus and oestrus are long and are accompanied by an obligatory long luteal phase (whether the bitch is pregnant or not). These are followed by anoestrus, which is not governed by season of the year; the average length of the cycle is approximately 7 months, although this may vary greatly between individuals. For the domestic queen, there is often (but not always) a seasonal period of (winter) anoestrus, which is followed in springtime by periods of repeated oestrous activity at intervals of approximately 14–21 days. However, the queen is an induced ovulator and, in the absence of mating,

there is no luteal phase so that oestrus is followed by a period of sexual inactivity (termed *interoestrus*), which is different endocrinologically from the luteal phase. The luteal phase does occur when the queen ovulates but may be associated with either pregnancy or non-pregnancy. The non-pregnant luteal phase is shorter than the pregnant luteal phase.

The hypothalamic–gonadal axis

In all domestic species the primary control of reproductive function is the interaction of the environment and brain with the hypothalamus and pituitary gland. The synthesis and release of gonadotrophin-releasing hormone (GnRH) is controlled by a complicated system that involves a variety of neurotransmitters, opiates, melatonin (associated with day length) and a number of neural and hormonal feedback loops. GnRH is synthesized in neurons of the hypothalamus and released from nerve endings at the median eminence. It is transferred via the hypophyseal portal system to the anterior pituitary gland. Here GnRH stimulates the release of the gonadotrophins, follicle-stimulating hormone (FSH) and luteinizing hormone (LH). These gonadotrophins act upon the ovaries, which during follicle growth, ovulation and the luteal phase produce, in a sequential order, oestrogen and progesterone (Figure 1.1). These steroidal hormones act upon specific target tissues, including the tubular

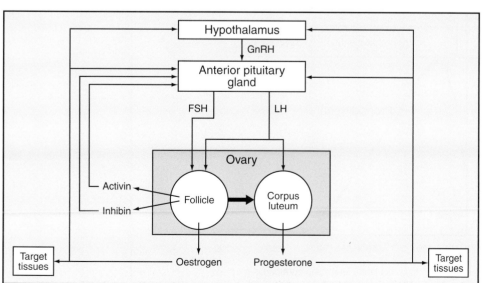

1.1

Hypothalamic–pituitary–gonadal axis, demonstrating the site of hormonal production and feedback loops. FSH = follicle-stimulating hormone; GnRH = gonadotrophin-releasing hormone; LH = luteinizing hormone.

reproductive tract and the brain behavioural centres. Basic mechanisms of negative and positive feedback are involved in the control of hormone secretion, and environmental factors such as light, social cues, duration of lactation and adequacy of food supply may have a modifying effect.

The bitch

The reproductive physiology of the bitch is intriguing; the non-seasonal mono-oestrous breeding pattern with obligatory anoestrus probably reflects the consequences of domestication of a species with a highly organized social and reproductive structure, which was initially aligned to seasonal breeding.

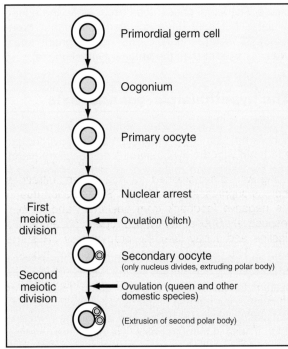

1.2 Different stages of oogenesis. The time of ovulation in the bitch and queen is shown.

Puberty

The onset of the first (pubertal) oestrus is very variable, and although this occurs at a mean age of approximately 9 months, it can vary between 6 and 14 months in a large proportion of bitches. In some bitches puberty may not occur until 24 or 30 months of age. The onset of puberty appears to be related to when the bitch reaches approximately 80% of adult bodyweight and, therefore, can be linked in some cases to breed because larger breeds reach adult bodyweight at a later age than smaller breeds.

In females, because only a small number of oocytes are ovulated at each oestrus, there is not the same requirement for mitotic proliferation of germ cells as there is in the male. As a result, development of the germ cells ceases when they reach the primordial germ cell stage (oogonia stage; equivalent to the spermatogonia stage in the male, see Chapter 2). Essentially, oogonia undergo mitotic division and the first stage of meiotic division and are then arrested (Figure 1.2) within the primordial follicle. The regular recruitment of primordial follicles into a pool of growing follicles commences at the first oestrus (puberty) and then continues thereafter at each oestrus.

Peak fertility

Peak fertility appears to be reached at approximately 2 years of age, and is generally maintained until 6 or 7 years of age. Within this period bitches can produce a litter at each mated oestrus, although this is not common breeding practice, and recommendations are normally to breed only at alternate oestrous cycles. In older bitches both the conception rate and the litter size decrease.

Cyclicity

The average interval between the onset of successive pro-oestrus periods is approximately 7 months. This is divided roughly into pro-oestrus (10 days), oestrus (10 days), luteal phase (pregnancy or non-pregnancy, 2 months) and anoestrus (4.5 months) (Figure 1.3). For individual bitches cyclicity may range from being highly variable to almost regular.

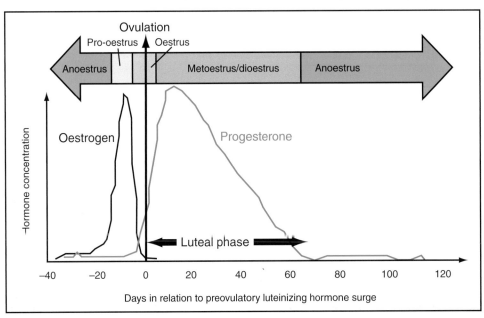

1.3 Different stages of the oestrous cycle in relation to changes in plasma hormone concentrations and ovulation.

Oestrous cycle

Anoestrus

Anoestrus is the period between the end of the luteal phase and the onset of the subsequent return to pro-oestrus. When the previous cycle has been a pregnant cycle, the early part of anoestrus encompasses lactation. The length of anoestrus exceeds the period of lactation such that bitches are not in pro-oestrus or oestrus whilst lactating.

The period of anoestrus is considered to be obligatory (following each luteal phase); it generally lasts a minimum of 7 weeks but averages 18–20 weeks. Throughout anoestrus the reproductive tract is quiescent and the internal and external genitalia, including the mammary glands, are at their smallest observed size. The vaginal wall is relatively thin and is easily traumatized by digital examination or collection of epithelial cells.

It is often assumed that basal hormonal concentrations persist throughout anoestrus, and whilst this is a convenient explanation, actually there are changes in the responsiveness of the pituitary gland as anoestrus progresses. Approximately 60 days before the next ovulation follicles may be detected within the ovaries. Relatively high concentrations of oestrogen can be detected in late anoestrus, from approximately 10–20 days prior to the onset of pro-oestrus. Although there is little detailed information on the endocrinology of late anoestrus, it is clear that it includes increased LH pulse frequency accompanied by FSH pulses and a less prominent increase in mean FSH concentration.

Pro-oestrus

Pro-oestrus is associated with stimulation of follicle development by FSH and LH, and the subsequent secretion of oestrogen from the granulosa cells of the follicle. Approximately two to eight follicles grow per ovary. These protrude above the margin of the ovary approximately 10 days before ovulation, when initially they are approximately 4 mm in diameter. Follicle diameter increases to between 6 and 9 mm just before the preovulatory LH surge. Follicular oestrogen promotes increased vascularity and oedema of the reproductive tract, as well as increased activity of the glandular epithelium. This leads to swelling of the external and internal reproductive tract, whilst within the uterus the mucosal capillaries leak at endothelial junctions resulting in passage of blood and plasma cells into the uterine lumen.

The clinical signs associated with pro-oestrus include enlargement and reddening of the vulval lips and the appearance of a serosanguineous vulvar discharge. There are also behavioural changes, including increased attractiveness to male dogs, increased urine marking and a tendency to roam. By definition the period of pro-oestrus includes sexual attractiveness but refusal to allow mating.

Clinical examination of the bitch may demonstrate enlargement and swelling of vaginal epithelial folds when viewed with an endoscope, and increased oedema of the uterus when imaged with ultrasonography. There is also substantial epithelial cell proliferation induced by oestrogen, and this is very significant within the vagina, where the mucosa changes from a cuboidal epithelium to a stratified squamous epithelium. Presumably this helps to prevent the vagina from being traumatized during mating, but it is extremely useful because it enables monitoring of the stage of the oestrous cycle (see Chapter 5).

Elevated LH and FSH concentrations are crucial for stimulating follicle growth. However, as follicles mature they produce the hormone inhibin, which is a selective inhibitor of FSH secretion, such that in late pro-oestrus FSH concentrations do not increase further and may decline. Nevertheless, FSH plays an important role in the maturation of the follicle and the equipping of cells for conversion into corpora lutea after ovulation.

Oestrus

Oestrus is a behavioural definition that encompasses sexual attractiveness and the acceptance of mating. It is characterized initially by elevated concentrations of oestrogen, which then decline prior to ovulation (Figure 1.4). During the period of oestrogen secretion the follicles enlarge to 9–12 mm in diameter; they usually reach their maximum size between the time of the LH surge and ovulation. Along with the increased oestrogen concentrations there is suppression of LH and FSH secretion due to the negative feedback from oestradiol and inhibin, respectively. Subsequently, oestrogen concentrations decline and 1 day later the preovulatory LH surge begins. The concentration of LH is elevated typically for between 1 and 3 days; although peak values may be reached on any of these days it appears that the initial surge rather than the peak is the event to which other physiological aspects are related in time.

The bitch will still allow mating in the early luteal phase (see Chapter 5) so that, technically, she enters the period of metoestrus whilst remaining in oestrus. Interestingly, the behaviour of the bitch appears to be primed by rising oestrogen concentrations, but full expression of oestrus requires the withdrawal of oestrogen in the presence of progesterone. These endocrine events occur close to the onset of standing oestrous behaviour (see later), but they are probably important changes at the transition from pro-oestrus (sexually attractive but unreceptive) to oestrus. During oestrus there is inviting receptive behaviour, including standing to be mated, movement of the tail to uncover the vulva and spinal lordosis. At this time pheromones are secreted under the influence of oestrogen within the reproductive tract discharge and urine. The pheromones are detected by the olfactory or vomeronasal organs of the dog and act to increase male reproductive activity. Pheromones also appear to have actions resulting in stimulation of GnRH centres because oestrogens may have significant effects on oestrus advancement or oestrus synchronization in other bitches.

The hormonal changes result in a characteristic appearance to the vaginal smear, and the reduction

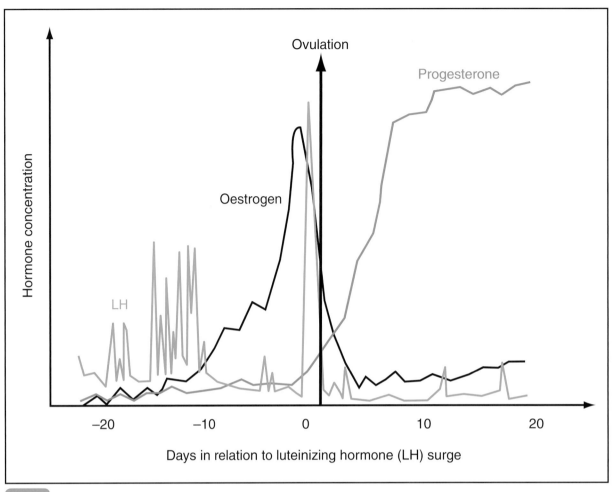

1.4 Periovulatory changes in plasma oestrogen, LH and progesterone concentrations in the bitch.

in oestrogen results in loss of the oedema and a resultant wrinkling and crenulation of the vaginal wall mucosa when viewed using an endoscope, together with a similar loss of turgidity of the vulva. These events often relate to the underlying endocrine changes and may be used when planning the timing of mating or insemination.

Ovulation: The interesting observation that a decline in oestrogen and rise in progesterone are required to facilitate full expression of behavioural oestrus fits well with the observed underlying endocrine changes. Ovulation is *spontaneous* in the bitch, in that it does not require stimulation by mating. As in other species, ovulation is stimulated by a surge in the release of LH that occurs a few days after peak concentrations of oestrogen are reached. Interestingly, luteinization of the follicle wall occurs prior to ovulation, and there is a significant increase in plasma progesterone prior to ovulation. Indeed the decline in oestrogen and increase in progesterone (leading to a significant change in the ratio of these hormones) appears to stimulate the LH surge, as well as being responsible for the significant change in sexual behaviour mentioned previously (see Figure 1.4). The concentration of progesterone appears to increase a few hours before or during the onset of the LH surge. Increased follicular proges-

terone is likely to be a critical factor in ovulation in the bitch, as it is in other species.

The preovulatory LH surge is often depicted as the central event of the cycle, and, because oestrous behaviour continues after ovulation, the stages of pregnancy are best described in relation to the day of the LH peak, rather than to the first day of oestrus or the day of mating, as is normal practice in other species.

As well as stimulating ovulation, the LH surge shifts steroid production in the granulosa cells from oestrogen to progesterone and helps the follicle to develop into a corpus luteum. In this process, the granulosa cells develop into large luteal cells by hypertrophy, and the theca cells develop into small luteal cells by proliferation, although the exact nature of this transformation has not been confirmed in the bitch.

Bitches normally have multiple ovulations, and histological and laparoscopic examinations show that although most ovulations occur between 48 and 60 hours after the LH surge, some follicles may not ovulate until as late as 96 hours after the LH surge. Interestingly, there is a very large variation (both between and within bitches) in relation to the day on which ovulation occurs. Whilst on average ovulation occurs 12 days after the onset of pro-oestrus, in some cases it can be as early as day 5 or as late as day 25.

Fertile and fertilization periods: The basic endocrinological events in the bitch are not unlike those of other species in that there is a preovulatory surge of LH that occurs approximately 2 days prior to ovulation. However, at the time of ovulation, the oocytes are immature and cannot be fertilized immediately. Fertilization can only occur after extrusion of the first polar body and completion of the first meiotic division to form the secondary oocyte (see Figure 1.2). This maturation takes approximately 48–60 hours and occurs within the distal oviduct. Interestingly, the oocytes remain viable within the reproductive tract for many days after they have become fertilizable, i.e. they do not begin to undergo degeneration until 9–10 days after ovulation (Figure 1.5). When compared with other species, this relative delay in the availability of oocytes for fertilization combined with their lengthened survival time has a significant impact upon the onset and duration of the fertilization period.

For all species, the 'fertilization period' is the time when oocytes are available to be fertilized. In the bitch this period commences 2 days after ovulation and extends until approximately 5 days after ovulation (Figure 1.5). The fertilization period is likely to be the approximate time of maximal fertility, which subsequently declines rapidly over the next few days owing to degeneration of the oocytes and closure of the cervix, which prevents sperm entering the female reproductive tract.

Whilst the fertilization period is extremely important in the bitch, it is not only during this period that mating can result in pregnancy. Intrauterine insemination after the end of the classically defined fertilization period has resulted in pregnancies of a small litter size (presumably by fertilization of ageing oocytes), whilst breeding prior to the fertilization period commonly results in pregnancy, by virtue of the survival of sperm in the female reproductive tract. Thus, the 'fertile period' can be defined as the time during which mating could result in conception. The fertile period therefore includes the fertilization period, but its start precedes the fertilization period by several days (Figure 1.5). Interestingly, in fertile stud dogs, survival of sperm for up to 7 days does not appear to be uncommon, thus enabling the fertile period to commence 5 days prior to ovulation (sperm surviving for 7 days to fertilize oocytes 2 days after ovulation).

Luteal phase

In many species there is an abrupt decline in sexual behaviour after ovulation such that oestrus ends almost coincidentally with the post-ovulation rise in plasma progesterone. Following oestrus the early maturation stage of the luteal phase is termed metoestrus, and the later stage is termed dioestrus. However, in the bitch standing oestrous behaviour continues for approximately 7 days after ovulation and therefore conventional terminology, which mixes behavioural and endocrine definitions, becomes confusing. Technically, the luteal phase commences after ovulation and during this time progesterone concentrations continue to increase. Progesterone

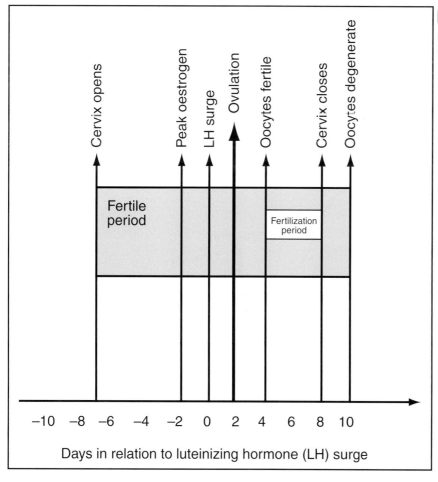

1.5 Onset and duration of the fertile and fertilization periods of the bitch in relation to periovulatory physiological events.

alters the characteristics of mucosal secretions and decreases smooth muscle excitability. It is also thought to slow the passage of ova in the uterine tubes, thereby delaying the entry of embryos, if present, into the uterus. Progesterone also 'closes' the cervix and prepares the uterine environment to facilitate support of the embryos.

The rise in progesterone begins a few hours before or during the preovulatory LH surge and continues to increase to reach 10–25 ng/ml by the end of standing oestrus (at approximately day 10). The progesterone profiles in the early luteal phase are identical in both pregnant and non-pregnant bitches, and indeed unlike other species there are few differences in the remaining luteal phase whether or not the bitch is pregnant (Figure 1.6). Details of methods for diagnosis of pregnancy are given in Chapter 11.

Within the luteal phase, prolactin, LH and to some extent progesterone are luteotrophic factors; both prolactin and LH have a clear luteotrophic action as early as week 2, and are required as luteotrophic factors from approximately day 25 onwards. Administration of prolactin inhibitors or suppressors of LH will terminate the luteal phase after day 25 in both pregnant and non-pregnant dogs. Although prostaglandin F (PGF) is luteolytic, it is not secreted in normal non-pregnant cycles. Exogenous administration of PGF or synthetic prostaglandins will terminate the luteal phase of pregnancy and non-pregnancy. Pharmaceutically, the luteal phase can therefore be ended by both anti-luteotrophic and luteolytic products. In clinical practice prolactin inhibitors (such as cabergoline and bromocriptine) and prostaglandins are used alone or in combination to terminate pregnancy and to treat conditions that may occur during the luteal phase (e.g. pyometra).

Pregnancy: Progesterone is produced solely from the corpus luteum throughout the duration of pregnancy. Progesterone concentrations are, on average, slightly higher in pregnant bitches and the progesterone plateau is broader during pregnancy, whilst the luteal phase is slightly longer in non-pregnancy (see Figure 1.6). It is likely that production of progesterone is much higher in pregnancy, but this is not readily demonstrated in the plasma because during pregnancy there is a substantial increase in plasma volume, which has a dilution effect on progesterone concentrations.

The pregnant luteal phase lasts approximately 63 days from ovulation to parturition, and the non-pregnant phase lasts approximately 66 days. During the second half of the luteal phase the progesterone concentration starts to decline and there is a concomitant increase in the plasma prolactin concentration (Figure 1.7). Prolactin is secreted by the pituitary gland and, by supporting the corpora lutea, tends to maintain an elevated progesterone concentration. Inhibition of prolactin secretion, or antagonism of its actions, results in an abrupt termination of the luteal phase and termination of pregnancy. Prolactin concentrations have been shown to be four times greater in pregnant bitches compared with non-pregnant bitches; however, in some non-pregnant individuals prolactin concentrations are high.

As previously mentioned, LH is a luteotrophic agent in the bitch and its inhibition may result in luteolysis. The pattern of LH secretion has not been characterized throughout the pregnant or non-

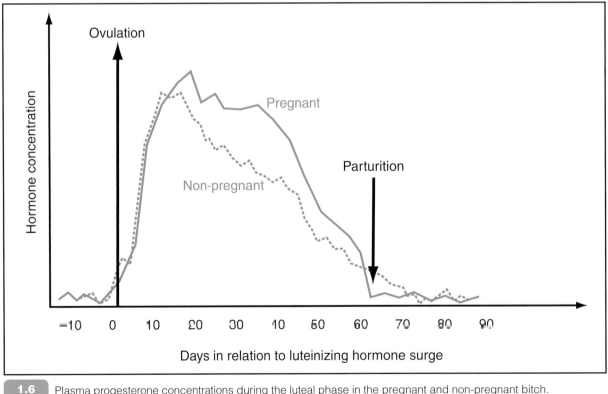

1.6 Plasma progesterone concentrations during the luteal phase in the pregnant and non-pregnant bitch.

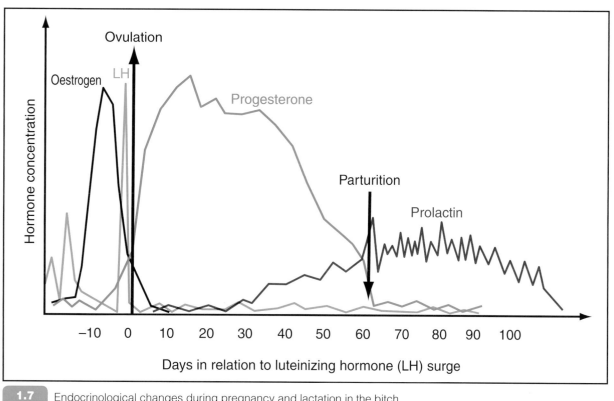

1.7 Endocrinological changes during pregnancy and lactation in the bitch.

pregnant luteal phase. Interestingly, the corpora lutea appear to be almost autonomous during the first 14 days after ovulation, and at this time have a need for relatively little luteotrophic support.

Relaxin is the only known hormone that is specific to pregnancy in the bitch. It is present in plasma 25 days after ovulation and peak values are reached at approximately day 50. Relaxin appears likely to be produced entirely from the placenta, and there is some evidence to demonstrate higher values in bitches with larger litter sizes because of the increased volume of the placentas. Although measurement of relaxin is widely used as a method of pregnancy diagnosis, the function of relaxin is to cause softening of the fibrous connective tissues prior to birth.

In addition to these hormonal changes associated with pregnancy, a variety of other physiological changes have been observed in the bitch. For example, there is a pregnancy-specific increase in blood volume that contributes to a normochromic normocytic anaemia. This is associated with a significant reduction in the percentage packed cell volume (PCV). In addition, pregnancy-specific increases in plasma fibrinogen and other acute phase proteins are found after approximately day 20 onwards.

The concentration of oestradiol does not appear to differ between pregnant and non-pregnant bitches, and tends to increase consistently during the luteal phase. However, there may be pregnancy-specific increases in the concentrations of oestrone and oestrone sulphate. Whilst one study demonstrated an increase in the total oestrogen concentration in urine during pregnancy, there have been no additional investigations, and the value of oestrogen

measurement as a method of pregnancy diagnosis remains to be elucidated.

Pseudopregnancy: The long luteal phase of the non-pregnant cycle has been called pseudopregnancy because of its similarity to pseudopregnancy in other species following a sterile mating. However, the long luteal phase in bitches occurs spontaneously and is not dependent upon mating-induced ovulation, or upon initial maternal recognition of pregnancy. The term 'physiological pseudopregnancy' may therefore be used, because in every non-pregnant bitch the luteal phase is long and there is some degree of mammary gland enlargement. Clinical pseudopregnancy (sometimes referred to as overt pseudopregnancy) is considered to be the development of extensive mammary gland enlargement, lactopoiesis and sometimes lactation, combined with behavioural changes typical of pregnancy and lactation. The incidence of clinical pseudopregnancy in the bitch is not known. In some breeds it has been reported anecdotally to affect up to 25% of individuals, whilst in other breeds it is not a recognized problem. The clinical signs of pseudopregnancy are mediated by elevated concentrations of prolactin, which occur at the same time as prolactin would be elevated in pregnancy. Whilst the timing of this event is similar, the magnitude of the increase is not as large in pseudopregnancy, but based on these observations it is not surprising that non-pregnant bitches commonly have clinical signs that are similar to those found during pregnancy, including lactation. In some cases pseudopregnancy may occur following iatrogenic termination of the luteal phase, including for example surgical neutering during the luteal phase.

The clinical signs of pseudopregnancy are variable and may include anorexia, nervousness, aggression, nest making, nursing of inanimate objects, lactation and occasionally false parturition. In the majority of non-pregnant bitches, mammary gland development is associated with the production of milk. In many cases pseudopregnancy undergoes spontaneous remission, especially if there is no stimulus for continued lactation (no suckling). However, in some cases either the physical or the psychological signs, or the effect of these signs upon the owner, may warrant treatment of the condition. In clinical practice, pseudopregnancy is most commonly treated by administration of prolactin inhibitors such as cabergoline or bromocriptine; treatment for 5–7 days usually produces rapid resolution of the clinical signs.

It is also noteworthy that, as well as clinical signs of pseudopregnancy, during the non-pregnant luteal phase the endometrium undergoes histological 'repair', returning to an appearance similar to that present before oestrus. Apoptosis of glandular basal epithelial cells appears to occur mainly during the early luteal phase, whilst degeneration of cells in the luminal epithelium occurs from the mid-luteal phase to early anoestrus.

End of the luteal phase and parturition: In the pregnant bitch there is a precipitous decline in plasma progesterone, which is widely accepted to be the final trigger for the initiation of parturition (see Figure 1.6). In other species the decline in progesterone occurs as a result of the production of cortisol by the fetus, which leads to increased levels of enzymes that convert progesterone to oestradiol. This endocrine shift removes the progesterone block to myometrial contractions and results in increased basal uterine contractions as well as increased reproductive tract secretions. During both the prepartum luteolysis and parturition there are increased concentrations of prostaglandin F metabolite (PGFM), which can be detected in the plasma. The PGFM appears to originate from the fetoplacental unit. It would appear, therefore, that in the dog the prepartum cascade of events is not dissimilar to those in other species.

There is no similar mechanism for the termination of the luteal phase in the non-pregnant bitch. As a consequence plasma progesterone concentrations decline slowly and the luteal phase of non-pregnancy is longer than the luteal phase of pregnancy (see Figure 1.6). This differs considerably from most other domestic species, in which the non-pregnant luteal phase is significantly shorter than the pregnant luteal phase.

Lactation: In the pregnant bitch after parturition the action of sucking by the neonates induces increased secretion of prolactin to values similar to those of late pregnancy (see Figure 1.7). Prolactin is required for normal lactation. Most commonly, lactation lasts for approximately 6 weeks after which prolactin concentrations decline progressively and milk production reduces until the puppies are fully weaned.

The queen

The reproductive physiology of the queen is unique among the domestic species and comprises a polyoestrous seasonal breeding pattern in which ovulation is stimulated by coitus and is not spontaneous. Queens are capable of achieving multiple pregnancies within a breeding season, making them one of the most prolific of the domestic species.

Puberty

The first oestrus in a queen may occur between 4 and 12 months of age. Whilst there are many factors that influence the onset of puberty, an important one is breed (many shorthaired breeds reach puberty earlier than longhaired breeds); particular breeds such as the Burmese reach puberty early (often at 4 months of age), whilst Persians reach puberty late (often at 12 months of age). A further critical event is the time of year at their birth; kittens born in summer or autumn often reach puberty at their first spring (when they are 5–6 months of age), whilst spring-born kittens often do not reach puberty until the subsequent spring (when they are 12 months of age). The timing of the first oestrus is also influenced by body condition, social environment and plane of nutrition. The average bodyweight at puberty is 2.5–3.0 kg (when approximately 80% of adult bodyweight has been reached). Interestingly, free-roaming cats tend to reach puberty earlier than pet cats.

Oogenesis in the cat follows the sequence described above, with oogonia being arrested after the first prophase of meiotic division. Oocyte development occurs within the follicle at each oestrus, similar to that in most other domestic species.

Peak fertility

In queens peak reproductive capacity appears to be reached between 2 and 8 years of age, when between two and three litters per year can be reared if the queen is allowed to breed naturally. Fertility tends to reduce after the age of 10 years when there are fewer cycles and smaller litters, although fertile queens over 20 years of age have been reported.

Cyclicity

The wild or free-roaming queen is a seasonal long-day breeder with anoestrus being observed over 3–4 months of winter when the day length is short; melatonin secretion from the pineal gland has been demonstrated to be important in regulating this event. However, in breeding colonies with controlled lighting regimes and in many domestic homes, queens may be non-seasonal and exhibit oestrus over winter, but this may be sporadic.

The breeding season for the average queen in the northern hemisphere lasts from February to October. During the breeding season the average interval from one ovulation to the next is between 14 and 21 days (it is often shorter in the oriental breeds). The cycle of the non-mated queen is divided into interoestrus (8–15 days), pro-oestrus (0–1 day) and oestrus (2–6 days). Queens are induced ovulators and when mating results in ovulation the luteal phase may follow oestrus. The non-pregnant luteal phase lasts approximately 25–45 days, whilst the pregnant luteal phase lasts approximately 65 days (Figure 1.8). On average queens will display oestrous behaviour every 2–3 weeks unless it is outside the breeding season, the queen has ovulated and is pregnant or pseudopregnant, the queen has been given depot progestogens, there is significant systemic illness or the queen has been neutered.

Oestrous cycle

Anoestrus
Anoestrus is the absence of cycling activity, which occurs naturally in periods of short daylight hours associated with increased secretion of melatonin. Interestingly, concentrations of both melatonin and prolactin are elevated during the period of short daylight hours, and are low during periods of high light intensity. During anoestrus queens will hiss or strike out when sexual advances are made by tom cats.

In the northern hemisphere, anoestrus often occurs between October and January. However, there may be breed variations in the response to different lighting regimes, with shorthaired cats more commonly continuing to cycle over winter. Presumably as day length increases and melatonin concentrations decline, there is increased activity within the hypothalamic–gonadal axis with increased secretion of LH and FSH, resulting in follicle growth, the production of oestrogen and a return to pro-oestrous and oestrous behaviour. In the laboratory environment, queens that are exposed to 10 hours of light per day are unlikely to enter anoestrus, but 14 hours of light per day are required to ensure good levels of fertility. Artificial light can be used to terminate anoestrus, although generally up to 2 months of supplementary lighting is needed to achieve this effect. It would appear that in queens the period of artificial lighting needed to terminate anoestrus can be shortened by exposure to a tom cat; in this instance return to oestrus may occur within 4–6 weeks.

Following parturition, queens enter anoestrus, during which time they are lactating. Whilst many queens do not cycle during lactation (presumably, as in other species, lactational anoestrus is mainly due to suppression of the mechanism regulating pulsatile secretion of GnRH in the hypothalamus, which is responsible for follicular development and steroid production), a large proportion return to oestrus during the second, third or fourth week of lactation (especially if this is in the middle of the breeding season). For queens that do not cycle during lactation, the return to cyclicity commonly occurs 2–8 weeks after weaning.

Interoestrus
The period between one oestrus and the next in queens that have not ovulated is called interoestrus. During this time oestrogen concentrations are low and no sexual behaviour is observed. The length of interoestrus is variable and although it is often 8–15 days it may be shorter, especially in oriental breeds. It is likely that towards the end of interoestrus there is increased secretion of LH and FSH, which results in stimulation of follicle growth and a return to pro-oestrous and oestrous behaviour.

Pro-oestrus and oestrus
Pro-oestrus and oestrus in the queen are associated with stimulation of follicle development by FSH and LH. As follicles enlarge there is subsequent secretion of oestrogen from the granulosa cells. In some

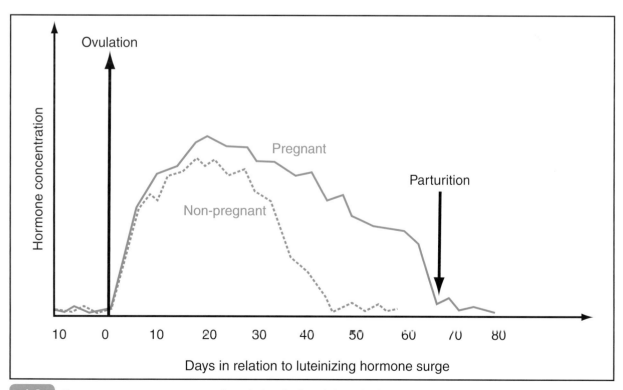

1.8 Plasma progesterone concentrations during the luteal phase in the pregnant and non-pregnant queen.

queens no obvious period of pro-oestrus can be detected, whilst in others there are similar signs to those present during oestrus, but the queen will not allow mating. During oestrus there may be slight swelling and reddening of the vulval lips, requiring close inspection to identify these changes. Unlike the bitch there is no obvious vulval discharge, although this may be because queens have fastidious grooming habits. During pro-oestrus the queen may become more affectionate and rub herself against other cats, humans and inanimate objects. During oestrus there is often increased activity, frequent vocalization with howling, and some queens have a reduced appetite. If stroked the queen may crouch with her front legs pressed to the ground, and her hindlegs elevated (spinal lordosis) with her tail turned to one side to present the vulva; often when in this position queens will make treading movements with their front legs. Occasionally queens in oestrus demonstrate urine spraying and marking. During oestrus the queen is passive to the male, but intromission causes significant vocalization and often the queen will then free herself aggressively from the male and may strike at him. This is usually followed by excessive licking of the perineal region and a short period of sexual rest followed by multiple further matings.

Some queens that have a nervous disposition or who are at the low end of the social scale in the hierarchy may not show behavioural oestrus, even though presumably they have elevated oestrogen concentrations. Other issues such as inexperience, distraction and male preference may also overcome normal physiology and result in this syndrome, which is often termed silent heat.

Throughout oestrus, follicles that have grown have a period of persistence during which time oestrogen concentrations remain elevated. Towards the end of oestrus (i.e. after 4 or 5 days) and in the absence of ovulation, the follicles appear to become atretic and oestrogen concentrations decline; consequently, oestrous behaviour wanes and the queen returns to interoestrus.

Ovulation: The queen is an *induced ovulator*, where mechanical stimulation of the vagina is required for neuronal triggering in the hypothalamus with a resultant release of GnRH, which induces the LH surge. Normally this stimulation is produced by mating, but it can be created artificially with a cotton swab placed into the vagina; this method may be used in clinical practice to induce ovulation, terminate oestrus and induce pseudopregnancy. The release of LH begins within a few minutes of normal mating and concentrations peak 2–4 hours later. Interestingly, ovulation does not always occur when there has been only a single mating because the magnitude of the LH surge appears to be related to the day upon which the queen is mated, together with the quality, interval and total number of matings. Generally, it is perceived that the LH surge is of greater magnitude and longer duration when multiple matings occur within a short period of time. Approximately 50% of queens ovulate after a single mating, and the optimal LH release to stimulate ovulation appears to be when four matings

occur within a 2–4 hour window of time. Greater coital stimulation does not appear to increase the LH response. Some studies have suggested that mating later in oestrus may require fewer matings to induce ovulation, possibly because higher plasma oestrogen at this time allows the anterior pituitary gland to respond better to copulatory stimulation and produce an LH surge of greater magnitude.

In a recent study, Tsutsui *et al.* (2009) found that 60% of cats mated once on day 1 of oestrus ovulated, whereas 83% of cats mated once on day 5 of oestrus ovulated. Interestingly, the low ovulation rate for queens mated on day 1 was not significantly increased even when queens were mated three successive times on day 1. In these queens only 70% ovulated, even though the mean plasma peak LH concentration was high. Interestingly, 100% of queens mated three successive times on day 5 ovulated. High plasma LH concentrations were observed in both groups subjected to multiple matings, and it would appear that LH levels were sufficiently high to induce ovulation in cats on day 5 but not on day 1. The fact that the ovulation rate in animals mated three times on day 1 of oestrus was low, but the plasma LH concentrations were high in the animals that ovulated, suggests that the response of the pituitary gland to the copulatory stimulation was insufficient. In the study by Tsutsui *et al.* (2009) it was also observed that even when ovulation occurred there were lower conception rates when mating happened only once compared with when there were three successive matings. This perhaps suggests that the number of sperm ejaculated was insufficient to achieve fertilization of all oocytes. Overall the results suggest that mating more than once in the middle of oestrus is required to ensure high ovulation and conception rates in queens.

In some queens ovulation occurs without mating. This is termed 'spontaneous ovulation' and may be seen in queens that are groomed frequently, or those that self-groom, and is also more common in older queens and those that are in the presence of a tom cat. In these cases presumably there is some other suitable, possibly pheromonal or visual, stimulation that causes the surge release of LH to trigger ovulation.

Fertile and fertilization periods: Following an LH surge of sufficient magnitude, ovulation is stimulated, probably via a mechanism similar to that observed in other species. Oocytes are ovulated after extrusion of the polar body (i.e. they are secondary oocytes) and are immediately fertilizable. Generally ovulation occurs within 24–48 hours after the peak of LH release. Oestrogen concentrations decline rapidly after ovulation and, as a result of the decline in oestrogen and rise in progesterone following ovulation, queens that ovulate frequently go out of oestrus within 24–48 hours of mating; anecdotally, the total duration of oestrus is shorter in queens that are mated and ovulate, compared with those that do not ovulate.

Although the cat is more similar to other common domestic species in that secondary oocytes are released at ovulation, other aspects of gamete biology are not dissimilar to those seen in the bitch. Fertilization occurs within the oviduct and the embryos migrate along the oviduct to enter the uterus as morulae approximately 3–4 days after ovulation.

Overlapping of follicle waves: The basic description of the oestrous cycle given above is that after a non-ovulating cycle the queen returns to interoestrus with basal plasma oestrogen concentrations for 8–15 days before the growth of the next follicular wave. However, in some queens the follicle waves follow one another so closely that there is either no gap or there is a very small gap between the observed oestrous behaviour: decreasing oestrogen produced by the atretic follicles is overlapped by increasing oestrogen from the next follicle wave. This is most commonly observed in some of the oriental breeds and results in a syndrome of an almost persistent oestrus, which can be inconvenient for the owner. This may need to be treated using either mechanical stimulation of the vagina to induce an LH surge, or the administration of an LH-like preparation such as human chorionic gonadotrophin (hCG).

Luteal phase

Most commonly there is a rapid end to oestrus after ovulation such that the beginning of the luteal phase is almost synchronous with the end of oestrus. It is likely that the LH surge is important not only for ovulation but also for formation of the corpora lutea.

Progesterone appears to be produced solely from the corpora lutea. Progesterone profiles and concentrations do not differ between pregnant and non-pregnant queens in the early luteal phase. Thereafter, however, there are significant differences in the progesterone profile and the duration of progesterone secretion, suggesting some mechanism for maternal recognition of pregnancy (see Figure 1.8).

Pregnancy: Plasma progesterone increases rapidly after ovulation; approximately 24 hours after the LH surge, progesterone concentrations have increased significantly and continue to rise to a peak on approximately day 25 after ovulation. Following a small decline, the concentrations remain stable over a broad plateau until approximately day 60 after ovulation (see Figure 1.8). Progesterone appears to be produced principally by the corpus luteum (the placenta produces little or no progesterone) and as a consequence ovariectomy or the administration of luteolytic compounds at any stage of gestation results in pregnancy loss. There is a clear difference in progesterone profiles between queens that ovulate and are pregnant and those that ovulate but are not pregnant; progesterone is produced for a longer period of time in pregnancy and progesterone concentrations are higher than in non-pregnancy (see Figure 1.8).

The progesterone profiles of pregnancy are probably associated with specific luteotrophic factors, which act as a signal to maintain the secretion of progesterone from the corpora lutea. It would appear most likely that the 'signal' originates from the fetus/placenta/uterus because pregnant queens that undergo hysterectomy in early pregnancy have a luteal phase similar to that seen in pseudopregnant animals. Potential hormonal signals include prolactin and relaxin, which are recognized as luteotrophic factors in the bitch.

Prolactin concentrations start to increase from approximately day 30 after ovulation (Figure 1.9). The

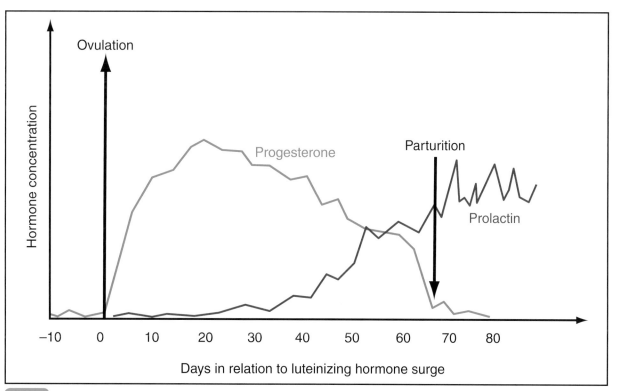

1.9 Endocrinological changes during pregnancy and lactation in the queen.

concentrations continue to rise, and reach a plateau on day 50. They then remain stable until a few days prior to parturition when concentrations increase again. It is clear that prolactin has an important luteotrophic role in pregnancy, because administration of prolactin inhibitors (such as cabergoline and bromocriptine) results in a rapid reduction in prolactin concentration, followed by cessation of progesterone secretion, and subsequent termination of pregnancy. This information may be used in clinical practice as a basis for pregnancy termination; prolactin inhibitors are often administered concurrently with prostaglandins for this purpose. After parturition, prolactin concentrations are elevated and remain so throughout the period of lactation (Figure 1.9).

Relaxin is another hormone with similar profiles in the pregnant bitch and queen. In both species relaxin is the only hormone specific to pregnancy. Relaxin is secreted by the placenta, although it is also possible that there is some ovarian production. In the queen, plasma relaxin concentrations increase after implantation and the establishment of the early placenta; plasma concentrations often increase from approximately day 25 after ovulation and continue to be elevated until a few days after parturition. In respect of its potential luteotrophic action, relaxin may act directly on the corpus luteum, may stimulate prolactin secretion from the pituitary gland indirectly, or may stimulate other unknown luteotrophic factors.

Plasma oestrogen levels decline after ovulation and remain low throughout pregnancy, but increase at approximately day 60 of pregnancy and then start to decline just before parturition. Up to 10% of cats may show behavioural oestrus when pregnant, most commonly between days 20 and 40 of pregnancy. This may reflect episodes of follicle growth and regression that appear to occur during the luteal phase (although in most cases plasma oestrogen concentrations remain low). Oestrous activity during pregnancy has led to the contention that queens may undergo 'superfetation' (where fetuses of different ages are present simultaneously, i.e. a second crop of fetuses develops from mating during pregnancy). However, it is not likely that oestrous activity during pregnancy results in mating, with transport of sperm and fertilization of a new crop of fetuses. More likely, superfetation does not occur but has been presumed when pregnant queens have been observed with oestrous activity and have coincidentally delivered one or more undersized kittens (at a normal parturition relating to the initial oestrus).

However, so called 'superfecundity' does occur in the queen; here several males mate with a queen during one oestrus and sperm from different males fertilizes oocytes from one ovulation, resulting in kittens with different fathers being born in the same litter.

Pseudopregnancy: In queens that have ovulated but are not pregnant, there is an initial increase in progesterone that is not distinguishable from that seen in pregnancy. Progesterone concentrations in the plasma reach a maximum approximately 25 days after ovulation and thereafter start to decrease, similar to the situation seen in pregnancy; however, in non-pregnant queens progesterone concentrations continue to decline gradually and reach basal values at approximately day 30–40 (see Figure 1.8). This slow decline of progesterone is thought to be the result of an absence of a pregnancy-specific luteotrophic factor; in non-pregnancy the corpora lutea seem to be programmed to have a lifespan of approximately 40 days, and this is only prolonged when there is pregnancy recognition resulting in an extended luteal phase. Prolactin concentrations do not increase in the pseudopregnant queen, nor is there any increase in relaxin.

Pseudopregnancy is not associated with any clinical signs with the exception of slight enlargement of the nipples. There is normally no mammary gland enlargement and no behavioural changes. In fact the only obvious clinical sign is the absence of cyclical activity when compared with a non-ovulating queen.

Interestingly, the non-pregnant luteal phase of the queen is different from that of the bitch because it is approximately half the length of the pregnancy, and it is not followed by an obligatory period of anoestrus. This allows a rapid return to cyclical activity and increased chances of a fertile mating within the breeding season.

End of the luteal phase and parturition: In the pregnant queen there is a decline in progesterone just prior to the onset of parturition. It is likely that the mechanism is similar to that seen in other species and involves fetal production of cortisol.

In non-pregnant queens there is a gradual reduction in plasma progesterone to basal values by approximately day 30–40 after ovulation. This slow decline of progesterone does not seem to require an active mechanism of luteolysis involving the uterus, because hysterectomy in early pseudopregnant cats does not influence the duration or profile of the luteal phase.

Lactation: Elevated prolactin concentrations are required for maintenance of lactation in the queen. After parturition, sucking by the kittens results in an elevated concentration of prolactin, and as sucking activity decreases throughout lactation towards weaning, so prolactin concentrations decline. After weaning prolactin concentrations decline slowly to basal values over approximately 2 weeks and lactation then ceases.

References and further reading

Tsutsui T, Higuchi C, Soeta M *et al.* (2009) Plasma LH, ovulation and conception rates in cats mated once or three times on different days on oestrus. *Reproduction in Domestic Animals* **11 (0),** 76 79

Physiology and endocrinology of the male

Gary C.W. England

Introduction

The male reproductive system is relatively complex in that it is required to perform four functions:

1. The production and distribution of male sexual hormones.
2. The production and maturation of sperm in the testes.
3. The maturation, transport and storage of sperm in the epididymal duct system.
4. The deposition of sperm in the female reproductive tract.

These functions are underpinned by the hypothalamic–gonadal axis, which is responsible for controlling maturation of sperm, enhancing male behaviour and ensuring the development of the masculine body form.

It is becoming increasingly common for veterinary surgeons to be consulted by both professional and amateur breeders to advise on normal breeding practice, undertake semen evaluation and the preservation of semen, investigate infertility and perform artificial insemination. Knowledge of the normal reproductive function, including basic anatomy, physiology, endocrinology and methods of clinical examination are important to enable the clinician to fulfil these roles.

Reproductive function

The paired testes are the primary organs of male reproduction, producing the male gametes (spermatozoa) and steroid hormones (androgens (testosterone) and oestradiol). The testes differ from ovaries in that all the potential gametes are not present at birth. Instead, germ cells undergo continual cell divisions, forming new spermatozoa throughout the reproductive life of the male.

Anatomy

In the fetus the developing testes cannot be differentiated initially from ovaries; however, during development a cylindrical, almost gelatinous, structure called the gubernaculum forms on the pole of the testis and by its effective contraction 'pulls' the testis through the inguinal canal into the scrotum within an outpouching of the peritoneum (the vaginal tunic). In the dog, the descent of the testes often occurs in the fetus so that the puppy is born with descended testicles, although this may take up to 8 weeks after birth in some individuals. There are reports of descent as late as 6–8 months after birth, and often these animals are considered to be abnormal and part of the cryptorchid syndrome (meaning that they are probably genetically affected). Kitten testicles are usually descended by the time of birth.

In both the cat and the dog, the testes are held within the scrotum covered by thin scrotal skin with little subcutaneous fat. There is close association of the testicular arterial supply and venous drainage via the pampiniform plexus, which provides an efficient system to allow cooling of the arterial blood. Together with the cremaster muscle (which is able to alter the distance of the testes from the body) and often a lack of hair on the scrotal skin, these mechanisms provide thermoregulatory control which can prevent increases in testicular temperature that are detrimental to sperm production.

In the dog the scrotum is pendulous and the testes are positioned almost horizontally, whereas in the tom cat they are held more closely to the body. Dog testes vary in size in relation to the body mass of the animal, but for a 15 kg medium-sized dog they are on average 3.0 x 2.0 x 1.5 cm. Generally there is a good correlation between body mass and testicular mass (Woodall and Johnstone, 1988). Tom cat testes measure approximately 13 x 8 x 6 mm for an average-sized animal.

The testes are composed of two tissue types: the seminiferous tubules and the interstitial tissue. The seminiferous tubules are the site of spermatogenesis and are U-shaped tubes, where both ends open into collecting ducts called the vasa efferentia, which themselves open into the epididymis. The latter structure is a highly convoluted tube that arises medially on the testis. It is located on the dorsolateral surface of the testis and terminates on the caudal pole. The epididymis can be considered to have three regions: head, body and tail. Sperm are moved along the epididymis, as the result of peristalsis, to the tail region, which acts as a store. Here sperm are maintained in a viable but non-motile state prior to ejaculation. The epididymal tail is pea-sized in the average-sized dog and can be palpated closely apposed to the caudal aspect of the testis within the scrotum. During ejaculation, sperm leave the epididymal tail and pass into the vasa deferentia, which run within the vaginal sac and convey the spermatozoa from the testes to the urethra. The vas deferens passes through the inguinal canal and approaches the urethra dorsally within the region of the prostate gland. A dilated area of the vas deferens close to the prostate gland (termed the ampullary region) is present but is not a significant structure in the dog, and is absent in the tom cat.

Surrounding the proximal urethra is the prostate gland, the sole accessory gland in the dog (in which it is large and contributes a significant volume of watery secretion to the ejaculate); although present, the prostate gland is not significant in the tom cat. In the normal dog the prostate gland is positioned within the pelvis and surrounds the cranial part of the pelvic urethra. The prostate gland is symmetrical and divided into two lobes, each of which is approximately 2 cm in diameter. The major accessory gland in the tom cat is the paired bulbourethral glands. These glands are positioned craniolateral to the base of the penis and are approximately 3 mm in diameter. The bulbourethral glands are absent in the dog. Vesicular glands are absent in both the dog and the tom cat.

The urethra passes into the penis, the root of which arises in the caudal perineum. A bone (os penis) is present within the penis of the dog and tom cat. This allows the male to achieve intromission before the attainment of a full erection. The os penis of the dog has a deep groove on its ventral surface which houses the urethra.

The glans penis comprises two parts (differentiated by the location of erectile tissue): (a) the bulbus glandis, which is composed of erectile tissue surrounding the os penis and urethra; and (b) the pars longa glandis, which has erectile tissue dorsal and longitudinal to the os penis and urethra only. The cat penis is unusual in that it is directed caudally, and under the influence of testosterone small spines or papillae are present along the glans penis. These spines become fully developed at puberty and are thought to help stimulate a pituitary-mediated release of luteinizing hormone (LH) in the female at the time of intromission.

Physiology

The male reproductive system has two principal functions: the production of sperm and the production of steroid hormones.

The production of spermatozoa (spermatogenesis) occurs within the seminiferous tubule. This tubule contains two cell types, the Sertoli cells (the somatic cells) and the germ cells. A cross-section of the testicular tubule shows two distinct zones: a basal zone, which contains spermatogonia dividing by mitosis; and an adluminal zone, which contains primary spermatocytes undergoing meiosis to produce secondary spermatocytes and spermatids. The basal and adluminal zones are separated by junctional complexes formed between Sertoli cells, creating an effective blood–testis barrier. This barrier prevents

blood macromolecules and interstitial fluid from the adluminal zone entering the basal zone, thus creating an environment in which meiosis can occur.

The production of hormones (steroidogenesis) occurs within the interstitial tissue compartment of the testis surrounding the seminiferous tubules. This region is composed of Leydig cells, which are closely associated with blood and lymphatic vessels. Leydig cells are the only testicular cells with receptors for LH. The LH (also previously known as interstitial cell-stimulating hormone, ICSH) binds to the receptors on the Leydig cells and stimulates steroidogenesis via a process mediated by cyclic adenosine monophosphate (cAMP). The androgens (testosterone) produced by the Leydig cells are essential for the development of the secondary male sexual characteristics, normal behaviour, function of the accessory glands, production of spermatozoa and maintenance of the male duct system. Given that the interstitial cell compartment surrounds the seminiferous tubule compartment, the latter is bathed in a fluid rich in testosterone.

Hormonal control

The production and secretion of gonadotrophin-releasing hormone (GnRH) from the hypothalamus is the principal method of control of the two gonadotrophins, LH and follicle-stimulating hormone (FSH) (Figure 2.1). Gonadotrophin secretion by the anterior pituitary gland is under the positive control of GnRH, which is released in an episodic manner. GnRH is synthesized in the neurons of the hypothalamus and released at nerve endings at the median eminence. It is transferred via the hypophyseal portal system to the anterior pituitary gland. Here GnRH stimulates the release of FSH and LH, which act upon the testes to stimulate and support spermatogenesis and the production of steroid hormones (androgens) (Figure 2.1).

There is an integrated feedback system for the control of hormone secretion. The basic mechanism of control is where a gland produces a hormone that stimulates a second gland to produce another hormone; usually the second hormone decreases the production of the hormone from the first gland. This is called negative feedback and is the mechanism by which GnRH release is controlled; there is a negative feedback loop that involves testosterone and the active metabolites oestradiol and dihydrotestosterone. Negative feedback is thought to occur at both the hypothalamic and the pituitary gland levels. This intricate system therefore allows control via the hypothalamus of the central nervous system (CNS), the anterior pituitary gland and the testes (Figure 2.1). There may also be some negative feedback control by oestradiol, which is produced via local aromatization of testosterone.

The negative feedback system that acts upon LH and FSH release is common to both gonadotrophins. This is an important consideration when agents are used clinically in an attempt to modify reproduction. Despite this linkage of the control, the concentrations of LH and FSH do not always rise in a parallel

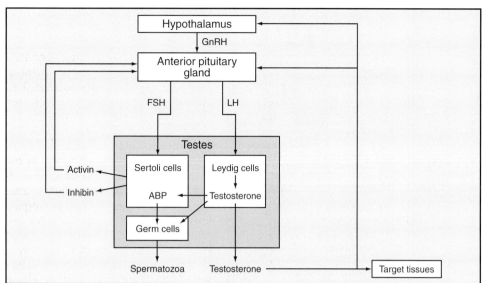

2.1

Hypothalamic–pituitary–gonadal axis demonstrating the site of hormone production and feedback loops. ABP = androgen-binding protein; FSH = follicle-stimulating hormone; LH = luteinizing hormone.

manner. As a consequence, the existence of an additional inhibiting factor called inhibin, which is solely responsible for the control of FSH secretion from the pituitary gland, has been proposed. There may also be other products of the Sertoli cell that have the opposite effect and stimulate FSH secretion (these have been termed activins) (Figure 2.1).

As well as its systemic effects, the testosterone produced by the Leydig cells has an important role in supporting spermatogenesis. Testosterone secretion therefore also occurs locally, where together with FSH it acts on the seminiferous tubules to stimulate Sertoli cell support of the germ cells, thus aiding spermatogenesis. This action is either a direct one, or occurs via an action on the peritubular cells, which promotes the production of a protein that modulates Sertoli cell function. The normal functioning of the Sertoli cells is dependent on the concentration of testosterone being considerably higher in the testis than in the peripheral circulation. These high concentrations of testosterone within the testis

are maintained partly by testosterone binding to androgen-binding protein (ABP), which is produced by the Sertoli cells. ABP may not be involved in the direct transport of androgens to spermatogenic cells, but may play a role in the transport of androgens throughout the whole of the male tract.

Interestingly, the hormone prolactin is thought to act synergistically with LH in regulating testosterone production by the Leydig cells; prolactin receptors are present on the Leydig cells. Once spermatogenesis reaches the stage of spermatid production, inhibin production occurs to create the negative feedback on FSH secretion.

The stimulatory and feedback control mechanisms result in a number of opportunities for testing specific components of the hypothalamic–gonadal axis (Figure 2.2). For example, if the hypothalamus is not functioning properly (Figure 2.2b) there will be reduced concentrations of GnRH, both the gonadotrophins and testosterone. However, if the pituitary gland is not functioning properly (Figure 2.2c) GnRH

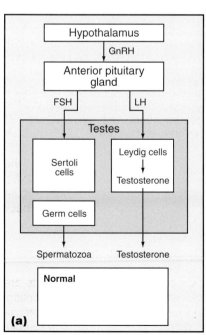

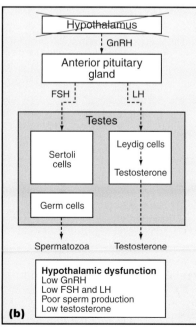

2.2 Changes in hormone concentrations associated with normal and pathological conditions of the hypothalamic–gonadal axis. **(a)** Normal. **(b)** Hypothalamic dysfunction. FSH = follicle-stimulating hormone; GnRH = gonadotrophin-releasing hormone; LH = luteinizing hormone. (continues) ▶

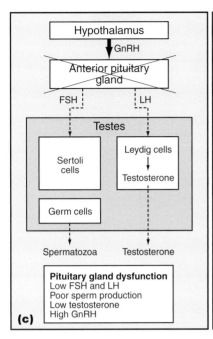

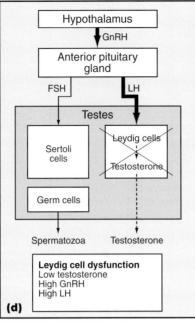

2.2 (continued) Changes in hormone concentrations associated with normal and pathological conditions of the hypothalamic–gonadal axis. **(c)** Pituitary gland dysfunction. **(d)** Leydig cell dysfunction (or absence of the testes). FSH = follicle-stimulating hormone; GnRH = gonadotrophin-releasing hormone; LH = luteinizing hormone.

concentrations will be elevated but both the gonadotrophins and testosterone will be reduced. In the situation of poor gonadal function or the absence of the testes (Figure 2.2d), GnRH and the gonadotrophins will be elevated but testosterone concentrations will be reduced.

Tests of the functioning of the pituitary gland and Leydig cells can be further confirmed by administration of GnRH and subsequent measurement of plasma LH and testosterone, respectively (Figure 2.3). In the normal functioning state administration of a given dose of GnRH will result in a specific pituitary gland response and release of LH, which will act upon the Leydig cells to produce a given rise in testosterone concentration (Figure 2.3b). If the

pituitary gland is not functioning properly a given dose of GnRH will not influence LH concentrations, which will remain low, as will concentrations of testosterone (Figure 2.3c). Where the pituitary gland is normal but Leydig cells are functioning abnormally, the LH response will be normal and LH concentrations will increase, but the testosterone response will be depressed and concentrations will remain low (Figure 2.3d).

A similar test can be used to examine just the gonadal component of the axis, by administering LH (or an LH-like substance, e.g. human chorionic gonadotrophin, hCG) and monitoring the release of testosterone (Figure 2.4). These tests are often called the GnRH and the LH stimulation test, respectively.

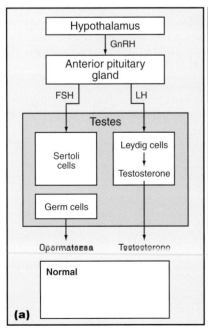

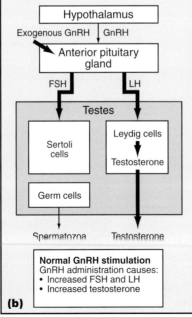

2.3 Changes in hormone concentrations following administration of exogenous GnRH in normal and pathological conditions of the hypothalamic–gonadal axis. **(a)** Normal. **(b)** Hypothalamic dysfunction. FSH = follicle-stimulating hormone; GnRH = gonadotrophin-releasing hormone; LH = luteinizing hormone. (continues) ▶

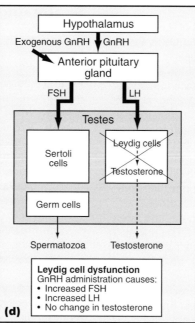

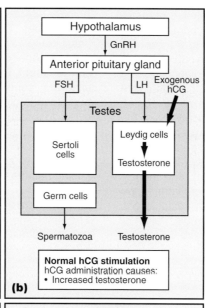

2.3 (continued) Changes in hormone concentrations following administration of exogenous GnRH in normal and pathological conditions of the hypothalamic–gonadal axis. **(c)** Pituitary gland dysfunction. **(d)** Leydig cell dysfunction (or absence of the testes). FSH = follicle-stimulating hormone; GnRH = gonadotrophin-releasing hormone; LH = luteinizing hormone.

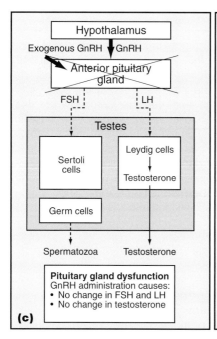

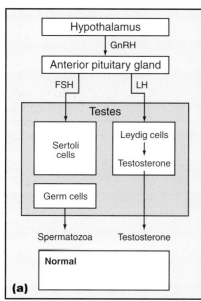

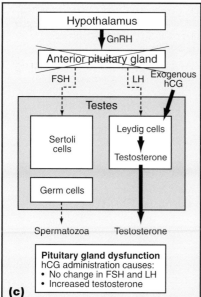

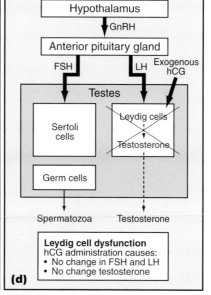

2.4 Changes in hormone concentrations following administration of exogenous hCG in normal and pathological conditions of the hypothalamic–gonadal axis. **(a)** Normal. **(b)** Hypothalamic dysfunction. **(c)** Pituitary gland dysfunction. **(d)** Leydig cell dysfunction (or absence of the testes). FSH = follicle-stimulating hormone; hCG = human chorionic gonadotrophin; LH = luteinizing hormone.

Hormone concentrations

The pulsatile nature of the release of LH makes measurement difficult, and a single sample is therefore virtually meaningless. Serial sampling provides a profile, which is a better reflection of the secretion of this gonadotrophin. Basal serum concentrations of LH in the dog are approximately 1.0–1.2 ng/ml, with surges reaching 3.8–10 ng/ml. Basal testosterone is usually between 0.5 and 1.5 ng/ml, rising to a peak of 3.5–6.0 ng/ml. Seasonal shifts have been reported in the concentrations of both LH and testosterone, yet despite the relationship between the mechanisms of secretion of the two hormones, the seasonal variations are independent of each other.

The concentration of LH in the tom cat has been found to range between 3 and 29 ng/ml. The pulsatile release of testosterone has been shown to result in a concentration that varies from 0.1 to 3.3 ng/ml within a 6-hour period.

As mentioned previously, administration of exogenous GnRH allows the clinician to test the pituitary gland release of LH and subsequent testicular release of testosterone, and the administration of hCG allows testing of the testicular release of testosterone. In normal males, GnRH administration causes an increase in LH secretion in 30 minutes and an increase in testosterone in 60 minutes. Administration of hCG is a common diagnostic test used to evaluate the presence of testes that are not within the scrotum (e.g. in cases of suspected cryptorchidism) and for testing the functioning of the Leydig cells. For the latter test it is normal to collect a blood sample for assay of testosterone immediately prior to the intravenous injection of hCG (200–500 IU in the dog; 100–500 IU in the tom cat), and then to collect a second blood sample 1–2 hours later. In the intact male a significant increase in testosterone occurs within 30 minutes such that samples taken at 1 or 2 hours are always diagnostic (Figure 2.5).

Spermatogenesis

During their development in the fetus, primordial germ cells differentiate into gonocytes which subsequently undergo mitosis during both late fetal and prepubertal life and differentiate into sperm-producing cells (spermatogonia). These stem cell spermatogonia are then arrested in the seminiferous tubules until the onset of puberty. This situation differs from that in the female, who at this stage has all of the eggs that she will ever possess. An important feature of spermatogenesis is the production of sperm whilst the number of sperm-producing cells (spermatogonia) is maintained (Figure 2.6). This process is divided into three phases:

- Spermatocytogenesis
- Spermiogenesis
- Spermiation.

Spermatocytogenesis

The stem cell spermatogonia located along the basement membrane of the seminiferous tubules multiply by mitosis. This allows both the cyclical production of *primary spermatocytes* and the maintenance of stem cell number. This initial division occurs within the basal compartment of the seminiferous tubule, whilst the remainder of the process occurs in the adluminal compartment. Primary spermatocytes undergo meiosis, and genetic material is exchanged between homologous chromosomes of the primary spermatocytes to produce *secondary spermatocytes*. These undergo a further meiotic division to produce *haploid spermatids*. Four spermatids therefore develop by meiosis from each primary spermatocyte, each containing half the normal chromosomal number.

Spermiogenesis

The final morphological transformation is termed spermiogenesis and involves differentiation into

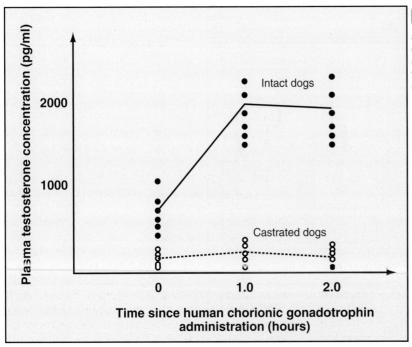

2.5 Concentrations of plasma testosterone prior to and after administration of 200 IU human chorionic gonadotrophin in six entire dogs (mean value = black line) and five castrated dogs (mean value = dashed line).

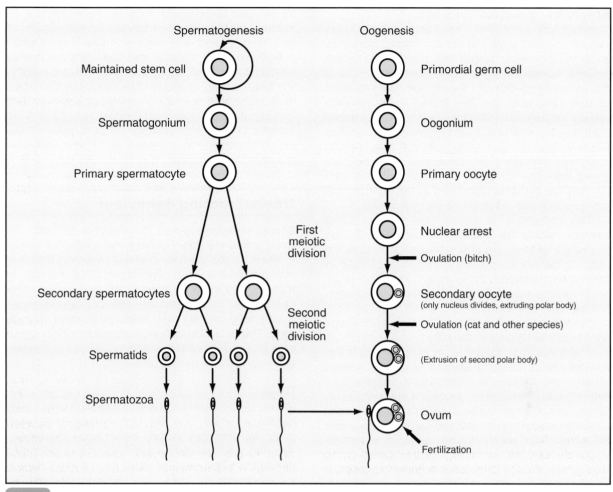

2.6 Stages of spermatogenesis compared with the respective stages of oogenesis.

mature spermatids, which are released into the lumen of the seminiferous tubules as spermatozoa. As spermatogenesis proceeds through these different stages, the developing gametes migrate from the basement membrane of the seminiferous tubules towards the lumen.

Spermiation

The process that involves the release of the germ cells into the lumen of the seminiferous tubule after spermiogenesis is known as spermiation. The released germ cells are now considered to be *spermatozoa*.

The germ cells are found close to, and associated with, Sertoli cells throughout their development. The role of the Sertoli cell includes providing support and nutrition, phagocytosing waste products and secreting essential luminal fluids. Throughout spermatogenesis, the germ cells are linked by cytoplasmic bridges and are in close contact with Sertoli cells. At spermiation, these links break and a remnant of the cytoplasmic link, known as the cytoplasmic droplet, is retained by the spermatozoa. These droplets are positioned just behind the sperm head, so they are termed proximal cytoplasmic droplets. The droplets move from the neck of the spermatozoa to the distal midpiece (here they are termed distal cytoplasmic droplets) as the sperm move along the epididymis. Sometimes sperm with distal cytoplasmic droplets can be identified microscopically in ejaculated samples, and usually this is considered to indicate ejaculation of immature spermatozoa.

Leydig cells also have a role in supporting spermatogenesis. The interstitial cell compartment surrounds the seminiferous tubule; which is consequently bathed in a fluid rich in testosterone secreted by the Leydig cells. In addition, myoid cells, which form the boundary tissue of the seminiferous epithelium, help in the propulsion of spermatozoa and fluid within the seminiferous tubules, moving them towards the vasa efferentia.

If a single seminiferous tubule is examined histologically, four to five types of germ cells can be identified arranged in specific layers, with the most advanced stages being present in the central luminal region. It should be noted that all seminiferous tubules do not contain germ cells at the same stages of maturation because this would result in all sperm being released at one moment in time and then there would be no more sperm available until another *spermatogenic cycle* had been completed. Therefore, examination of an entire histological section of the testis demonstrates individual seminiferous tubules each containing germ cells at different stages of the cycle. The duration of spermatogenesis is approximately 62 days in the dog (data are not available for the tom cat).

Sperm transport and storage

Spermatozoa pass from the seminiferous tubules through the rete testis into the epididymis. Here they undergo the final stages of maturation, which includes specific membrane changes and the loss of the cytoplasmic droplet. Sperm with an immature plasma membrane are also known to be immotile and during passage along the epidiymis maturational changes occur providing the potential for motility. Although they have motility potential they remain essentially immotile until ejaculation, when they are mixed with secretions from the accessory glands (in the dog the prostate gland and in the tom cat the bulbourethral glands) and then become motile.

The maturation of sperm that occurs during transport along the epididymis also includes the acquisition of fertilizing ability; sperm taken from the head of the epididymis are generally incapable of fertilization, whereas those taken from the distal body and the tail are able to achieve normal interactions with the oocyte.

Epididymal transit is thought to take on average approximately 12 days in the dog.

It is convenient, therefore, to consider the tail of the epididymis to be a storage organ for sperm, although the situation is slightly more complex than this because there must also be a mechanism to remove very old sperm should ejaculation not occur.

Further changes occur to the spermatozoa within the female tract, which include capacitation and the acrosome reaction, giving the sperm the ability to change the characteristics of its swimming behaviour and ultimately to fertilize the oocyte (see below).

Sperm production

Given that there is a pool of sperm stored within the epididymides, the number of spermatozoa that can be found within an ejaculate is related to the frequency of ejaculation. Sexually rested males have greater numbers of sperm in the ejaculate, but with repeated ejaculation the number of sperm in the ejaculate falls to reach the number that are 'produced' in the interval between ejaculations. In this way, if daily ejaculates are collected, after 4–5 days the number of sperm present in the ejaculate reflects the 'daily sperm production'. In normal dogs a good correlation exists between daily sperm production, the volume of the testes and the width of the scrotum. Interestingly, in the dog daily sperm production appears to be lower than in some other domestic species; this is probably related to the long duration of spermatogenesis in the dog. Large dogs in general ejaculate greater numbers of spermatozoa than small dogs. Values for sperm production in the tom cat are not available.

Puberty

Leydig cells appear to mature at approximately 5 months of age and plasma testosterone concentrations are elevated from this time onwards. However, there are some breed differences, with smaller dogs demonstrating secondary sexual characteristics earlier than large-breed dogs. Spermatozoa can be found in the seminiferous tubules by 6–7 months of age, and overall the average male dog reaches puberty at 10–12 months and can ejaculate normal sperm at this time. Often over the next few months the total number of sperm in the ejaculate continues to increase and reaches a plateau at approximately 2 years of age. The tom cat reaches puberty at 8–10 months of age. Interestingly, for both species, the onset of puberty in the male is slightly later than observed in the female.

Normal mating behaviour

Sexual behaviour differs considerably between the tom cat and the dog. Furthermore, mating behaviour is often complicated by the way in which breeders manage their animals. For example, on the day of mating, bitches and queens are frequently transported over long distances, are introduced to the male briefly, and are then expected to mate immediately. This situation can eliminate the normal courtship phase associated with pro-oestrous behaviour and may result in mating problems.

Dogs

In the normal situation, the dog and bitch exhibit play behaviour when they are first introduced to each other and this may include mounting of the female. The dog may ejaculate a small volume of clear fluid during this play period or whilst he is trying to locate the vulva with his penis. This fluid is the first fraction of the ejaculate and does not contain sperm; it originates from the prostate gland and its function is to flush any urine or cellular debris from the urethra. The dog will continue to mount, thrust and dismount, until his position allows the penile tip to enter the vagina. This is known as intromission (Figure 2.7a). The dog will then achieve a full erection during which time the thrusting movements increase rapidly and the second fraction of the ejaculate is produced. This fraction is sperm rich. Once the thrusting has subsided the dog will turn through 180 degrees and dismount the bitch whilst his penis remains within the vagina. The dog and bitch will now stand tail-to-tail – this is called the copulatory tie (Figure 2.7b). The tie is associated with the dog ejaculating the third fraction of ejaculate, which is again a clear fluid and prostatic in origin; its purpose is to flush the sperm forwards through the cervix into the uterus. The tie lasts an average of 20 minutes but varies considerably between dogs and can be as short as 5 minutes or more than 60 minutes in duration.

Cats

In the queen, the period of sexual introduction and play is variable, depending upon the experience and aggression of the male. The normal sequence of events occurs rapidly compared with the dog. The male usually approaches the female from the side or back and grasps her neck in his mouth. Whilst maintaining this grasp he mounts the female and positions himself to align the genital regions. The queen

2.7 Normal mating in the dog. **(a)** Intromission. **(b)** Copulatory tie.

normally lowers her chest and elevates the pelvic region whilst deviating her tail to one side. Pelvic thrusting and ejaculation occur rapidly. During intromission the queen often emits a cry and attempts to end mating by rolling, turning and striking at the male. The female then exhibits a marked postcoital reaction that consists of violent rolling and excessive licking of the vulval and perineal region. The queen will usually not allow further mating at this time, but commonly returns to being sexually receptive within 10–20 minutes.

Semen collection and evaluation

Semen quality is assessed to evaluate the fertilizing capability of the male animal. Semen evaluation is based upon the assumption that certain specific characteristics of the semen may reflect the ability of the sperm to fertilize an oocyte.

Semen collection in the dog has been performed using the artificial vagina technique, although most commonly semen is collected by manually stimulating the penis in the presence of a teaser bitch. Using this method, the three fractions can be collected separately into test tubes via funnels. The volume of the canine ejaculate varies according to the size of the dog, but for medium-sized breeds such as Labrador Retrievers, total volumes up to 40 ml have been recorded: the first fraction often averages 0.5–1.0 ml; the second fraction is 0.5–3.0 ml; and the third fraction comprises the remaining 5–35 ml.

It is normal to evaluate the motility of sperm microscopically using a subjective assessment of undiluted sperm maintained at body temperature. The percentage of sperm with normal progressive forward motility is usually between 80 and 90%. Stained samples are used to evaluate sperm morphology and vital (live–dead) staining characteristics. In normal fertile dogs there are <20% morphologically abnormal spermatozoa. Finally, sperm concentration is evaluated in killed diluted samples

placed into a haemocytometer counting chamber, and sperm concentration is multiplied by the volume of the ejaculate to calculate the 'total sperm output' for the ejaculate. The total sperm output is commonly between 300 and 400 million sperm per ejaculate in medium-sized breeds of dog. Further details of semen collection and evaluation for the purposes of investigating infertility and for performing artificial insemination are given in Chapters 8 and 9, respectively.

Semen can be collected successfully from the tom cat using an artificial vagina if the tom cat has been trained previously. Feline semen can also be collected by electroejaculation under anaesthesia, but although a larger volume may be obtained than when using an artificial vagina, there is often contamination of the sample with urine. Typical characteristics of the ejaculate include a total volume of 0.02–0.12 ml and normal forward progressive motility of between 60 and 95%; there are generally <10% morphologically abnormal spermatozoa. Commonly the total sperm output is between 0.15 and 13 million spermatozoa. Further details of semen collection and evaluation for the purposes of investigating infertility are given in Chapter 8.

Spermatozoa
Structurally, a spermatozoan (Figure 2.8) is divided into:

- The head – contains the nucleus and the acrosome (contains the acrosomal enzymes)
- The midpiece – contains the mitochondria required for metabolism
- The tail – propels the sperm via the movement of flagella.

Canine and feline spermatozoa both have round heads that are two-dimensionally flattened. In the dog, the total sperm length is approximately 70 μm, with the head comprising approximately 10% of the length, the midpiece 20% of the length and the tail the remaining 70%.

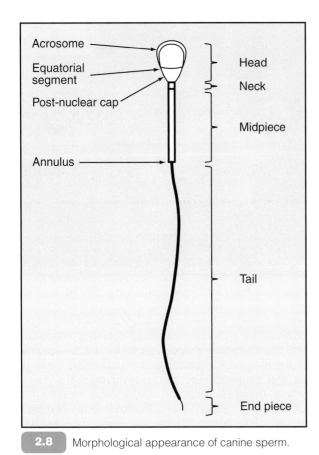

2.8 Morphological appearance of canine sperm.

Labels: Acrosome, Equatorial segment, Post-nuclear cap, Annulus, Head, Neck, Midpiece, Tail, End piece

Fertilization

Once sperm have been ejaculated they become motile but are not fertile. Movement of the sperm through the female reproductive tract is caused by flushing mechanisms (associated with the volume of the ejaculate) combined with contractions of the female reproductive tract and the sperm flagella activity. During this journey, the opportunity arises to form a sperm reservoir within the female reproductive tract (in the bitch this is thought to be via binding to epithelial cells at the tips of the uterine horns), and for the sperm to achieve fertilizing potential when they ultimately come into close association with oocytes. There are three major events that occur during the maturation of sperm to enable fertilizing ability:

• Capacitation
• Acrosomal exocytosis
• Expression of hyperactive motility.

Capacitation is a preparatory phase that appears to be reversible; after a period of time in the female tract a cohort of sperm will be capacitated, awaiting progression to the acrosome reaction, but if suitable triggers are not present they de-capacitate. There is a regular turnover of capacitated sperm so that generally there is always a population available to respond to appropriate triggers. While it is not possible to observe morphological changes associated with capacitation, there are substantial changes in the movement of ions (notably calcium and sodium) across the sperm membranes and it may be possible to track these changes (at least *in vitro*) using particular labelling techniques.

The acrosome reaction appears to be a progression from capacitation and involves an influx of calcium ions and a series of events that culminate in fusion and vesiculation of the outer acrosomal membranes with the plasma membrane, which results in the release of the acrosomal contents, principally an enzyme called acrosin. It is believed that there are mechanisms that regulate the time sequence of events of the acrosome reaction, notably by slowing the possibility of the reaction during sperm transport in the female tract, and enhancing the possibility in the presence of particular triggers such as the zona pellucida or elevated concentrations of progesterone (at least in the dog).

Hyperactive motility normally occurs as sperm approach and attach to the zona pellucida, although a transitional motility type (between normal fast forward progressive motility and hyperactive motility) has been observed, and may be involved in detachment of sperm from epithelial binding if this has occurred. Hyperactive motility is characterized by violent whip-like lateral movement of the sperm tail, resulting in marked side-to-side movement of the sperm head without significant forward motion.

In sperm that have been capacitated and have then undergone the acrosome reaction, the release of acrosin starts the process of 'digesting' the zona pellucida. Combined with the significant head movement that occurs during hyperactive motility, this allows the sperm to enter the perivitelline space of the oocyte and to be available for fertilization. The inner acrosomal membrane is exposed, and fusion between the oolemma and the sperm head begins at the equatorial segment of the sperm head. As the sperm head is incorporated within the oocyte, the sperm chromatin undergoes decondensation within its nucleus. The male and female pronuclei then move towards each other and the nuclear envelopes break down so that male and female chromosomes are released into the ooplasm, and the first cell division is initiated.

References and further reading

Woodall PF and Johnstone IP (1988) Dimensions and allometry of testes, epididymides and spermatozoa in the domestic dog (*Canis familiaris*). *Journal of Reproduction and Fertility* **82**, 603–609

3

Prevention of breeding in the female

Stefano Romagnoli and Hasan Sontas

Introduction

Temporary or permanent control of reproduction in the bitch and the queen is a relatively common request. Owners may require oestrus prevention or suppression for a variety of reasons related to hormone-dependent behavioural or physiological changes occurring in their pets during pro-oestrus or oestrus. These include escape behaviours, roaming, aggression towards other animals and humans, disobedience (working or show dogs), reduced performance (racing dogs), attractiveness to males, vulvar bleeding in the bitch and hypervocalization in the queen. Furthermore, many owned but unsupervised pets are at risk of mismating and producing multiple litters per year, thus increasing the pet overpopulation problem.

Current legislation in some countries (such as the United States) requires that stray animals are euthanased within a short period of time following their arrival at a shelter if they fail one or more key behavioural tests. This regulation results in the destruction of several million animals each year, which is why breeding prevention programmes (such as to spay or neuter as soon as possible) have been started in many countries during the last two or three decades (Kutzler and Wood, 2006). However, public opinion in many European countries and in the USA is becoming more sensitive to welfare issues for companion animals, and as a consequence the usefulness and practical consequences of gonadectomy are currently being questioned by pet owners and veterinary surgeons alike (Society for Theriogenology, 2009).

In dogs and cats, female reproduction can be controlled using surgical or non-surgical methods. Ovariectomy and ovariohysterectomy are the most common surgical approaches; other approaches such as uterine tube ligation have been reported but are not practised as a routine. Minimally invasive methods such as intravaginal or intrauterine devices have been attempted with little if any success. Parenteral administration of steroidal or non-steroidal agents such as progestogens, androgens and gonadotrophin-releasing hormone (GnRH) agonists to obtain a prolonged (albeit temporary) control of reproduction remains the preferred method in most countries. Vaccines against components of the reproductive system such as the zona pellucida are under study and may be available in the future.

The most common approach to oestrus prevention differs among countries. The spectrum of invasive and non-invasive methods allows for different treatment modalities depending on the cultural background of the pet owner, the clinician's preference or factors related to the animal such as age, breed, temperament, intended use, household environment or the social, ethical, economic and regulatory features of each country. For example, in Norway, gonadectomy is considered mutilation under the Animal Welfare Act and only allowed to be performed for medical reasons or in animals in official service. In contrast, surgical sterilization is the most commonly recommended contraceptive method in the USA. In many European countries, both surgical and medical methods are performed and a number of drugs approved for this purpose are commercially available.

The most effective, cheap and non-invasive method for prevention of breeding is confinement of the female during oestrus. However, this is not practical for most owners and generally has already been ruled out when a veterinary surgeon is consulted. In the past, intravaginal devices to block copulation were developed for the bitch. Owing to difficulties in fitting and retention, as well as to reported side effects including perforation of the vaginal wall and vaginal infections, and a high failure rate, these are no longer recommended in bitches. A 'Y'-shaped metallic intrauterine device was marketed commercially for canine contraception in some South American and European countries during the last decade of the twentieth century. The contraceptive effect of such a device is due to the chronic inflammatory reaction induced and to spermicidal activity of the metallic ions released by the electrolytic copper. Although it was proven effective in experimental trials, this 'canine intrauterine coil' has not withstood the test of time. It has been withdrawn from the market in most European countries because of problems such as failure to accomplish removal of the device and uterine perforation. Conception and pregnancy have been reported with the device *in situ*, causing failure to whelp and therefore necessitating a Caesarean operation and ovariohysterectomy.

Medical approach

In small animals, the medical approach to prevention of breeding is based on administration of synthetic analogues of progesterone (progestins or progestogens), synthetic analogues of testosterone (androgens) or long-acting GnRH agonists. Progestogens and androgens are commonly employed for their actions on the canine and feline female reproductive system, leading to suppression of ovarian activity due to the negative feedback on the hypothalamic–pituitary–gonadal axis. A number of progestogens are commercially available; androgens are also available, although the choice is more limited and their use less common. Drugs currently being studied include GnRH antagonists and melatonin.

Progestogens

Synthetic analogues of progesterone (progestins or progestogens) are commonly used in dogs and cats for temporary (starting the treatment shortly before onset of pro-oestrus) or prolonged (starting the treatment in anoestrus) postponement of oestrus, or for suppression of oestrus (starting the treatment after pro-oestrus onset). The following compounds are currently available in different countries:

- Medroxyprogesterone acetate (MPA)
- Megestrol acetate (MA)
- Delmadinone acetate (DMA)
- Chlormadinone acetate (CMA)
- Proligestone (PGS).

From a clinical point of view all these products act in the same way, through a block of the production and/or release of GnRH from the hypothalamus (Romagnoli and Concannon, 2003). Figure 3.1 details the suggested dosages of the most commonly used progestogen-based compounds in the bitch and queen.

Medroxyprogesterone acetate

MPA is a widely available long-acting progestogen. Two types of formulation are available: oral (dogs and cats, for oestrus suppression) and depot (predominately dogs, for prolonged postponement). Given that MPA is metabolized slowly by the liver, effective circulating concentrations can be maintained for 3–6 months after a single intramuscular injection. Various treatment regimens have been recommended for bitches, with the minimum effective parenteral (prolonged postponement) dose being approximately 2.5–3.0 mg/kg, and the minimum effective oral (oestrus suppression) dose being 1.0–2.0 mg/kg/day for 4 days then 0.5–1.0 for 12 days. In queens the minimum effective oral dose is 2.5 mg/week (see Figure 3.1). It should be noted that, while a relevant amount of scientific and clinical data are available for parenteral dosages, little data are available for oral dosages of MPA in bitches and queens. Higher dosages have been suggested, but should be discouraged because of the increasing risk of side effects at MPA doses greater than 3.0 mg/kg in both bitches and queens. MPA should be administered in anoestrus, and the duration of treatment should not exceed 2 years (or four injections at 6-month intervals). In bitches, the interval from treatment to the first spontaneous oestrus ranges from 1.5 to 26 months following the last injection. In some countries MPA is not recommended for use in cats.

Megestrol acetate

MA is a short-acting progestogen which is better suited than MPA to temporary postponement and suppression of oestrus. It is typically available as an oral formulation (although parenteral formulations are available) with dosages varying in the bitch depending on the stage of the oestrous cycle: from 0.55 mg/kg/day for 32 days in anoestrus, up to a maximum dosage of 2.2 mg/kg/day for 8 days in early pro-oestrus. In anoestrous queens MA

Drug	Dosage for bitches	Dosage for queens
Medroxyprogesterone acetate	Parenteral: 2.5–3.0 mg/kg i.m. q6months Oral: 1.0–2.0 mg/kg/day for 4 days then 0.5–1.0 mg/kg/day for 12 days	2.0 mg/kg i.m. q5months
Megestrol acetate	**In pro-oestrus:** ≤2.0 mg/kg/day orally administered for ≤2 weeks **In anoestrus:** 0.5 mg/kg/day orally administered for up to a maximum of 40 days A typical dosage for oestrus suppression is 2.0 mg/kg/day for 8 consecutive days A typical dosage for temporary postponement of oestrus is 0.5 mg/kg/day in late anoestrus	**In pro-oestrus:** 5 mg/cat/day orally for 4 days, then 5 mg/cat orally q2wks **In anoestrus:** 5 mg/cat orally q2wks or 2.5 mg/cat orally q1wk (better if divided into two doses q3.5d)
Delmadinone acetate	0.25–5.0 mg/kg orally q24h <10 kg bodyweight: 1.5–2.0 mg/kg i.m., s.c. q6months 10–20 kg bodyweight: 1.0–1.5 mg/kg i.m., s.c. q6months >20 kg bodyweight: 1.0 mg/kg i.m., s.c. q6months	0.25–0.7 mg/kg orally q1wk 2.5–5.0 mg/kg s.c. q6months For oestrus suppression: 0.5–1.0 mg/kg s.c. for 6 days or 2.5–6.75 mg/kg, one or two injections 24 hours apart
Proligestone	10–33 mg/kg s.c. repeated after months 3 and 4, and thereafter q5months	10 mg/kg s.c. repeated after months 3 and 4, and thereafter q5months

3.1 Suggested dosages of the most commonly used progestogen compounds in bitches and queen for the control of oestrus.

should be given at a dose of 5.0 mg/cat q2wks, and can be increased to 5.0 mg/cat/day for up to 4 days in cases of breakthrough heat. Dosing of injectable formulations of MA should be done carefully based on bodyweight, keeping in mind that no dose–response data are available to demonstrate the most appropriate protocol for injectable MA in small animals. In bitches the interval from last treatment to first spontaneous oestrus ranges from 1 to 7 months.

Delmadinone and chlormadinone acetate
DMA and CMA are synthetic progestogens whose action on the reproductive and endocrine systems is similar to that of MPA. Their effectiveness is reportedly good for prolonged postponement of the oestrous cycle in bitches if injected every 4 months, with occasional breakthrough heats. Suggested dosages for bitches are based on bodyweight (<10 kg = 1.5–2.0 mg/kg; 10–20 kg = 1.0–1.5 mg/kg; >20 kg = 1.0 mg/kg), and for queens the suggested dose is 2 mg/cat (see Figure 3.1). Published information on both drugs is scant. Post-treatment fertility is reportedly normal for both species. The reported interval from last treatment to first spontaneous oestrus is about 8 months for DMA, and is variable but up to 2 years for CMA.

Proligestone
PGS is the most recent type of progestogen available for use in dogs and cats. It has been developed to act selectively on the hypothalamic–pituitary–gonadal axis and has less progestational activity than other synthetic progestogens. However, a range of side effects have been reported for this compound in studies performed in the Netherlands. The recommended dosage regimen is 10–33 mg/kg s.c. repeated after 3 and 4 months, and then every 5 months. Breakthrough heats may occur during treatment, although their incidence has not been studied. In bitches the interval from last treatment to first spontaneous oestrus ranges from 3 to 9 months.

Androgens
In a similar way to progestogens, administration of androgens in the female will suppress gonadotrophin release due to negative feedback upon the hypothalamic–pituitary axis. Furthermore, androgen receptors have been identified in oestrogen target tissues, and androgen administration may cause a decrease in the functional response of the tissues to oestrogen. Natural and synthetic androgens have been used for the control of oestrus in bitches and queens. As with progestogens, treatment with androgens should be started when the animal is in the first half of anoestrus, because delaying their administration until late anoestrus may prove ineffective (Kutzler and Wood, 2006).

Testosterone
Testosterone and its esters, when administered in weekly intramuscular injections (e.g. testosterone propionate at a dose of 110 mg/dog), have been used to prevent oestrus in Greyhounds. A common protocol consists of administering mixed testosterone esters at a dose of 25 mg/kg i.m. q4–6wks. Oral methyltestosterone may suppress oestrus when administered at a daily dose of 0.25–0.5 mg/kg or at a weekly dose of 25 mg/dog. Prolonged anoestrus (up to 1–2 years) may be observed after treatment withdrawal.

Mibolerone
Mibolerone is a synthetic androgen that has been used for prevention of oestrus in the bitch; it is administered orally, once daily. Although it may provide long-term oestrus prevention, for up to 5 years, the use of this compound is not recommended for more than 2 years. The dosage of mibolerone depends on breed and bodyweight. A dose of 30–180 μg/day is recommended for bitches of all breeds with a bodyweight of 12–50 kg, except for German Shepherd Dogs (and their cross-breeds), which should be treated with the highest dose. In the bitch the interval from last treatment to first spontaneous oestrus ranges from 7 to 200 days.

Gonadotrophin-releasing hormone agonists
A recent development in the control of reproduction in the bitch is the use of long-acting GnRH agonists such as deslorelin. The clinical efficacy of deslorelin for inhibiting reproduction in the bitch was evaluated recently using a subcutaneous implant of 4.7 mg deslorelin in a total of 10 adult bitches (healthy or with mammary gland neoplasia for which the owners were requesting suppression of cyclicity without performing gonadectomy) (Romagnoli *et al.*, 2009). The bitches were implanted in anoestrus or dioestrus, and the treatment was repeated every 5 months. Some of the bitches implanted in anoestrus came into heat within 4–15 days following treatment, whilst none of the bitches implanted in dioestrus showed heat during treatment. Suppression of reproductive cyclicity was achieved successfully in 6/10 bitches for 1–4 years. No behavioural, local or generalized side effects were observed in any of the treated bitches. The 4.7 mg deslorelin implant may work well for suppression of cyclicity, provided that it is administered in dioestrus and at intervals of 4.5 months. Based on ongoing studies in adult queens, the 4.7 mg deslorelin implant seems to be effective in inhibiting reproductive cyclicity for at least 12 months, and good efficacy has also been observed in delaying puberty in both tom cats and queens implanted at 2–4 months of age. Deslorelin is currently authorized in Europe for use in male dogs only; therefore, its use in bitches and queens should be regarded as 'off-label'.

Side effects
The side effects of progestogens, androgens and GnRH agonists are summarized in Figure 3.2.

Side effect	Progestogens	Androgens	GnRH agonists
Increased frequency of GnRH peaks			√ [a]
Decreased frequency of GnRH peaks	√	√	√ [b]
Suppression of ovarian activity	√	√	√ [b]
Suppression of uterine motility	√	√	√ [b]
Suppression of adrenocortical axis	√		
Increased secretion of prolactin and growth hormone	√		
Insulin resistance	√		
Endometrial proliferation and secretion	√		√ [c]
Atrophy of endometrial lining		√	
Cervical closure	√	√	√ [d]
Proliferation of mammary parenchyma	√		√ [c]
Atrophy of mammary parenchyma		√	√ (possible)
Lactational arrest	√	√	√ (possible)
Fetal developmental defects	√	√	
Delayed parturition	√		√ [e]
Increased appetite and bodyweight	√	√	
Vaginitis		√	
Increased libido/aggressiveness/mounting behaviour		√	
Anabolic effects		√	
Decreased libido/aggressiveness	√		
Clitoral hypertrophy		√	
Urinary incontinence/urine spraying		√	
Skin discoloration	√	√	
Growth of anal hepatoid glands		√	
Thickening of cervical dermis		√	
Epiphora		√	
Physeal closure in prepubertal animals	√	√	

3.2 Summary of the most relevant side effects of progestogens, androgens and GnRH agonists on the reproductive system, mammary glands, bodyweight and behaviour of bitches and queens. [a] Present only during the first 2–4 weeks following treatment, after which it disappears. [b] Present only after the end of the first 2–4 weeks following treatment. [c] Present only during the first 2–4 weeks following treatment in bitches implanted with deslorelin during dioestrus. [d] This effect is maintained in bitches implanted in anoestrus, while it appears at the end of the induced oestrus in bitches treated in anoestrus. [e] Present if implanted during the last trimester of pregnancy.

Progestogens

The incidence of side effects of progestogens appears to be associated with overdosing, the wrong choice of patient and/or administration during stages of the oestrous cycle other than anoestrus. For instance, while a dose of 2.5 mg/kg of MPA in a young, healthy bitch can probably be used safely for 18–24 months (or for 3–4 injections at 6-month intervals), a dose of 6.0 mg/kg is at risk of causing side effects well before 18 months, and a dose of 10 mg/kg may cause acromegaly and pyometra after an equal or shorter treatment period. Similarly, a dose of 3.0 mg/kg will cause development of benign and malignant mammary gland tumours in bitches if administered every 3 months for 12 consecutive treatments (Romagnoli and Concannon, 2003).

Potential side effects of progestogens

A 'safe' dosage may cause side effects if it is:

- Administered in dioestrus (because of the coincident endogenous secretion of progesterone adding to exogenous administration)
- Given to an animal with a pre-existing disease (such as subclinical cystic endometrial hyperplasia or diabetes)
- Administered to a pregnant female (because it will delay parturition).

The side effects of a normal dose in healthy females include:

- Delayed onset of parturition
- Skin reaction at inoculation sites
- Masculinization of female fetuses (if given inadvertently during pregnancy)
- Behavioural/metabolic changes (e.g. decreased libido, increased appetite and bodyweight, polyuria/polydipsia)
- Reversible adrenocortical suppression in queens (occurs with MA at a dose of 5.0 mg/cat for 7–14 days).

The side effects of a normal dose in females with subclinical disease, a high dose, or prolonged treatment with a normal dose include:

- Benign and malignant mammary gland nodules
- Cystic endometrial hyperplasia and/or pyometra
- Acromegaly
- Diabetes mellitus
- Glucocorticoid effect and Cushing's syndrome.

Although side effects have been reported more commonly with MPA, CMA, DMA and MA than with PGS, it should be noted that the clinical use of gonadal steroidal compounds in dogs and cats started almost 50 years ago, at which time very little was known about the reproductive and endocrine effects of steroids. In contrast, PGS has been introduced to the veterinary market more recently, and is perhaps used with more caution and attention to dosages. Reported side effects of PGS include pain at the injection site and discoloration of the hair. Although the early clinical trials did not report development of uterine disease or mammary gland tumours when PGS was used at the suggested dose regimen, no large clinical studies have been performed in recent years. Therefore, PGS can probably be regarded as slightly safer than MPA. However, small animal clinicians should use the same degree of caution when using any progestogen. Fertility is generally not affected by treatment with progestogens, although the endometrial proliferation induced by these drugs may alter endometrial status, particularly with prolonged treatment.

Androgens

In a similar way to progestogens, androgens may cause side effects if administered at excessively high dosages, for too long or to the wrong patient. Reported side effects of androgens following chronic treatment at normal dosages include:

- Clitoral hypertrophy and vaginitis (most common)
- Virilization
- Masculinization of female fetuses (if given inadvertently during pregnancy)
- Musky body odour
- Urinary incontinence
- Urine spraying
- Mounting behaviour

- Thickening of the cervical dermis
- Epiphora.

Furthermore, overdosing may result in anal gland inspissation and resulting odour. Androgens are known to be much less harmful to fertility owing to the lack of endometrial proliferation, although endometrial atrophy may become chronic following prolonged treatment.

Reproductive side effects following chronic use of mibolerone at efficacious dosages include ovarian fibroma, endometrial atrophy and vaginal mucosa atrophy. Beagles treated with mibolerone appear to have a reduced response to challenge with adrenocorticotropic hormone (ACTH). In general, the use of androgens is not recommended in the following cases:

- Bitches intended for future breeding
- Bitches ≤7 months old, owing to early epiphyseal closure
- Bedlington Terriers, owing to their genetic predisposition to chronic progressive hepatitis
- Pregnant animals
- Bitches with androgen-dependent neoplasms
- Animals with a history of liver or kidney disease.

Mibolerone is not recommended in cats because the effective dose (50 µg/day) is close to the toxic dose (60 µg/day).

Gonadotrophin-releasing hormone agonists

Agonists of GnRH appear to be much less harmful to the reproductive and general health of bitches and queens compared with gonadal steroids, because their action is based on the removal of endogenous steroids from the general circulation, thereby eliminating any potential risk of hormone-related conditions. However, these compounds may be a source of problems in terms of both owner convenience and health risks for the animal being treated.

Agonists of GnRH provide a powerful stimulus for the pituitary gland, which is forced to release all its reserves of gonadotrophins until downregulation occurs. The interval from treatment to the onset of downregulation may vary from 2 to 4 weeks. During this period ovarian activity and secretion are typically stimulated, which may cause induction of oestrus if the female is treated in anoestrus, or a surge in progesterone secretion if treatment occurs in dioestrus. Oestrus induction can be a major inconvenience for an owner who is requesting control of oestrus for their pet. The oestrus is normally fertile; therefore, a bitch may conceive and become pregnant, but spontaneous abortion can occur in some cases when pituitary gland downregulation occurs. The surge in progesterone production may exacerbate a subclinical uterine or ovarian condition, as well as delay parturition if treatment is performed during the last trimester of pregnancy. Induction of heat is not observed when bitches are treated in dioestrus or re-treated in anoestrus (when the previous implant is still functioning), although breakthrough heats have been observed within the last month prior to re-treatment.

As with all treatments that are supposed to be repeated at regular intervals, owner compliance is an important limiting factor. Failure of owners to return their bitches or queens for successive treatments often accounts for breakthrough oestrus. Treatment during a breakthrough oestrus should be avoided, because administration of any reproductive drug during the follicular phase is likely to carry a risk for the female.

New drugs

Recently, GnRH antagonists and melatonin have been experimentally used with some success to control reproduction in dogs and cats. Although promising, clinical efficacy of these drugs is still being investigated, and therefore their use in the prevention of female breeding cannot currently be advocated.

Melatonin

In seasonal breeding species such as the queen, melatonin secreted from the pineal gland is responsible for the timing of reproductive activity. Secretion of melatonin from the pineal gland is regulated by light through a nervous pathway connecting the retina to the pineal gland. Leyva *et al.* (1989) demonstrated that the cat is very responsive to changes in photoperiod, with a longer duration of darkness causing increased and prolonged release of both melatonin and prolactin. Whilst a longer photoperiod has a positive effect on folliculogenesis in the queen, such an effect can be temporarily suppressed by melatonin administration. A short photoperiod is highly suppressive of ovarian activity in queens. Secretion of melatonin and prolactin are strictly related as both hormones tend to increase during darkness (Leyva *et al.*, 1989).

Recently, a subcutaneous implant containing 18 mg of melatonin, inserted following sedation and local anaesthesia, has been tested for the control of reproduction in queens (Gimenez *et al.*, 2009). Three to eleven days after implantation 33–78% of the queens displayed behaviour and vaginal cytological changes consistent with oestrus. However, none of the queens displayed receptiveness to the tom cat. The duration of the interoestrous interval was approximately 2 and 4 months when the queens were implanted during interoestrus and oestrus, respectively.

Gonadotrophin-releasing hormone antagonists

GnRH antagonists are substances that block GnRH receptors in a competitive manner at the level of the pituitary gland. Unlike agonists, GnRH antagonists display an immediate action following treatment. A single dose of a GnRH antagonist causes a decrease to very low concentrations of gonadotrophins and gonadal steroids within two hours (Vickery *et al.*, 1989). Due to their immediate inhibitory effect on the gonadal axis, GnRH antagonists have been used for oestrus control in both cats and dogs.

Clinical considerations for safe drug use

The ideal candidate for progestogen or androgen treatment is an adult postpubertal female in anoestrus, whilst the ideal candidate for treatment with a GnRH agonist is an adult postpubertal female in dioestrus. Before a female is treated with long-acting compounds she should be evaluated for normal uterine and mammary gland anatomy and physiology, as well as glucose metabolism. A minimum database of clinical information should be gathered prior to administering a long-acting compound and include:

- A thorough reproductive history to assess whether or not oestrus occurred within the last 1–2 months
- A complete clinical examination
- Palpation of the mammary glands to rule out the presence of mammary nodules
- A vaginal smear to rule out the presence of oestrus (Figure 3.3)
- Serum biochemistry to check for liver function and glucose metabolism.

Dosing should always be based on bodyweight, avoiding those guidelines which divide dogs into two categories (<15 kg and >15 kg) or consider all cats as weighing 5 kg, as lighter animals always end up receiving higher dosages.

In general, progestogen or androgen treatment for 12–18 months is considered adequate in most patients, although longer treatments can also be 'safe' provided that the female is given a thorough clinical check prior to each treatment. Whilst most bitches and queens may tolerate treatment periods

Stage	Pro-oestrus	Oestrus	Early dioestrus	Late dioestrus/anoestrus
Vaginal smear	All vaginal cell types present: cornified cells (<50%), neutrophils, blood cells Figure 3.3a	Cornified cells only present Figure 3.3b	Large intermediate vaginal cells in abundance along with neutrophils Figure 3.3c	Isolated groups of small intermediate cells present Figure 3.3d
Treatment	**Do not treat**	**Do not treat**	Treatment with GnRH agonists	Treatment with androgens (first half of anoestrus only) Treatment with progestogens Gonadectomy
Duration of stage	2 weeks	2 weeks	1 week	7 weeks (late dioestrus) 12–20 weeks (anoestrus)

3.3 Timing for use of reproductive drugs or performing ovariectomy in the bitch, based on endocrine and vaginal cytology patterns. (continues) ▶

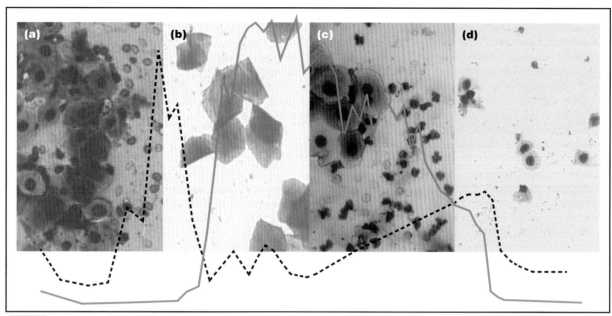

3.3 (continued) Timing for use of reproductive drugs or performing ovariectomy in the bitch, based on endocrine (oestradiol = black dashed line; progesterone = green line) and vaginal cytology patterns.

Condition	Causes	Comments
Prepubertal queens (depot compounds)	Mammary gland hyperplasia	It is best to use a short-acting oral compound (e.g. MA) initially for 1–2 weeks and then change to a long-acting compound once potential side effects have been ruled out
Pregnancy	Fetal development defects Delayed parturition	
Pseudopregnant bitches	Pseudopregnancy to relapse and become chronic	
Dioestrus	Side effects due to overdosing	The stage of the oestrous cycle should always be identified using vaginal cytology/serum progesterone assay; the bitch/queen is best treated during anoestrus. Dioestrus should be ruled out in queens: 30% of queens ovulate spontaneously, maintaining thereafter a dioestrus lasting 30–45 days
Vaginal haemorrhage	Worsening of underlying condition	Prolonged sanguineous vulvar discharge following parturition in the bitch can be a critical problem which should be treated either with a uterine contractive drug (e.g. ergonovine) or surgery. Mild bloody vulvar discharge can be caused by uterine neoplasia, cystic endometrial hyperplasia with superimposed endometrial inflammation, pyometra or metritis; none of these conditions will benefit from progestogen administration
Diabetes	Worsening of underlying condition	Recommended that blood glucose levels are measured before/after a prolonged treatment to confirm health status with regard to glucose metabolism
Bitches with prolonged heat	Worsening of underlying condition	Prolonged heat may be caused by ovarian cysts or granulosa cell tumours; may also be a split heat or a misinterpretation of normal signs of oestrus by the owner. For none of these is progestogen treatment indicated

3.4 Conditions (and potential side effects) for which progestogens are not recommended.

of more than 12–18 and up to 24 months, animals with a pre-existing disease such as subclinical diabetes, small mammary gland nodules or cystic endometrial hyperplasia may have their condition worsen rapidly as a result of a progestogen treatment, or may experience a temporary worsening if treatment with a GnRH agonist is given in dioestrus. Conditions for which progestogens are not recommended are summarized in Figure 3.4.

Surgical approach

Surgical gonadectomy refers to the removal of the gonads under general anaesthesia. In the female,

gonadectomy (spaying) is generally performed by removal of both ovaries (ovariectomy) or of the ovaries and uterus (ovariohysterectomy) via an abdominal approach. Whether or not bitches and queens should be spayed, the choice of procedure, and the best age at which to perform surgery are questions that need to be addressed, and each option has advantages and disadvantages.

Advantages of gonadectomy

Ovarian removal is associated, in both bitches and queens, with a reduced risk of mammary gland and uterine diseases (neoplasia, pyometra, mastitis), as well as the absence of ovarian diseases (tumours, cysts), progesterone-related diseases (false pregnancy, feline mammary gland hypertrophy), oestrogen-related diseases (vaginal hyperplasia/prolapse, persistent oestrus, bone marrow aplasia), pregnancy-related diseases (unwanted pregnancy, pregnancy complications, abortion, dystocia, uterine prolapse, subinvolution of placental sites) and parturition-related diseases (dystocia, uterine prolapse, subinvolution of placental sites).

Bitches that are gonadectomized prior to puberty have a 95% reduction in the risk of developing mammary gland tumours compared with bitches spayed after the first heat (8% risk), after the second heat (26% risk), or with bitches spayed after 2.5 years of age or left intact (100% or full risk) (Egenvall *et al.*, 2002). A similar reduction in the risk of developing mammary gland neoplasia has been confirmed for spayed queens (Overley *et al.*, 2005). From a behavioural point of view it is commonly believed that spayed bitches and queens, apart from not showing oestrus-related behaviour, have a more relaxed, somewhat lazy attitude. Gonadectomy may also contribute to reducing pet overpopulation.

Disadvantages of gonadectomy

Surgical gonadectomy is an irreversible technique, which may be a problem for some owners who would like to have progeny from their pets at a later date. The disadvantages of gonadectomy in bitches and queens include generic surgical risk, a few behavioural abnormalities, obesity and, in bitches, urinary incontinence and osteoporosis (Burrow *et al.*, 2005) (Figure 3.5).

Generic surgical risk

A number of complications of gonadectomy have been well documented, and their incidence varies depending on whether ovariectomy or ovariohysterectomy is performed (Figure 3.6). There are some unsubstantiated associations between the incidence of complications and: the age of the animal; the ability of the veterinary surgeon; the presence of reproductive diseases; the breed. The complication observed most frequently seems to be vaginal or intra-abdominal haemorrhage, which is much more common in bitches of large (80% of all complications) as opposed to small (20% of all complications) size.

Side effect	Percentage occurrence in the bitch	Percentage occurrence in the queen
Generic surgical risk	7–27	33
Obesity	25–50	50–75
Urinary incontinence	5–12	–
Osteoporosis	Reported (under experimental conditions; has not been observed to occur spontaneously)	–
Behavioural changes	Reported	Reported

3.5 Side effects and percentage occurrence as a result of gonadectomy in the bitch and queen. The incidence of each condition is calculated for the total number of spayed animals.

	Procedure	Incidence	Reference
Short-term complications			
Anaesthetic problems	Ovariohysterectomy	3–41% 3% cats	Dorn and Swist, 1977; Berzon, 1979
Haemorrhage of ovarian/uterine pedicle due to incomplete ligature or rupture of blood vessels (bitch, especially if spayed during heat)	Ovariohysterectomy [a] Ovariectomy	2–79% 4% cats	Berzon, 1979; Janssens and Janssens, 1991
Vaginal haemorrhage	Ovariohysterectomy [a] Ovariectomy	2–15% 1% cats	Pearson, 1973; Berzon, 1979
False pregnancy	Ovariohysterectomy Ovariectomy	13%	Okkens *et al.*, 1981b
Suture dehiscence/infection/abscess/oedema	Ovariohysterectomy Ovariectomy	2.7–25% 45% cats	Dorn and Swist, 1977; Berzon, 1979; Janssens and Janssens, 1991
Peritonitis/evisceration	Ovariohysterectomy Ovariectomy	0.3–2% 1% cats	Berzon, 1979
Long-term complications			
Ovarian remnant syndrome	Ovariohysterectomy Ovariectomy	17–43%	Pearson, 1973; Okkens *et al.*, 1981b
Ligature of ureters/ureterovaginal fistula	Ovariohysterectomy	2–3%	Okkens *et al.* 1981a
Granuloma of the uterine/ovarian pedicle	Ovariohysterectomy Ovariectomy	6–28%	Pearson, 1973; Okkens *et al.* 1981a
Stump granuloma	Ovariohysterectomy	15–51%	Pearson, 1973; Okkens *et al.* 1981b

3.6 Incidence of short- and long-term complications attributable to ovariectomy or ovariohysterectomy. [a] The risk of complications is higher following ovariohysterectomy than ovariectomy. (continues) ▶

	Procedure	Incidence	Reference
Long-term complications (continued)			
Fistula due to use of non-absorbable suture material	Ovariohysterectomy	38%	Pearson, 1973
Cotton gauze in abdomen	Ovariohysterectomy	Reported	Mai *et al.*, 2001
Infection of the uterine stump	Ovariohysterectomy	35%	Okkens *et al.* 1981b
Obesity	Ovariohysterectomy	0–50% 14% cats	Dorn and Swist, 1977; Edney and Smith, 1986; Howe *et al.*, 2000
Urethral sphincter mechanism incompetence	Ovariohysterectomy	3–20%	Okkens *et al.*, 1997
	Ovariectomy	9–21%	Arnold *et al.*, 1989; Okkens *et al.* 1997
Intestinal and peritoneal adhesions	Ovariohysterectomy	5.5%	Pearson, 1973
Ovarian remnant syndrome and granulosa cell tumour	Ovariohysterectomy	Reported	Pluhar *et al.* 1995
Ovarian remnant syndrome and uterine leiomyoma	Ovariohysterectomy	Reported	Sontas *et al.* 2010
Ovarian remnant syndrome and vaginal neurofibroma	Ovariohysterectomy	Reported	Sontas *et al.* 2009

3.6 (continued) Incidence of short- and long-term complications attributable to ovariectomy or ovariohysterectomy. [a] The risk of complications is higher following ovariohysterectomy than ovariectomy.

Vulvar blood loss may occur following ovariohysterectomy (blood from the uterine pedicle, the suspensory ligament or the broad ligament) or following ovariectomy (blood from the ovarian pedicle). Some complications may be due to the stage of the reproductive cycle in which the surgery is performed. In bitches intraoperative bleeding is more common during pro-oestrus and oestrus (due to high oestrogen concentrations), while false pregnancy may result when surgery is performed during dioestrus (due to rising prolactin concentrations following a sudden fall in progesterone). The onset of oestrus after gonadectomy in bitches and queens is a well know surgical complication, and is attributable to an ovarian remnant. It may occur more frequently following ovariohysterectomy than ovariectomy because the abdominal incision tends to be more caudal with ovariohysterectomy, therefore making it more difficult to reach the ovaries.

Ureteral obstruction, resulting from the inclusion of the ureters in the ligature, or the development of a ureterovaginal fistula have been reported only following ovariohysterectomy. Granulomas of the ovarian (less common) or uterine (more common) pedicle may be due to the use of non-absorbable suture material, while inflammation of the uterine pedicle may occur if the most caudal part of the uterus is accidentally caught in the suture during ovariohysterectomy performed because of pyometra. Pyometra may develop following ovariectomy if an ovarian remnant is left in place and the bitch or queen resumes cycling following surgery, or if progestogens are administered later in life.

Obesity

An increase in bodyweight following gonadectomy is observed in bitches and is particularly common in queens. Gonadectomized bitches have twice the risk of obesity of intact bitches. Food intake increased significantly in bitches during the first 90 days following ovariectomy/ovariohysterectomy in comparison with bitches who received a sham laparotomy. Appetite increases significantly in gonadectomized bitches 6 months, and in gonadectomized cats 3 days, after surgery.

Urinary incontinence

Decreased capacity of the external urethral sphincter is observed following gonadectomy in bitches, but not in queens. The average urethral closure pressure in intact bitches is 18.6 ± 10.5 cmH$_2$0; 12 months following gonadectomy it is 10.3 ± 6.7 cmH$_2$0 in continent bitches, and it may fall to 4.6 ± 2.3 cmH$_2$0 in incontinent animals. Signs of urinary incontinence (urine loss during sleep or recumbency) may appear from 2 weeks to 9 years after surgery, and may typically be corrected by the administration of oestrogen- or alpha adrenergic based drugs. In those cases that do not respond to medical treatment, surgery may be considered.

Orthopaedic disease

There are some data which suggest that there is an increased risk of rupture of the cranial cruciate ligament in animals that are neutered at an early age. In addition, surgical neutering has been identified as a risk factor for developing clinical signs of hip dysplasia (this may be mediated by increased weight in the neutered animal). Loss of trabecular bone is the most important complication of menopause in women and is thought to be due to a lack of oestrogen stimulation, causing reduced secretion of calcitonin. Loss of trabecular bone has been observed in Beagle bitches 11 months following ovariohysterectomy, although at present it is unclear whether this bears any clinical significance, presumably because of the short canine lifespan.

Behavioural changes

Recent data from behavioural studies indicate that gonadectomy may exacerbate dominance in bitches, regardless of their attitude prior to surgery. A significant increase in the degree of reactivity has been observed in German Shepherd Dog bitches 5 months following ovariectomy. Owners should be advised about the importance of evaluating the behaviour of their bitches prior to deciding whether to neuter them (Overall, 2007).

Other risks

Some additional risks may be associated with surgical neutering, although the mechanism of action has not been elucidated. For example, there appears to be an increased risk of certain haemangiosacromas and of bladder and urethral tumours, although some of these risks may be balanced by other tumours having a greater risk in entire animals. Some endocrine diseases may also be more prevalent in neutered animals, and whilst the specific risk of hypothyroidism is not proven, there is reasonable evidence of an increased risk of diabetes mellitus, although this may be mediated by obesity (as noted previously for hip dysplasia).

Timing of surgery

As with the administration of progestogens and androgens, the ideal stage of the oestrous cycle at which to perform gonadectomy is anoestrus. This is particularly true for bitches, because when such surgery is performed in pro-oestrus or oestrus there is an increased risk of short-term postsurgical complications, whilst when it is performed in dioestrus there is an increased risk of false pregnancy. In bitches gonadectomy is best performed 2.5–5.5 months following the onset of pro-oestrus; an assay of serum progesterone may be of help in ruling out dioestrus when an appropriate history is not available. Although no specific study has ever looked at the best stage for performing gonadectomy in cats, the queen is not known to experience any such problems in relation to the stage of the oestrous cycle in which she is gonadectomized.

Ovariectomy or ovariohysterectomy?

Whether to remove only the ovaries or also the uterus is a true dilemma for many veterinary surgeons. One school of thought recommends the removal of the ovaries and uterus, based on the concept that what is removed cannot cause disease. However, another approach is to remove only the ovaries because the uterus quickly undergoes atrophy following ovariectomy, and the risk of developing urinary incontinence may be lower for anatomical reasons.

Research performed in Utrecht (Okkens *et al.*, 1997) in which 138 bitches that had undergone ovariohysterectomy and 126 that had undergone ovariectomy were followed up over 8–11 years, has clearly shown that, between the two groups of dogs: a) there is no difference in the short-term or long-term surgical complications; b) there is no difference in the incidence of cystic endometrial hyperplasia, pyometra or any other uterine disease; and c) there is no difference in the incidence of urinary incontinence. In fact the only differences between the two approaches are the degree of invasiveness and the duration of the surgery (and therefore the duration of anaesthesia),

all of which are increased in the case of ovariohysterectomy. This causes a higher risk of surgical complications and greater stress for the animal, as well as having cost implications for the owner.

Okkens *et al.* (1997) conclude that 'there is no indication to remove also the uterus in elective castration procedures of healthy bitches, and therefore ovariectomy is to be considered the procedure of choice.' This view is shared by a number of authors (Johnston *et al.*, 2001; Van Goethem *et al.*, 2006; Whitehead, 2006), and many clinicians currently consider that it is unethical to perform ovariohysterectomy instead of ovariectomy, unless there are specific health reasons (Society for Theriogenology, 2009).

Although such data are available only for bitches, it is likely that the same situation is also true for queens. In patients with specific indications for ovariohysterectomy, it is advisable not to remove the cervix because of its important role in isolating the abdominal cavity from the external environment, even in the neutered female. Hysterectomy alone should not be performed because bilateral development of ovarian cysts has been reported anecdotally, with bitches having to undergo repeat laparotomy to remove the cystic ovaries. In addition, if a previously hysterectomized bitch of small size is mated by a large dog, the vaginocervical suture may rupture, with consequent development of coital peritonitis (Becker *et al.*, 1974).

Prepubertal gonadectomy

Prepubertal gonadectomy is defined as the surgical sterilization of immature animals aged from 6–14 weeks. Apart from ensuring that the animal will never be able to reproduce after puberty, gonadectomy is a less invasive, less traumatic surgical procedure when performed prior to puberty. The incidence of short-term postsurgical complications is lower in immature than in adult animals, and most of the complications are minor problems such as swelling of the abdominal suture. Complications seem to be less common the younger the age at which ovariectomy is performed (Howe, 1997).

Long-term complications of prepubertal gonadectomy include immaturity of the external genitalia, urinary incontinence and cystitis, abnormal growth, and behavioural changes. The presence of an infantile vulva is reported in bitches, with infolding vulvar lips increasing the risk of vulvar dermatitis. The risk of obesity does not change in animals that are gonadectomized before puberty. It is unknown whether there is a predisposition to obesity, but if this is the case gonadectomy seems to have the same stimulus regardless of age, although neutering as an adult may be less of a stimulus for obesity compared with neutering early in life. The risk of urinary incontinence increases in bitches gonadectomized prior to 5 months of age, and even more so if it is performed prior to 3 months of age. In addition, the risk of cystitis increases in bitches neutered prior to 5 months of age (Spain *et al.*, 2004a).

An increase in the length of the long bones occurs in all dogs and cats that undergo prepubertal gonadectomy. In addition, a decrease in the diameter of the pre-pelvic urethra and a 30% decrease in basal metabolic rate were observed in queens castrated at 7 weeks *versus* 7 months. The effects of prepubertal gonadectomy on behaviour vary depending on species and sex. In dogs there is increased aggressiveness and barking towards family members and strangers, increased fear of noise and increased sexual behaviour, but decreased separation anxiety and decreased urination in the home. In cats castrated prior to 5 months of age, there is a decrease in activity and an increase in shyness towards strangers (Spain *et al.*, 2004ab).

References and further reading

Arnold S, Hubler M and Reichler I (2009) Urinary incontinence in spayed bitches: new insights into the pathophysiology and options for medical treatment. *Reproduction in Domestic Animals* **44** (Suppl. 2), 190–192

Becker RL, Giles RC and Hildebrandt PK (1974) Coitally induced peritonitis in a dog. *Veterinary Medicine Small Animal Clinician* **10**, 53–54

Berzon JL (1979) Complications of elective ovariohysterectomies in the dog and cat at a teaching institution: clinical review of 853 cases. *Veterinary Surgery* **8**, 89–91

Burrow R, Batchelor D and Cripps P (2005) Complications observed during and after ovariohysterectomy of 142 bitches at a veterinary teaching hospital. *Veterinary Record* **157**, 829–833

Dorn AS and Swist RA (1977) Complications of canine ovariohysterectomy. *Journal of the American Animal Hospital Association* **13**, 167–171

Edney ATB and Smith PM (1986) Study of obesity in dogs visiting veterinary practices in the United Kingdom. *Veterinary Record* **118**, 391–396

Egenvall A, Bonnett P, Ohagen P *et al.* (2002) Incidence of and survival after mammary tumors in a population of over 80,000 insured female dogs in Sweden from 1995 to 2002. *Preventive Veterinary Medicine* **69**, 109–127

Gimenez F, Stornelli MC, Tittarelli CM *et al.* (2009) Suppression of estrus in cats with melatonin implants. *Theriogenology* **72**, 493–499

Howe LM (1997) Short-term results and complications of prepubertal gonadectomy in cats and dogs. *Journal of the American Veterinary Medical Association* **211**, 57–62

Howe LM, Slater MR, Boothe HW *et al.* (2000) Long-term outcome of gonadectomy performed at an early age or traditional age in cats. *Journal of the American Veterinary Medical Association* **217**, 1661–1665

Jackson EKM (1984) Contraception in the dog and cat. *British Veterinary Journal* **140**, 132–137

Janssens LAA and Janssens GHRR (1991) Bilateral flank ovariectomy in the dog: surgical technique and sequelae in 72 animals. *Journal of Small Animal Practice* **32**, 249–252

Johnston SD, Kustritz MVR and Olson PNS (2001) Prevention and termination of canine pregnancy. In: *Canine and Feline Theriogenology*, ed. SD Johnston *et al.*, pp. 168–179. WB Saunders, Philadelphia

Keskin A, Yilmazbas G, Yilmaz R, Ozyigit MO and Gumen A (2009) Pathological abnormalities after long-term administration of medroxyprogesterone acetate in a queen. *Journal of Feline Medicine and Surgery* **11**(6) 518–521

Kutzler M and Wood A (2006) Non-surgical methods of contraception and sterilisation. *Theriogenology* **66**, 514–525

Leyva H, Madley T and Stabenfledt G (1989) Effect of light manipulation on ovarian activity and melatonin and prolactin secretion in the domestic cat. *Journal of Reproduction and Fertility Supplement* **39**, 125–133

Mai W, Ledieu D, Venturini L *et al.* (2001) Ultrasonographic appearance of intra-abdominal granuloma secondary to retained surgical sponge. *Veterinary Radiology and Ultrasound* **42**, 157–160

Martin LJM, Siliart B, Dumon HJW and Nguyen P (2006) Spontaneous hormonal variations in male cats following gonadectomy. *Journal of Feline Medicine and Surgery* **8**, 309–314

Okkens A, Dielman S and van de Gaag I (1981b) Gynaecological complications following ovariohysterectomy in dogs, due to partial removal of the ovaries or inflammation of the uterocervical stump. *Tijdschrift voor Diergeneeskunde* **106**, 1142–1158

Okkens AC, Kooistra HS and Nickel RF (1997) Comparison of long-term effects of ovariectomy *versus* ovariohysterectomy in bitches. *Journal of Reproduction and Fertility* (Suppl) **51**, 227–231

Okkens AC, van de Gaag I, Biewanga WJ, Rothuizen J and Voorhout G (1981a) Urological complications following ovariohysterectomy in dogs. *Tijdschrift voor Diergeneeskunde* **106**, 1189–1198

Overall KL (2007) Working bitches and the neutering myth: sticking to science. *The Veterinary Journal* **173**, 9–11

Overley G, Shofer FS, Goldschmidt MH *et al.* (2005) Association between ovariohysterectomy and feline mammary carcinoma. *Journal of Veterinary Internal Medicine* **19(4)**, 560–563

Pearson H (1973) The complications of ovariohysterectomy in the bitch. *Journal of Small Animal Practice* **14**, 257–266

Pluhar GE, MA Memon and LG Wheaton (1995) Granulosa cell tumor in an ovariohysterectomized dog. *Journal of the American Veterinary Medical Association* **207**, 1063–1065

Romagnoli S and Concannon PW (2003) Clinical use of progestins in bitches and queens: a review. In: *Recent Advances in Small Animal Reproduction*, ed. PW Concannon *et al.*, International Veterinary Information Service (www.ivis.org), Ithaca, NY. Document number A1206.0903

Romagnoli S, Stelletta C, Milani C *et al.* (2009) Clinical use of deslorelin for the control of reproduction in the bitch. *Reproduction in Domestic Animals* **44**, 36–39

Society for Theriogenology (2009) Position statement on mandatory spay-neuter; Basis for position on mandatory spay-neuter in the canine and feline. http://www.therio.org/displaycommon. cfm?an=1&subarticlenbr=191 (Accessed April 2009)

Sontas BH, Altun D, Guvenc K, Arun S and Ekici H (2009) Vaginal neurofibroma in a hysterectomized poodle dog. *Reproduction in Domestic Animals* doi: 10.1111/j.1439-0531.2009.01497.x

Sontas BH, Özyogurtcu H, Turna Ö, Arun S and Ekici H (2010) Uterine leiomyoma in a spayed poodle bitch: a case report. *Reproduction in Domestic Animals* doi: 10.1111/j.1439-0531.2008.01289.x

Spain CV, Scarlett JM, Houpt KA (2004a) Long-term risks and benefits of early-age gonadectomy in dogs. *Journal of the American Veterinary Medical Association* **3**, 380–387

Spain CV, Scarlett JM, Houpt KA (2004b) Long-term risks and benefits of early-age gonadectomy in cats. *Journal of the American Veterinary Medical Association* **3**, 372–379

Van Goethem D, Shaefers-Okkens A and Kirpensteijn J (2006) Making a rational choice between ovariectomy and ovariohysterectomy in the dog: a discussion of the benefits of either technique. *Veterinary Surgery* **35**, 136–143

Vickery BH, McRae GI, Goodpasture JC and Sanders LM (1989) Use of potent LHRH analogues for chronic contraception and pregnancy termination in dogs. *Journal of Reproduction and Fertility Supplement* **39**, 175–187

Whitehead M (2006) Ovariectomy versus ovariohysterectomy. (Letter) *Veterinary Record* **159**, 723–724

4

Prevention of breeding in the male

Michelle Kutzler

Introduction

There are an estimated 200 million cats kept as pets worldwide and in many countries, including the United States, the United Kingdom and China, pet cats outnumber pet dogs. It is estimated that within the UK, feral cats kill approximately 100 million birds and mammals each year. Unwanted dogs and cats act as reservoirs or vectors of transmissible diseases to humans and to economically valuable domestic species. Dogs are the major vector of rabies worldwide and account for more than 55,000 human deaths associated with disease each year. Free-roaming dogs are a source of ecological and social problems, attacking other animals and people, causing road accidents, frightening the public and creating fecal and urine contamination. Although it is generally accepted that females are more important from a population control point-of-view, a single intact male may produce more offspring in one year than a single female is capable of producing in a lifetime. The Humane Society of the United States estimates that, each year, between 8 and 10 million dogs and cats enter US shelters of which 4–5 million are euthanased as no homes can be found.

In 1992, the American Humane Association put forth a resolution, stating 'no dog or cat adopted from a shelter should be allowed to reproduce'. The American Veterinary Medical Association put forth a similar resolution in the following year, voting to endorse the concept of prepubertal gonadectomy surgery in dogs and cats. The American Animal Hospital Association and the American College of Theriogenologists have also approved position statements supporting early spay–neuter surgery. Despite this, in the USA, the number of new litters of fertile dogs and cats born each day vastly exceeds the delivery system for surgical sterilization methods, resulting in excess numbers of unwanted animals.

Castration is also used in many pet households to suppress certain male behaviours, such as urine marking in tom cats and territorial aggression in dogs. In the UK, castration of tom cats before the age of 6 months is recommended to all owners. The opinions on the routine castration of dogs are more divided. In some countries (e.g. Scandinavia) where stray dogs are not a problem, routine surgical neutering is considered mutilation and is illegal. In Norway only 7% of dogs are neutered and canine overpopulation is not a problem.

The health benefits from neutering are that testicular tumours and many prostatic diseases are prevented. However, there are less well known benefits for entire animals such as a decreased risk of haemangiosarcomas, osteosarcomas, transitional cell carcinomas and prostatic adenocarcinomas. Intact dogs also have fewer obesity problems and associated diseases such as diabetes mellitus.

Castration is performed far more frequently in tom cats than in dogs. The pungent urine odour from intact tom cats may account for the higher percentage of castrated cats. An alternative to surgery is treatment with gonadotrophin-releasing hormone (GnRH) agonists or reproductive steroid hormones (progestogens or androgens), which results in negative feedback, effectively shutting down the hypothalamic–pituitary–testicular axis. Immunocontraception via vaccination against GnRH and luteinizing hormone (LH) are also possible. The current problem with hormonal downregulation and immunocontraception is that both of these methods are reversible and have to be given repeatedly, making them less suitable for large-scale population control. Intratesticular, intraepididymal or intra-vas deferens injections, or other methods for mechanical disruption of fertility (including testicular ablation using ultrasound waves), provide methods for non-surgical sterilization of the male dog and cat. New methods of contraception are currently under development that inhibit either the production of sperm (spermatogenesis) or its function by targeting motility, orientation, binding or fusion with the ovum (Figure 4.1). These new technologies may be a panacea for pet overpopulation. However, currently within the USA, castration (orchidectomy) remains the procedure of choice for permanent sterilization of male companion animals.

Medical approach

Hormonal downregulation

GnRH acts as the master reproductive hormone through regulation of the release of LH and follicle-stimulating hormone (FSH) from the pituitary gland. In males, LH regulates testosterone synthesis; FSH is necessary for the initiation and maintenance of spermatogenesis. Testosterone is needed for spermatogenesis and for the development of secondary sexual characteristics, including behavioural characteristics such as territorial marking (spraying),

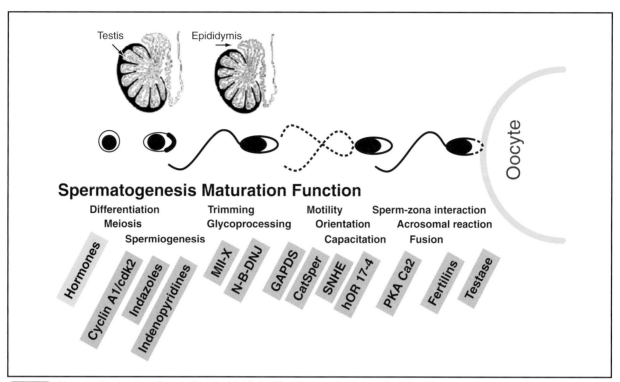

Spermatogenesis Maturation Function

Differentiation	Trimming	Motility	Sperm-zona interaction
Meiosis	Glycoprocessing	Orientation	Acrosomal reaction
Spermiogenesis		Capacitation	Fusion

Hormones · Cyclin A1/cdk2 · Indazoles · Indenopyridines · MII-X · N-B-DNJ · GAPDS · CatSper · SNHE · hOR 17-4 · PKA Ca2 · Fertilins · Testase

Oocyte

4.1 New methods of contraception that inhibit either the production or function of sperm are currently under development. Many potential targets have been identified along the pathway from spermatogonial stem cells to fertilization. A partial list associated with the putative stage of the pathway is shown. (Reproduced from Blithe (2008) with permission from *Contraception*)

mounting and aggressiveness. Hormonal downregulation is a temporary method for suppressing fertility in breeding animals (Figure 4.2). Sustained exposure to GnRH reduces GnRH-stimulated gonadotrophin secretion through GnRH receptor downregulation, internalization and signal uncoupling. Exogenous steroid hormones suppress fertility indirectly via inhibition of pituitary gland gonadotrophin secretion and release, mainly that of LH. A summary of hormones that are commercially available for downregulation of the hypothalamic–pituitary–testicular axis is listed in Figure 4.3.

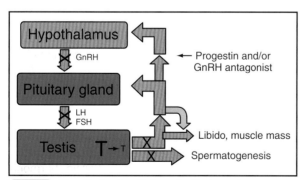

4.2 Hypothalamic–pituitary–testicular axis. Spermatogenesis requires a high intratesticular concentration of testosterone (T). Addition of a progestogen or a GnRH agonist results in cessation of GnRH production, which inhibits production of LH and FSH. In the absence of gonadotrophins, intratesticular and serum testosterone decline to levels that are insufficient to support spermatogenesis and libido. (Redrawn after Blithe (2008) with permission from *Contraception*)

Contraceptive	Delivery
Progestogens	
Megestrol acetate	Oral
Medroxyprogesterone acetate	Oral, i.m.
Melengestrol acetate	Implant
Levonorgestrol	Implant
GnRH analogues	
Deslorelin	Implant
Leuprolide acetate	i.m. depot
Chemical	
Bisdiamine WIN 18,446	Oral

4.3 Products that are commercially available in some countries. (Data from Munson (2006) with permission from *Theriogenology*)

Gonadotrophin-releasing hormone agonists

Several studies have examined the use of the GnRH superagonist, deslorelin, as a male contraceptive in dogs. Subcutaneous administration of a 4.7 mg slow-release implant of deslorelin reduces plasma LH and testosterone concentrations to undetectable levels within 4 weeks and produces subsequent infertility within 6 weeks. The treatment-induced effects on fertility are completely reversible. Testosterone and LH concentrations and semen quality return to normal by 60 weeks after implant administration. The threshold concentration of deslorelin necessary for suppressing spermatogenesis in male dogs is >0.25 mg/kg of bodyweight. A long-acting deslorelin implant is available commercially in the UK and most European countries. The manufacturer claims that this product will result in contraception for at least 6 months in 98% of male dogs. Serial administration of multiple implants at 6-month intervals did not result in adverse effects or diminished efficacy. Deslorelin has also proven to be safe and effective in male cats, resulting in spermatogenic arrest and reduction of testosterone concentration.

In addition to deslorelin, daily subcutaneous injections of nafarelin (2 µg/kg/day) decrease basal testosterone concentrations and result in infertility within 3 weeks after the onset of treatment. Within 8 weeks of treatment discontinuation, normal fertility is restored. Administration of a single subcutaneous injection of leuprolide acetate (1 mg/kg) to intact male dogs decreases ejaculate volume, increases morphologically abnormal spermatozoa and significantly decreases serum testosterone and LH concentrations for 6 weeks. Return to normal spermatogenesis occurs 20 weeks after treatment discontinuation. Leuprolide treatment is also effective in male cats. Subcutaneous administration of buserelin implants (6.6 mg) to intact male dogs decreases testosterone concentrations to basal values and results in infertility within 3 weeks of implant administration. Male contraception occurs for an average of 233 days.

Progestogens

Spermatogenesis is regulated by FSH and LH. Using the principles of negative feedback, gonadotrophin secretion should be suppressed in males given exogenous progestogens, resulting in a disruption in spermatogenesis. Intact tom cats treated with the oral progestogen megestrol acetate (MA) have a decreased libido and reduced urine spraying and aggression.

MA is metabolized rapidly when given orally, and has a half-life of 8 days in the dog. In male dogs, daily oral treatment with MA (2 mg/kg) for 7 days produced no change in semen quality, whereas higher doses (4 mg/kg) produced only minor secondary sperm abnormalities. It is important to note that a daily dose of 2.2 mg/kg bodyweight orally is effective at suppressing oestrus in bitches when treatment is initiated at the beginning of pro-oestrus. The reason for the different response occurring between genders is not clear.

Reported side effects of prolonged MA treatment in tom cats include fibroepithelial mammary gland hyperplasia and mammary gland carcinoma. In one study, over 33% of tom cats with mammary gland neoplasia had a history of progestogen therapy. Additional side effects in dogs and cats include increased appetite leading to weight gain, lethargy or restlessness, and clinical and pathological changes typical of diabetes mellitus. MA is not approved for use in cats.

Medroxyprogesterone acetate (MPA) is a long-acting injectable progestogen that has also been used to suppress oestrus in the bitch and queen, but to a more limited extent compared with MA owing to the high incidence of side effects. Subcutaneous administration of MPA (20 mg/kg) to male dogs produced a rapid response (within 3 days) and significant decrease in sperm motility, morphology and output. Because of the rapidity of the response, the author (England, 1997) postulated that the effect was mediated by the direct action of progestogens on spermatozoal maturation and transport in the epididymis. Semen quality was not adversely affected when MPA was given at dosages of 4 mg/kg or 10 mg/kg. Subcutaneous administration of MPA in dogs has resulted in clinical signs consistent with adrenocortical suppression (e.g. alopecia, hair discoloration, thinning of the skin and mobilization of subcutaneous fat). Although licensed in some countries MPA is generally not recommended for use in tom cats.

Prolactin

Prolactin is a protein hormone secreted by the anterior pituitary gland. In humans, hyperprolactinaemia resulting from a pituitary gland adenoma leads to oligo- or azoospermia. Intramuscular administration of exogenous prolactin (600 mg/kg of bodyweight q1wk for 6 months) to male dogs results in severe asthenozoospermia, teratozoospermia and oligospermia or azoospermia within 6 weeks of treatment. At the end of 3 months of treatment in dogs, degenerative changes within the seminiferous tubules are evident on testicular biopsy. Within 3 months of drug withdrawal, the sperm count returns to normal, and mating produces normal pregnancy. It is interesting to note that serum concentrations of testosterone, LH and FSH do not change significantly, which suggests that prolactin treatment may be having a direct effect on the testes rather than functioning via hormonal downregulation.

Androgens

Subcutaneous administration of 5 mg/kg of testosterone esters (testosterone proprionate = 0.6 mg/kg, testosterone phenylpropionate = 1.2 mg/kg, testosterone isocaproate = 1.2 mg/kg, testosterone decanoate = 2.0 mg/kg) to male dogs results in a significant decline in spermatozoal motility within 3 weeks after treatment, which persists for 3 months. Daily oral administration of 50 mg of methyltestosterone to male dogs for 90 days resulted in decreased daily output of spermatozoa (England, 1997). Chronic administration of danazol, a synthetic derivative of 17-alpha-ethinyl testosterone, resulted in loss of the spermatogenic elements and azoospermia (Dixit, 1977). These effects were reversible within a period of 60 days. Other than a transient elevation in serum aminotransferase, hepatic function was unaltered during danazol treatment. Another androgen in development is MENT™ (7-alpha-methyl-19-nortestosterone), which can be delivered continuously via a subcutaneous implant. This drug has a high binding affinity for androgen receptors but does not undergo 5-alpha reduction in the prostate gland, so there is no risk of severe hyperplastic changes as can occur with other androgen treatments. It is important to mention that testicular responsiveness to exogenous androgens may vary with genetics or geography.

Anti-androgens

Anti-androgens are compounds that are able to block the effect of androgens on their target cells, by inhibiting binding to the androgen receptor. However, in contrast to androgens, the receptor–anti-androgen complex is unstable and short-lived, and androgen-dependent gene transcription and protein synthesis are not stimulated. Currently available anti-androgens include both steroidal (cyproterone acetate and chlormadinone acetate) and non-steroidal (flutamide and nilutamide) compounds. Flutamide binds weakly to the androgen receptor. In contrast to flutamide, nilutamide has a long plasma half-life and thus high concentrations can be achieved in the target tissue, leading to a prolonged inhibition of androgen binding. In dogs, administration of either steroidal or non-steroidal anti-androgens results in only a slight, transient influence on spermatogenesis.

Immunosterilization

Over the past two decades, efforts have been made to develop a vaccine that can suppress fertility in male canids and felids. Several targets of immuno-contraception have been identified, such as GnRH, LH and sperm antigens. The physiological effects of antibody titres against GnRH include suppression of reproductive behaviour, suppression of synthesis and secretion of gonadotrophins and steroid hormones, gonadal atrophy and the associated arrest of gametogenesis. Antibody production against LH lowers serum testosterone concentrations in males, whereas antibody production against spermatozoal antigen prevents fertilization. However, the individual immune response is pivotal to ensuring an immunosterilant event. An ideal immunocontraceptive vaccine would have a high margin of safety for treated animals and the environment, be effective in a high percentage of treated animals, have a rapid onset and long duration of activity following a single treatment, inhibit sex hormone production, be efficacious in both dogs and cats, and be simple to deliver in the field.

Gonadotrophin-releasing hormone immunization

Development of GnRH vaccines for immunocontraception is problematic for several reasons. Native GnRH is not naturally immunogenic: it is a small decapeptide hormone that is well conserved throughout all mammalian species. Under normal conditions, GnRH is recognized by the immune system as self (allogenic). Consequently, administration of a vaccine derived from native GnRH results in no antibody production or a short-lived, weak response because the animal is tolerant to its own hormones. However, if GnRH is altered in such a way that induces its recognition as a foreign material (e.g. by coupling it with another molecule with many antigenic determinants), an immunoglobulin G (IgG) response will occur. Conjugation to keyhole limpet haemocyanin (KLH), ovalbumin, more potent adjuvants or other amplifying agents is essential to produce durable titres. Depending on the presentation and the adjuvant, GnRH vaccines can last for 1–2 years.

Molecules of GnRH have been conjugated to a number of different antigens to mobilize T-helper cells.

Vaccination with a fusion protein composed of canine GnRH and the T-helper cell epitope p35, which originates from canine distemper virus F protein, is strongly immunogenic and results in loss of testicular function in dogs. Vaccination of male dogs with a fusion protein of GnRH conjugated to tetanus toxoid has similar effects, which are reversible once the antibody titres wane. However, vaccination of male dogs with N-terminal modified GnRH conjugated to tetanus toxoid does not result in infertility.

A GnRH vaccine was commercially available in the USA for use in male dogs until 2009. The authorized application of this vaccine was for the treatment of benign prostatic hyperplasia in dogs. Both testosterone concentrations and prostate gland size decreased following vaccination. No significant systemic reactions or adverse events were observed over a 14-day post-vaccination observation period. This vaccine is currently being tested in male cats by the author. One experimental GnRH vaccine developed for male dogs (a fusion protein of GnRH conjugated to tetanus toxoid) resulted in a similar antibody response in cats to that seen dogs but did not result in infertility in cats.

Other experimentally derived GnRH vaccines have demonstrated efficacy in male cats. In one study, a glycine moiety was added to the native GnRH molecule at the C terminus as a spacer, and a cysteine molecule was added to ensure consistent alignment of the peptide to the activated protein carrier KLH. The aqueous-based GnRH–KLH conjugate was combined in a 1:1 ratio by volume with a novel adjuvant (AdjuVac®), which is an oil-based modified USDA-licensed Johne's disease vaccine that contains small quantities of killed *Mycobacterium avium*. No inflammation or tenderness was detected at the injection site in any of the vaccinated cats (Levy *et al.*, 2004). The GnRH vaccination resulted in sterility in male cats that exceeded the length of the study period (6 months) (Figure 4.4). This GnRH/KLH/AdjuVac® vaccine currently has an APHIS/USDA patent-pending status.

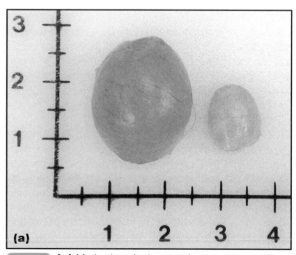

4.4 **(a)** Marked testicular atrophy was present 6 months after GnRH vaccination of a cat (right) compared to sham-treated cat (left). The scale is in centimetres. (Reproduced from Levy *et al.* (2004) with permission from *Theriogenology*) (continues) ▶

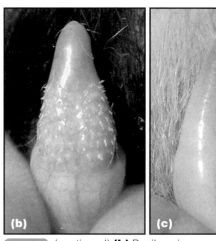
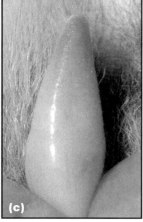

4.4 (continued) **(b)** Penile spines were still well developed in the sham-treated cat. **(c)** Penile spines were absent in the GnRH vaccinated cat. (Reproduced from Levy *et al.* (2004) with permission from *Theriogenology*)

Luteinizing hormone immunization

Reproductive function is severely impaired in male dogs immunized against LH for up to 1 year. However, this has not been tested in tom cats. A direction that has not yet been explored is immunization against human chorionic gonadotrophin (hCG), which has many structural similarities to LH. Treatment of dogs and cats with hCG results in antibody formation. Utilizing this humoral response to suppress fertility immunologically is an area that deserves further investigation.

Spermatozoal antigen immunization

Sperm antigens are an excellent target for contraceptive vaccines because these proteins are viewed as 'foreign' by the immune system. However, the entire spermatozoan cannot be used for vaccine development because it shares several antigens with other somatic cells. Therefore, research has focused on delineating appropriate sperm-specific epitopes that would increase vaccine immunogenicity specifically within the reproductive tract and efficacy. Lactate dehydrogenase (LDH-C4) sperm-specific antigen (SP-10) and acrosin have been isolated as two of the main sperm-specific antigens. To date, immunization against sperm antigens has not resulted in satisfactory control of fertility.

Chemical castration

Chemical castration (intratesticular, intraepididymal or intra-vas deferens injections) is another non-surgical approach to male contraception. Two reasons cited by many dog owners for opposing surgical and other non-surgical methods of castration are anthropomorphic empathy regarding emasculation, and a fear that desired behaviours such as protection and hunting instinct will be decreased following androgen withdrawal. Preservation of testicular tissue and testosterone production in conjunction with sterilization can be obtained using chemical castration, which may increase the cultural acceptance. Chemical agents injected into the testis, epididymis

or vas deferens cause infertility by inducing azoospermia. The technique is inexpensive, relatively easy to perform, and suitable for large-scale sterilization programmes. Sedation is typically used for this procedure in dogs and general anaesthesia is necessary for cats. The scrotal area is first clipped and disinfected. Then, a 22-gauge (dogs) or 27-gauge (cats) 1.25 cm needle is inserted percutaneously into the structure of interest (testis, epididymis or vas deferens). However, the tail of the epididymis and vas deferens in cats is smaller and more difficult to locate than that of the dog.

Injection into the testis

Intratesticular injections have been investigated as a method of inducing aspermatogenic orchitis and male contraception for more than five decades. The procedure for intratesticular injection involves inserting the needle at the caudal pole of the testis and gently pushing it towards the other pole, depositing the injected material homogenously as far as possible through the tissue. Injection of an adjuvant agent, such as Freund's complete adjuvant (FCA) or bacillus Calmette Guérin (BCG), directly into the testis incites a local inflammatory reaction which enables lymphoid cells to gain access to the testicular tissue, resulting in an autoimmune response. A single intratesticular injection of FCA or BCG (10–25 IU) results in severe oligospermia or azoospermia without granuloma formation or the development of circulating anti-sperm antibodies. The few spermatozoa that may be present are immotile. Infertility occurs within 6 weeks and lasts for several months. Intratesticular injection of high doses of FCA or BCG (>75 IU) results in a severe granulomatous reaction. In addition to using bacterial cell wall products, reproductive toxins can be injected directly into the testis. Intratesticular administration of a 100 mg solution (0.5 ml total volume per testis) of methallibure, dexamethasone, metopiron, niridazol, alpha-chlorohydrin or danazol causes testicular and epididymal atrophy and azoospermia in dogs.

In 2003 the US Food and Drug Administration (FDA) approved a zinc gluconate solution neutralized to pH 7 by arginine for chemical castration via intratesticular injection in the male dog. Zinc is considered to be non-mutagenic, non-carcinogenic and non-teratogenic. It is authorized for use only in prepubertal dogs (3–10 months old) with testes measuring 10–27 mm in width, but has been used off-label in large adult dogs. The procedure involves injecting a predetermined amount of zinc solution (based on scrotal width) into each testis. The needle should be changed following filling of the syringe and then inserted into the dorsocranial portion of the testis beside the head of the epididymis. The centre of the testis is the indicated target for deposition of zinc gluconate. Histopathological findings within 2.5 months of injection demonstrate almost complete fibrosis of the seminiferous tubules and Leydig cells.

It is important to note that intratesticular injection of a 70% glycerol solution does not result in azoospermia and sterility in dogs. Scrotal swelling and tenderness are common in the first few days following injection and usually resolve without treatment. However, scrotal ulcers or draining tracts in the scrotum or preputial area can occur 4–6 days after treatment. In one study, necrotizing injection site reactions occurred in 3.9% (4/103) of dogs following intratesticular administration of the zinc gluconate solution (Levy *et al.*, 2008). One of these dogs chewed away a portion of his scrotum and testis. All four dogs with severe reactions were large mature dogs that had received a dose of the drug at the upper end of the range (three dogs received 0.8 ml/testis and one dog received 1.0 ml/testis). Proper injection technique is critical when zinc gluconate is administered intratesticularly because leakage or injection into non-target tissues can result in severe tissue damage. The expense of the product is another reason limiting its use in large sterilization campaigns.

Injection into the epididymis
As an alternative to intratesticular injection, zinc arginine can be injected into the tail of the epididymis. Intraepididymal injection of 50 mg of zinc arginine (0.5 ml per testis) results in azoospermia within 90 days of injection. Histological examination of the testes reveals normal seminiferous tubules with atrophy of the rete testes, an absence of sperm within the epididymis and ductus deferens and no sperm granuloma formation. It is important to note that intraepididymal injection of saline does not induce azoospermia.

Injection of a sclerosing agent (3.5% formalin solution in phosphate buffered saline or 1.5% chlorhexidine gluconate in 50% DMSO) into the tail of both epididymides in dogs results in irreversible azoospermia via chemical occlusion, with secondary testicular atrophy. However, intraepididymal treatment with formalin alone induces only temporary azoospermia or oligospermia in treated dogs. A single bilateral intraepididymal injection of chlorhexidine in DMSO in male dogs results in the development of azoospermia by 91 days after treatment. Two bilateral intraepididymal injections of chlorhexidine in DMSO result in the development of azoospermia within 35 days of treatment.

Studies using these models of male contraception report no or minimal signs of discomfort observed following injection, with variation depending on the route of administration and agent injected. Afferent nerve endings associated with pain sensation are located on the scrotal skin and in the capsule of the testis rather than within the testicular and epididymal parenchyma. A transient increase in testicular diameter may follow within 24 hours of the injection, resulting in pain secondary to swelling, and this may persist for up to 7–15 days. Additional local and systemic reactions reported after intratesticular injections include scrotal ulceration and dermatitis, scrotal self-mutilation, preputial swelling, vomiting, diarrhoea, anorexia, lethargy and leucocytosis. It is important to note that, depending on the treatment method, dogs may remain fertile for up to 60 days post-injection owing to the reserves of sperm present in the epididymis. Also, unlike surgical castration, this kind of chemical sterilization does not eliminate gonadal sources of testosterone.

Intraepididymal injection induces chemical vasectomy in tom cats. Injection of a 4.5% solution of chlorhexidine digluconate into the tails of both epididymides results in azoospermia or severe oligospermia. Unlike the dog, sperm granulomas and spermatocoeles are consistently observed in cats following intraepididymal injections. In addition, transient scrotal swelling and pain may persist for up to 2 weeks following intraepididymal injection in cats.

Injection into the vas deferens
A single injection directly into the vas deferens with sclerosing chemical agents (10% silver nitrate, 3.6% formaldehyde in ethanol, 5% potassium permanganate, 100% ethanol or 3.6% formaldehyde) results in irreversible infertility, similar to that induced by intraepididymal injections of the same agents.

Occlusive silicone casts of the vas deferens have been tested in cats as an alternative to surgical vasectomy. A clear polymer gel made of styrene maleic anhydride mixed with DMSO has been injected into the vas deferens; the gel then solidifies. Sperm outflow is blocked and local inflammatory reactions caused by the presence of exuded sperm at the site of the plug placement lead to irreversible scarring.

Miscellaneous

Ultrasound
Ultrasound has been used to suppress spermatogenesis through a combined effect of heat and mechanics. The testes of male dogs and cats were treated with a high-intensity focused ultrasound beam. Each treatment consisted of 1–2 W/cm^2 for 10–15 minutes administered 1–3 times at intervals of 2–7 days. Ultrasound wave treatment suppressed spermatogenesis significantly without affecting testosterone concentrations. In separate studies, a burst (20–120 seconds) of ultrasound energy (3–19 W) was focused on to the vas deferens or epididymides of anaesthetized dogs. The ultrasound waves induced thermal coagulative necrosis of subsurface structures, resulting in luminal occlusion within 2 weeks after treatment. However, skin burns occurred in approximately 20% of cases (Roberts *et al.*, 2002ab).

Reproductive toxins
In addition to targeting GnRH using immunocontraception, toxins conjugated to GnRH can be used to disrupt the hypothalamic–pituitary–gonadal axis. Toxins must be carefully selected to be safe in other tissues without GnRH receptors. The internalization of GnRH receptors following ligand binding localizes the cytotoxic effects to pituitary gland gonadotroph cells. Pokeweed antiviral protein has been conjugated with GnRH and administered to intact male dogs as three daily injections (100 µg/kg). Serum concentrations of testosterone and LH and testicular volume decrease after treatment, and the effects of male contraception persist for 5–6 months. No adverse effects were noted other than transient (<24 hour) arthralgia in a few dogs (Sabeur *et al.*, 2003).

Other reproductive toxins used for male contraception include ketoconazole, embelin and alpha-chlorohydrin. Ketoconazole is an inhibitor of cell division and has been shown to exert spermatostatic effects in several species, including dogs, rabbits, monkeys and humans. Within 4–24 hours of oral administration of ketoconazole (50–246 mg/kg) to male dogs, the motility of ejaculated sperm declined at an accelerated rate compared with control samples from the same animals (Vickery *et al.*, 1985). The decline in sperm motility was correlated to the presence of ketoconazole in the seminal plasma. In addition, serum testosterone concentrations were profoundly suppressed following oral ketoconazole treatment. At high dosages, ketoconazole is poorly tolerated by the gastrointestinal tract and can cause hepatoxicity. However, similar spermatostatic effects occur following treatment with other orally administered 1-substituted imidazole compounds in dogs, without gastrointestinal and hepatic side effects.

Embelin (*Embelia ribes*) is an indigenous benzoquinone plant used in Asia for the prevention of pregnancy. Oral treatment with embelin (80 mg/kg q48h for 100 days) in male dogs caused a significant decrease in testicular weight and variable degrees of spermatogenic arrest, mainly at the spermatocyte state (absence of post-meiotic cells) (Dixit and Bhargaua, 1983). Within 8 months following embelin ingestion, normal spermatogenesis returns. Treatment with embelin did not result in any adverse effects on serum biochemistry or liver histology.

Alpha-chlorohydrin is an alkylating agent that causes depletion of spermatogenic elements from the seminiferous tubules. A single high dose (70 mg/kg) of alpha-chlorohydrin or chronic administration (8 mg/kg bodyweight for 30 days) inhibits spermatogenesis within 33 days in dogs. These effects are reversible within 100 days following treatment.

Bisdiamine compounds (amoebicidal drugs) target the male germinal epithelium and appear to have no harmful systemic effects. A safety and efficacy trial in tom cats demonstrated that 150 mg/kg of the bisdiamine WIN 18,446 administered daily in food causes complete and reversible spermatogenic arrest in all treated animals without damage to the spermatogonia (Munson, 2006). Although bisdiamines appear to be safe and effective in male cats, they can induce serious teratogenic effects in pregnant females. The cost of bisdiamines is currently prohibitive for most veterinary uses.

Surgical approach

For permanent irreversible fertility control with minimal side effects, surgical sterilization is recommended. Methods of anaesthetic management and surgical sterilization of male dogs and cats vary depending upon patient age (paediatric *versus* mature), testis location (cryptorchid *versus* descended) and technique (e.g. orchidectomy, vasectomy).

Although neutering is one of the oldest surgical procedures described in domestic animals, there is very little information in the literature that establishes the ideal age at which the surgery should be performed. Veterinary surgeons are rightfully concerned about the long-term health risks associated with early gonadectomy. Using voiding cystograms, no difference has been found in the diameters of the preprostatic or penile urethra in cats neutered at 7 weeks or 7 months old when compared with age-matched intact cats. However, male dogs gonadectomized at 7 weeks old had smaller penile diameters, decreased radiodensity and size of the os penis, and immature preputial development compared with male dogs gonadectomized at 7 months old or left intact. The immature preputial development increases the risk of paraphimosis secondary to a shorter prepuce relative to the length of the penis (personal observation). Obesity is a multifactorial, long-term consequence of orchidectomy. However, age at gonadectomy does not increase the predisposition towards obesity. The benefits and disadvantages associated with orchidectomy, irrespective of age at surgery, are listed in Figure 4.5.

	Incidence (%)	Substantial morbidity?	Specific breeds at risk
Benefits			
Reduced incidence of testicular neoplasia	0.9	No	None
Reduced incidence of benign prostatic hyperplasia or prostatitis	75–80 (by 6 years of age)	No	None
Adverse consequences			
Complications of surgery	6.1	No	None
Prostatic neoplasia	0.2–0.6	Yes	None
Transitional cell carcinoma	<1	No	Airedale Terrier; Beagle; Collie; Scottish Terrier; Shetland Sheepdog; West Highland White Terrier; Wire Fox Terrier
Osteosarcoma	0.2	Yes	Dobermann; Great Dane; Irish Setter; Irish Wolfhound; Rottweiler; St Bernard
Haemangiosarcoma	0.2	Yes	Boxer; English Setter; German Shepherd Dog; Golden Retriever; Great Dane; Labrador Retriever; Pointer; Poodle; Siberian Husky

4.5 Potential consequences of gonadectomy in male dogs. (Data from Root-Kustritz (2007) with permission from the *Journal of the American Veterinary Medical Association*) (continues) ▶

	Incidence (%)	Substantial morbidity?	Specific breeds at risk
Adverse consequences continued			
Cranial cruciate ligament rupture	1.8	Yes	Akita; American Staffordshire Terrier; Chesapeake Bay Retriever; German Shepherd Dog; Golden Retriever; Labrador Retriever; Mastiff; Neapolitan Mastiff; Newfoundland; Poodle; St Bernard
Obesity	2.8	Yes	Beagle; Cairn Terrier; Cavalier King Charles Spaniel; Cocker Spaniel; Dachshund; Labrador Retriever
Diabetes mellitus	0.5	No	Airedale Terrier; Cocker Spaniel; Dachshund; Dobermann; Golden Retriever; Irish Setter; Miniature Schnauzer; Pomeranian; Shetland Sheepdog

4.5 (continued) Potential consequences of gonadectomy in male dogs. (Data from Root-Kustritz (2007) with permission from the *Journal of the American Veterinary Medical Association*)

Routine castration

Dogs

In dogs castration may be performed by either an open or a closed tunic technique; with both techniques the testis is displaced cranially and accessed using a midline prescrotal skin incision. Male dogs are considered to be 'scrotal conscious', so the scrotum itself is not clipped or prepared, and the scrotum is draped out of the surgical field to avoid postoperative self-mutilation. The open tunic technique is preferred in larger dogs because ligatures may be placed directly around the vascular pedicle, resulting in more secure ligation. The advantages of the closed tunic technique are that it is simpler to perform, and the parietal vaginal tunic has not been opened, thereby minimizing the risk of peritoneal contamination via the communication between the abdomen and the parietal vaginal tunic. The disadvantage of the closed tunic technique is that the ligatures are less secure because the vessels are ligated whilst being surrounded by the tunic and the cremaster muscle, rather than being ligated directly.

Puppies: Paediatric puppy castration is performed with modifications to the techniques used in adult dogs. The basic principles of paediatric surgery include minimizing surgery time and tissue handling as well as controlling intraoperative bleeding meticulously. As puppy testes are mobile and can be difficult to identify, careful palpation must be used to determine whether both testes have descended into the scrotal region before inducing anaesthesia. Clipping and surgical preparation of the scrotum does not result in scrotal irritation in puppies because the paediatric scrotum is not as well developed. However, to reduce the risk of hypothermia, clipping of hair at the surgery site should be minimized and warmed surgical scrub solution should be used. Puppies may be castrated using a single midline prescrotal or scrotal incision. With a midline incision, the testes must be held securely underneath the incision site to prevent iatrogenic penile trauma. The incision may be closed using one or two buried interrupted sutures in the subcuticular layer, or incisions may be left open to heal by secondary intention. However, closure of the incision prevents postoperative contamination with urine and faeces,

and prevents extrusion of fat from the incision. Polydioxanone suture material should be avoided because it may cause calcinosis circumscripta.

Cats

Cats may also be castrated via an open or closed tunic approach. The closed technique is generally preferred because of the inability to create and maintain a high degree of surgical asepsis in the tom cat. The spermatic cord may be ligated using suture material or, preferably, may be ligated using the spermatic cord itself by placing an overhand knot in the cord. This avoids placement of exogenous foreign material into a potentially contaminated environment. Often after either technique the skin is not routinely closed.

Kittens: Castration in kittens is performed using similar techniques to those in the adult cat, including the use of two separate scrotal incisions to approach the testes. However, care must be taken if tying the spermatic cord to itself due to the fragility of the paediatric tissues. As with puppies, the testes of kittens are extremely small, highly mobile and occasionally difficult to stabilize in the scrotal region in preparation for the incision.

Postoperative complications

Postoperative complications vary with technique but include scrotal bruising and swelling, haemorrhage and infection at the surgery site. In the dog, swelling and bruising of the scrotum are more commonly seen after open castration. Severe haemorrhage after castration could result in a scrotal haematoma or intra-abdominal haemorrhage, requiring intensive supportive care and emergency laparatomy to locate and ligate the spermatic cord. Scrotal haematoma, if severe, may necessitate scrotal ablation.

Anaesthesia

Anaesthetic complications are a risk associated with any surgical sterilization technique. Anaesthetic concerns unique to paediatric animals include:

- Altered metabolism and excretion of drugs due to immature hepatic enzyme systems, decreased protein binding of drugs and decreased glomerular filtration rate

- Predisposition to hypoglycaemia with fasting due to decreased glycogen stores secondary to smaller liver size and skeletal muscle mass
- Decreased ability to maintain body temperature due to a low percentage of body fat and decreased ability to shiver.

Readers are referred to the *BSAVA Manual of Canine and Feline Anaesthesia and Analgesia* (Holden, 2007) for further information.

Laparascopic cryptorchid castration

Cryptorchidism is only occasionally seen in cats but is relatively common in dogs, with the reported range of incidence in both species being 0.8–9.8%. Cats of certain breeds may be predisposed to cryptorchidism, and Persian cats are over-represented. The overall incidence of right-sided compared with left-sided unilateral cryptorchidism and of unilateral cryptorchidism compared with bilateral cryptorchidism is not well documented in dogs and cats. Approximately half of cryptorchid testes have an abdominal location. However, fat or inguinal lymph nodes are often mistakenly diagnosed as inguinal testes on palpation. Inaccurate localization of a cryptorchid testis can lead to increased complications in the patient caused by unnecessary exploration of the inguinal region and abdomen.

When performing reproductive surgery, laparoscopy is a minimally invasive alternative to laparotomy. Prior to insufflation of the abdomen with carbon dioxide, the urinary bladder should be emptied via manual expression or catheterization. For cryptorchid castrations, a trocar (10–12 mm for large dogs, 3–5 mm for small dogs and cats) is inserted, through a small stab incision in the skin, into the abdominal cavity through a right para-umbilical approach. Cryptorchid testes are located by tracing the ductus deferens cranially from the inguinal ring to the testis, or by tracing the testicular vessels from the caudal aspect of the kidney to the testis. After the testis is identified, an additional trocar is placed ventral and slightly lateral to the testis. Transillumination of the abdominal wall can be used to avoid blood vessels during trocar placement. Haemostasis of the spermatic cord in large dogs is achieved with endoscopic clips or an Endoloop™. In small dogs and cats, haemostasis can be achieved with bipolar cauterization because the vessels and ductus deferens are smaller. The testis is then removed from the abdomen through one of the cannulae.

A variation of this technique is a combined laparoscopy–laparotomy. The testis is identified laparoscopically (as described above) but the deferent duct is grasped with forceps or a spay hook and brought to the abdominal wall. A small laparotomy incision is made over the testis and it is removed from the abdomen before ligation (Figure 4.6). This technique removes some of the need for special laparoscopy equipment and may also reduce surgery time. Compared with laparotomy alone, laparoscopy results in less postoperative pain, fewer intra-abdominal adhesions, fewer wound complications and reduced tissue trauma. However, laparoscopy requires expensive surgical equipment and special training. Also, there are anaesthetic risks from complications caused by intraperitoneal pressures over 15 mmHg.

Vasectomy

Vasectomy involves bilateral removal and/or occlusion of a portion of the vas deferens, rendering the animal infertile by preventing sperm from being ejaculated during copulation. Male secondary sex characteristics and behaviours, as well as androgen-dependent diseases, are not prevented as androgens are still produced. Vasectomy of dominant males has been suggested as a method to control the feral cat population because vasectomized dominant tom cats may prevent submissive, intact tom cats from inseminating non-spayed females. In addition, vasectomized males that mate with non-spayed females induce a 45-day pseudopregnancy, which

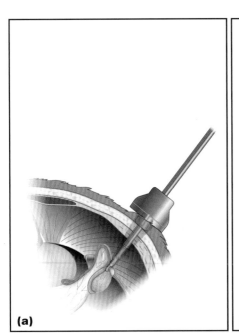

(a)

(b)

4.6 Laparoscopic-assisted cryptorchidectomy. The animal has been tilted slightly to the contralateral side to facilitate visualization of the retained testis. **(a)** The testis retained in the abdomen is grasped with 5 mm toothed grasping forceps and elevated to the abdominal wall. **(b)** The testis is exteriorized for ligation and transection of the vascular supply and ductus deferens. (Redrawn after Miller *et al.* (2004) with permission from the *Journal of the American Veterinary Medical Association*) decreases the number of oestrous cycles per year in the female. In the UK vasectomy is rarely performed as it is an elective surgical procedure with no clear benefit to the animal itself.

Following induction of general anaesthesia, the inguinal region is prepared for surgery. A 1–3 cm incision is made over the spermatic cord as it traverses the scrotum to the inguinal canal. Caudal traction placed on the scrotum will put tension on the spermatic cord, facilitating identification. The cord is dissected from the surrounding fat and fascia and a small incision is made through the parietal vaginal tunic. The testicular vessels lie in one fold of the visceral vaginal tunic, whilst the ductus deferens and its accompanying vessels lie in the other fold. The ductus deferens is isolated and the deferential artery and vein are dissected from the surface of the duct. The duct is clamped, ligated and divided. Recanalization has been reported when the duct is simply severed. The parietal vaginal tunic is closed with a fine absorbable suture and the skin is closed with an absorbable suture in a subcuticular pattern. The surgery is repeated on the opposite side.

Laparascopic intra-abdominal vasectomy may be performed, with occlusion of a segment of the ductus deferens using bipolar forceps and electrocoagulation. Vasectomy in the dog has also been described using a Vasocclude® device to apply a clip through a small scrotal puncture site. In the dog and cat, low numbers of sperm have been reported for up to 21 and 49 days post-vasectomy, respectively. Irrespective of the vasectomy method used, following the procedure the epididymis will distend with sperm, resulting in a spermatocoele. Extravasated epididymal fluid may induce a chronic granulomatous inflammatory reaction (sperm granuloma). Increased intraluminal pressure within the rete results in irreversible damage to the seminiferous cell population and testicular degeneration. Up to 33% of vasectomized humans experience chronic scrotal discomfort, and recurrent scrotal dermatitis (presumably associated with discomfort) has been reported in dogs following vasectomy.

References and further reading

Blithe D (2008) Male contraception: what is on the horizon? *Contraception* **78**, S23–S27

Dixit VP (1977) Chemical sterilization: effects of danazol administration on the testes and epididymides of male rabbit. *Acta Biologica et Medica Germanica* **36**, 73–78

Dixit VP and Bhargava SK (1983) Reversible contraception like activity of embelin in male dogs (*Canis indicus Linn*). *Andrologia* **15**, 486–494

England GC (1997) Effect of progestogens and androgens upon spermatogenesis and steroidogenesis in dogs. *Journal of Reproduction and Fertility* **Suppl. 51**, 123–138

Fayrer-Hosken R (2008) Controlling animal populations using anti-fertility vaccines. *Reproduction in Domestic Animals* **43**(Suppl. 2), 179–185

Holden D (2007) Paediatric patients. In: *BSAVA Manual of Canine and Feline Anaesthesia and Analgesia, 2nd edition*, ed. C Seymour and T Duke-Novakovski, pp. 296–302. BSAVA Publications, Gloucester

Hotston Moore A and England G (2008) Rigid endoscopy: urethrocystoscopy and vaginoscopy. In: *BSAVA Manual of Canine and Feline Endoscopy and Endosurgery*, ed. P Lhermette and D Sobel, pp.143–174. BSAVA Publications, Gloucester

Howe LM (2006) Surgical methods of contraception and sterilisation. *Theriogenology* **66**, 500–509

Jung MJ, Moon YC, Cho IH *et al.* (2005) Induction of castration by immunization of male dogs with recombinant gonadotrophin-releasing hormone (GnRH)-canine distemper virus (CDV) T-helper cell epitope. *Journal of Veterinary Science* **6**, 21–24

Kim JK, Jeong SM, Yi NY *et al.* (2004) Effect of intratesticular injection of xylazine/ketamine combination on canine castration. *Journal of Veterinary Science* **5**(2), 151–155

Ladd A, Tsong YY, Walfield AM *et al.* (1994) Development of an antifertility vaccine for pets based on active immunization against luteinizing hormone-releasing hormone. *Biology of Reproduction* **51**, 1076–1083

Levy JK, Crawford C, Appel L *et al.* (2008) Comparison of intratesticular injection of zinc gluconate versus surgical castration to sterilize male dogs. *American Journal of Veterinary Research* **69**, 140–143

Levy JK, Miller LA, Crawford PC *et al.* (2004) GnRH immunocontraception of male cats. *Theriogenology* **62**, 1116–1130

Miller NA, VanLue SJ and Rawlings CA (2004) Use of laparoscopic-assisted cryptorchidectomy in dogs and cats. *Journal of the American Veterinary Medical Association* **224**(6), 875–878

Munson L (2006) Contraception in felids. *Theriogenology* **66**, 126–134

Naz RK (2009) Status of contraceptive vaccines. *American Journal of Reproductive Immunology* **61**, 11–18

Nolen RS (2007) Nonsurgical alternatives explored as possible answer to dog and cat overpopulation. *Journal of the American Veterinary Medical Association* **230**(2), 169–170

Ortega-Pacheco A, Bolio-González ME, Colin-Flores RF *et al.* (2006) Evaluation of a Burdizzo castrator for neutering of dogs. *Reproduction in Domestic Animals* **41**, 227–232

Pérez-Marin CC, López R, Domínguez JM *et al.* (2006) Clinical and pathological findings in testis, epididymis, deferens duct and prostate following vasectomy in a dog. *Reproduction in Domestic Animals* **41**, 169–174

Roberts WW, Chan DY, Fried NM, *et al.* (2002a) High intensity focused ultrasound ablation of the vas deferens in a canine model. *Journal of Urology* **167**, 2613–2617

Roberts WW, Wright EJ, Fried NM, *et al.* (2002b) High-intensity focused ultrasound ablation of the epididymis in a canine model: a potential alternative to vasectomy. *Journal of Endourology* **16**, 621–625

Robertson SA (2008) A review of feral cat control. *Journal of Feline Medicine and Surgery* **10**, 355–375

Root-Kustritz MV (1999) Early spay-neuter in the dog and cat. *Veterinary Clinics of North America: Small Animal Practice* **29**(4), 935–943

Root-Kustritz MV (2007) Determining the optimal age for gonadectomy of dogs and cats. *Journal of the American Veterinary Medical Association* **231**, 1665–1675

Sabeur K, Ball BA, Nett TM, *et al.* (2003) Effect of GnRH conjugated to pokeweed antiviral protein on reproductive function in adult male dogs. *Reproduction* **125**, 801–806

Shafik A (1994) Prolactin injection, a new contraceptive method: experimental study. *Contraception* **50**, 191–199

Skorupski KA, Overley B, Shofer FS *et al.* (2005) Clinical characteristics of mammary carcinoma in male cats. *Journal of Veterinary Internal Medicine* **19**, 52–55

Tivers MS, Travis TRD, Windsor RV and Hotston Moore A (2005) Questionnaire study of canine neutering technique taught in UK veterinary schools and those used in practice. *Journal of Small Animal Practice* **46**, 430–435

Vickery BH, Burns J, Zaneveld LJ, Goodpasture JC and Bergstrom K (1985) Orally administered ketoconazole rapidly appears in seminal plasma and suppresses sperm motility. *Advances in Contraception* **1**, 341–353

White RAS (2005) The male urogenital system. In: *BSAVA Manual of Canine and Feline Abdominal Surgery*, ed. JM Williams, pp.270–297. BSAVA Publications, Gloucester

5

Determining breeding status

Angelika von Heimendahl and Gary C.W. England

Introduction

Understanding the basic physiology of reproduction is important to ensure that breeding of fertile males to fertile females occurs at the correct time in order to maximize the number of pregnancies as well as the number of offspring per litter.

The bitch, a spontaneous ovulator, shows marked variability in the time that ovulation occurs in relation to the onset of pro-oestrus and the signs of standing oestrus; the ability to determine the time of ovulation is therefore critical in this species. The queen on the other hand is an induced ovulator, and there is considerable variation in the likelihood of ovulation occurring depending upon the frequency and day of mating. The ability to determine the onset of oestrus and ensure appropriate detection of mating is therefore equally critical in the queen.

Breeders have developed different methods to determine breeding status, with varying degrees of success over the years. This chapter reviews the biological principles that underpin the reproductive management of bitches and queens during oestrus, and recommends practical methods to optimize veterinary input at the time of mating.

Bitch

Determination of the optimal breeding time of a bitch can be difficult because there is significant individual variation relating to the day on which ovulation occurs, and poor correlation between the time of ovulation and behavioural oestrus. The situation is often complicated by dog breeders 'choosing' the day on which to breed, commonly using arbitrary criteria. These include having set days (e.g. 11 and 13 days after the onset of pro-oestrus), measuring electrical conductivity in the vaginal mucosa (using the Draminski ovulation detector) or looking at 'ferning' in the saliva under the microscope. Other methods, such as monitoring the onset of vulvar softening or the change in the characteristics of vaginal discharge, can be more useful. The methods used in veterinary practice to determine the optimum mating time include measurement of plasma progesterone concentrations, exfoliative vaginal cytology and vaginoscopy.

Reproductive physiology

The reproductive physiology of the bitch is unusual in that the oocytes are immature at the time of ovulation and cannot be fertilized until 2 days afterwards. Once they have completed maturation they remain fertilizable for 2–3 days. This results in a 3-day 'fertilization period' that commences 2 days after ovulation and extends to up to 5 days after ovulation (see Chapter 1). Fertile matings may occur before the onset of the fertilization period since sperm can survive within the female reproductive tract for 7 or more days. The 'fertile period' is the time during which mating could result in conception and therefore commences 5 days prior to ovulation and extends until approximately 5 days after ovulation.

Optimal breeding time

The time of maximum fertility appears to be from the day of ovulation until 4 days after ovulation (Figure 5.1). There is very little difference in pregnancy rate or litter size when bitches are mated on any one of these days, whereas breeding earlier or later results in lower pregnancy rates and smaller litters. It is thought that the time of peak fertility commences just

Time period	Days in relation to ovulation
Fertile period	–5 to +5
Fertilization period	+2 to +5
Peak fertility	–1 to +4
Preferred time for natural service or fresh semen insemination	0 to +4
Preferred time for frozen semen insemination or breeding where semen quality is not optimal	+2 to +4

5.1 The timing of peak fertility in relation to the day of ovulation in the bitch.

before the onset of the fertilization period, since sperm need to capacitate in the female reproductive tract, and this process takes about 6 hours. The best way to predict the optimum breeding time is to pinpoint the day of ovulation, although this is not always easy. The most common cause of infertility in bitches is mating at the wrong time.

Number of days from the onset of pro-oestrus
Many breeders rely on counting the number of days from the onset of pro-oestrus, and believe that bitches ovulate on a set number of days from this event. Although the majority of bitches will ovulate between 9 and 14 days after the onset of pro-oestrus, ovulation may occur as early as 5 days and as late as 30 days after the onset. Some of the inaccuracies in the timing of mating are compensated for by the long survival time of both sperm and oocytes, but the results are best if ovulation is monitored.

Onset of oestrous behaviour
In some species oestrous behaviour is a reliable indicator of ovulation. However, in the bitch there is often poor correlation between hormonal and behavioural events. In laboratory Beagles it has been shown that the onset of standing oestrus occurs approximately 2 days before ovulation, such that mating 3 or 4 days after the onset of standing oestrus would be optimal. However, interestingly, these findings are not repeatable across many breeding bitches subject to different management regimes. This may be due to the fact that these management regimes do not allow natural courtship responses, or it may simply be that there is greater variation than was originally thought from the laboratory studies. Nevertheless, in the average breeding bitch, estimation of the time of standing oestrus cannot be relied upon to demonstrate the best time to breed.

Onset of vulvar softening
One assessment that may be useful in establishing the optimal time for breeding is the change in the texture and/or tone of the vulva. During pro-oestrus there is increasing turgidity of the vulva in response to increasing concentrations of oestrogen. A few days before ovulation, the oestrogen concentration drops rapidly, with a concomitant rise in progesterone. This results in a reduction in oedema and a consequent distinct softening of the vulva. Subjective examination of the vulva once or twice daily will easily demonstrate when this event has occurred; ovulation generally occurs 2 days later and breeding should therefore commence 3 days after this event.

Hormone concentrations
Measurement of the plasma or serum concentration of various hormones can be used to determine ovulation in the bitch, including:

- Oestrogen
- Luteinizing hormone (LH)
- Progesterone (Figure 5.2).

However, the level of oestrogen drops off too far ahead of ovulation to be useful. The concentration of LH surges approximately 2 days prior to ovulation in the bitch, and the ability to detect this event could be used to predict ovulation (Figure 5.3). In addition, luteinization of the follicle wall prior to ovulation results in a significant increase in plasma (or serum) concentrations of progesterone, which can be measured easily.

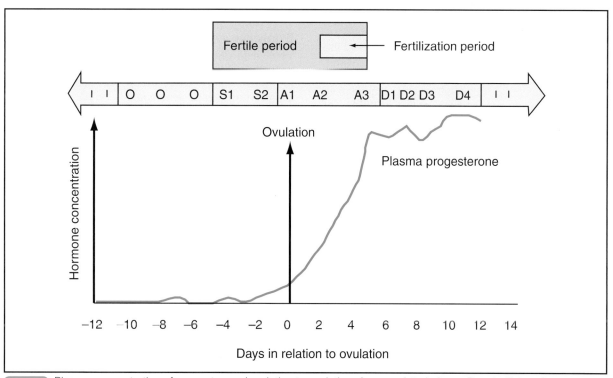

5.2 Plasma concentration of progesterone in relation to ovulation. A = angulated phase; D = declining phase; I = inactive phase; O = oedematous phase; S = shrinkage phase (see text for details).

Oestrous cycle stage	Time between events	Event
Pro-oestrus		Oestrogen peak
	+ 2 days	Luteinizing hormone peak
Oestrus	+ 2 days	Ovulation
	+ 2 days	Fertilization period
Metoestrus	+ 3 days	End of fertile period

5.3 Occurrence and timing of hormones around the period of ovulation (e.g. 2 days after the oestrogen peak is the LH peak).

Luteinizing hormone: Detection of an increase in plasma LH concentrations is a reliable and accurate method for determining the time of ovulation. However, in most countries there is no readily available commercial assay for canine LH and measurement relies upon radioimmunoassay techniques, which are commonly expensive and result in a delay in obtaining the results because samples have to be assayed in batches. A further potential problem is the requirement to take daily blood samples because the duration of the LH surge is relatively short. If an LH assay is available, breeding should commence 3 or 4 days after the LH surge has been detected.

Progesterone: The concentration of progesterone begins to increase rapidly from baseline levels approximately 2 days before ovulation and reaches a distinctly higher level at ovulation. This increase can be detected by serial monitoring, which allows for the anticipation of ovulation, confirmation of ovulation and detection of the onset of the fertilization period. Given that the initial rise in progesterone is progressive, it is only necessary to collect blood samples every second or third day, unlike the daily requirement for sampling to detect the LH surge. Progesterone concentrations double roughly every 2 days so that if the sampling interval is increased the accuracy of detection of ovulation is decreased. Progesterone concentrations may be measured by radioimmunoassay, quantitative or qualitative enzyme-linked immunosorbent assay (ELISA), or immunochemiluminescence assay. Many veterinary diagnostic laboratories offer measurement of progesterone with reporting of the results the same day.

Progesterone concentration
The concentration of progesterone may be reported either in ng/ml or nmol/l.

1 ng/ml = 3.17 nmol/l

Several qualitative or semiquantitative progesterone ELISA test kits have become commercially available for use in the veterinary practice. The results are usually obtainable within 30–45 minutes of sample collection. One of the pitfalls of testing in-house is the 'window' defined by the resultant colour change. Once the blood progesterone concentration has exceeded 10 ng/ml the test will interpret the result as 'mate immediately' (this will remain so for the next 2 months of the bitch's cycle). The in-house ELISA will only give relevant results when run at least once before ovulation with subsequent tests related to the preovulation results. If the first test confirms that ovulation has already taken place, other methods (such as vaginal cytology) will have to be used to make sure the bitch is not already in metoestrus.

Commercial laboratory results can give a more accurate indication of plasma progesterone concentration, and because mating does not have to be performed until several days after ovulation the possible delay caused by postage is not a problem. Observation of the progressive rise of progesterone and noting of the important values makes it possible to plan blood sampling as well as mating (Figure 5.4).

Exfoliative vaginal cytology
Collection, staining and microscopic examination of exfoliated vaginal epithelial cells is a simple method for monitoring the stage of the oestrous cycle, especially when serial examination is performed. The increase in plasma oestrogen during pro-oestrus and oestrus causes thickening of the vaginal wall, probably as a mechanism to protect the otherwise delicate mucosa at the time of mating. Oestrogen causes thickening of the vaginal mucosa predominantly via an increase in the number of cell layers. The mucosa changes from a cuboidal epithelium (in anoestrus), through a transitional phase (during pro-oestrus) into a stratified, keratinized squamous epithelium (during the fertile period). After the end of

Event	Progesterone level
Luteinizing hormone surge (36–48 hours before ovulation)	1.5–2.5 ng/ml (4.5–7.5 nmol/l)
Ovulation	5–8 ng/ml (15–24 nmol/l)
Fertile period	10–25 ng/ml (30–75 nmol/l)

5.4 Important progesterone concentrations.

the fertilization period, as progesterone concentrations increase, there is rapid sloughing of the newly developed epithelium and uncovering of a simple cuboidal epithelium similar to that observed during anoestrus. With the onset of metoestrus a large influx of polymorphonuclear leucocytes takes place.

Collection: Cells may be collected by aspiration of the vaginal cavity using a plastic catheter or by use of a saline-moistened cotton swab gently rolled over the surface of the vaginal mucosa. Swabs should be introduced and removed using a small speculum or guard.

Examination: Once collected the cells should be placed on a glass microscope slide by lightly rolling the cotton swab, or by application of the aspirated fluid, which is then spread into a thin film. Smears can be stained using either a simple Wrights–Giemsa stain ('Diff-Quick') or a modified trichrome stain. The former is readily available and has the advantage that sample preparation may take only minutes; the latter has the advantage of identification of keratinized cells, but the staining technique is laborious.

Interpretation: During pro-oestrus, as plasma oestrogen concentrations increase, the surface epithelial cells alter in shape, size and staining character. They change from small circular cells with little cytoplasm ('parabasal cells') to larger, irregularly shaped, flat (squamous) nucleated cells ('intermediate cells') (Figure 5.5a).

During oestrus the cells develop into anuclear cornified squamous cells ('superficial cells') and are characterized as having no nucleus, or a faint and/or small pyknotic nuclear remnant (Figure 5.5b).

After the end of the fertilization period, during sloughing of the epithelium, the superficial cells disappear and small parabasal cells reappear in the vaginal smear. Polymorphonuclear leucocytes are generally present in small numbers in the anoestrous vaginal smear, but disappear during the fertile period because the thickened mucosa is a barrier to their migration to the surface. These polymorphonuclear leucocytes reappear, often in large numbers, at the end of the fertilization period because epithelial sloughing results in a thinner mucosa at the same time as there is a significant chemoattractant effect within the vaginal lumen (Figure 5.5c).

The relative proportions of the different types of epithelial cell can be used as broad markers of the endocrine environment (Figure 5.6). Several indices of cornification and keratinization have been used. In general, the fertile period can be predicted crudely by calculating the percentage of epithelial cells that appear anuclear when using a modified Wrights–Giemsa stain. Breeding should be attempted throughout the period when >75% of the epithelial cells are anuclear, because in most bitches this coincides with the fertile period.

However, there is great variation between bitches, and in some cases the percentage of cornified cells reaches nearly 100% as early as 9 days before ovulation or as late as 2 days before ovulation. Conversely, in some cases, the peak value of anuclear cells may reach only 60%. Therefore, changes in vaginal cytology cannot be used prospectively to time ovulation accurately. Nevertheless, vaginal cytology permits monitoring of the normal progress of pro-oestrus, and waiting for significant cornification avoids unnecessary and costly testing, transportation or breeding until the pro-oestrous rise in oestrogen is complete. At the other end of the fertile period, vaginal cytology is extremely valuable for demonstrating the end of the fertilization period, an event which is often not clear when using qualitative progesterone ELISA tests.

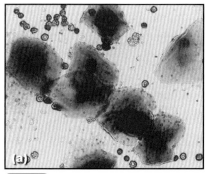 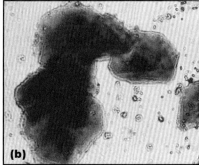 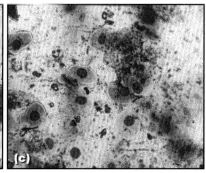

5.5 Vaginal cytology. The appearance of exfoliated cells during **(a)** pro-oestrus, **(b)** oestrus and **(c)** metoestrus.

Oestrous cycle stage	Hormone/event	Cytology
Anoestrus	Prolactin	Parabasal and small intermediate cells; occasional leucocytes
Pro-oestrus	Oestrogen peak	Large nucleated intermediate cells; large numbers of red blood cells
	Luteinizing hormone peak	Increase of cornified and keratinized cells; red blood cells decreased
Oestrus	Ovulation	Anuclear cells >75%; not many other cells
Metoestrus	End of fertile period	Large influx of polymorphonuclear leucocytes

5.6 Vaginal cytology at different stages of the oestrous cycle.

Vaginoscopy

Examination of the vagina using a rigid endoscope can be useful to document the stage of the oestrous cycle and determine the optimal time to breed. The vulval lips should be cleaned of any discharge prior to insertion of the endoscope. The use of a dilute disinfectant solution is not necessary but it may be useful for wetting the surrounding hairs and keeping them away from the vulva. Lubricating gel may be applied to the outer sheath but will not generally be necessary for bitches that are in oestrus. When lubricants are used it is important not to place material close to the tip of the endoscope because this quickly obscures the field of view.

Anoestrus: For bitches in anoestrus the vestibule and vaginal mucosal folds have a low height and a relatively thin appearance such that the underlying vasculature can be seen, and overall the folds appear a red or pink/red. The mucosa is dry or relatively tacky in appearance, and lubrication is warranted in most cases. When the vaginal folds are examined in profile they have a small, round and flaccid appearance. If there is significant insufflation of the vagina with air the folds may be compressed and become difficult to see.

Pro-oestrus: During pro-oestrus increased plasma oestrogen concentrations cause thickening and oedema of the vaginal mucosa. The mucosal folds therefore appear greatly enlarged, thickened and oedematous. Serosanguineous fluid can be observed within the lumen and may be seen to exit from the cervix. This is a stark change in appearance from the relatively thin, flat and dry mucosa noted in anoestrus. Furthermore, there is a considerable change in colour of the epithelium, from red with clearly visible vessels to pink or pink/white. As pro-oestrus progresses the mucosal surface becomes progressively less pink and typically appears white, because the thickened mucosa prevents the underlying capillaries (which were visible during anoestrus) from being seen.

Oestrus: In very late pro-oestrus or early oestrus, at approximately the same time as the LH surge, there is a progressive shrinkage of the folds which is accompanied by further pallor. These effects are the result of an abrupt withdrawal of oestrogen following the LH surge. Subsequently, over the next few days, mucosal shrinkage is accompanied by gross wrinkling of the mucosal folds and they now develop a distinctly angulated appearance but remain dense cream/white. The epithelial peaks become sharp-tipped and irregular in appearance. The mucosa also starts to look drier, and in many bitches there is a change in colour to clear/creamy of any discharge originating from the cervix. There is progressive flattening of the epithelium, which develops a concertina-like appearance.

End of fertile period: At the end of the fertile period there is rapid shedding of the epithelial surface. Sometimes passage of the endoscope lifts large sheets of cells from the vaginal wall. Complete shedding of the epithelium is usually achieved within 48 hours. The mucosal folds become less distinct and are clearly softer and more flaccid. The surface of the mucosa at this time becomes variegated in colour with white patches mixed with areas of thin reddish epithelium. Finally, as the luteal phase progresses (whether the bitch is pregnant or not) there is a reduction in the diameter of the lumen, similar to that seen in anoestrous. Initially there is apparent contraction of the folds, producing a rosette appearance. Thereafter, the flat, dry, red epithelium is similar to that observed during anoestrus.

Scoring system: A specific scoring system, which was modified by Hotston Moore and England (2008), has been used to determine the optimum time for breeding (Figure 5.7).

Phase	Mucosal appearance
Inactive phase (I)	Thin, red, dry mucosa with low and flattened mucosal folds (Figure 5.8a)
Oedematous phase (O)	Thickened and oedematous mucosa which appears swollen, rounded and grey/white (Figure 5.8b)
Shrinkage phase (S)	Thickened mucosa which is white with reduced turgidity, progressive furrowing, wrinkling and indentations. The mucosal fold profile still rounded rather than angular. The progressive nature of this phase means that the early changes are designated 'S1' and the later changes 'S2'
Angulated phase (A)	Thickened mucosa which is white, significantly reduced in turgidity. In profile the mucosal folds show progressive shrinkage and angulation such that the peaks are sharp-tipped and irregular in appearance. The mucosa is wrinkled and shrunken in appearance. The progressive nature of this phase means that the early changes are designated 'A1' and the later changes 'A2' and 'A3' (Figure 5.8c)
Declining phase (D)	Progressive decline in the size of the mucosal fold profile. Early in the phase (designated 'D1') there is a flaccid appearance to the mucosal folds. Subsequently, the folds become more rounded ('D2') and there is sloughing of the cornified layers of the epithelium ('D2' and 'D3'), resulting in a thin mucosa of variegated colour with flattened folds and a rosette appearance ('D4') (Figure 5.8d)
Inactive phase (I)	The declining phase is followed by a return to a phase characterized by a thin, red, dry mucosa with low and flattened mucosal folds. This appearance is also designated 'I', although there may be more debris present than found in phase I prior to the onset of pro-oestrus (Figure 5.8e)

5.7 Scoring system and related appearance of the vaginal mucosa in the bitch.

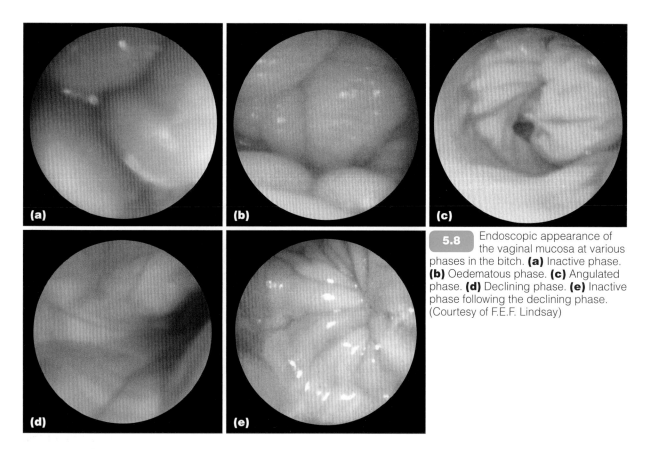

5.8 Endoscopic appearance of the vaginal mucosa at various phases in the bitch. **(a)** Inactive phase. **(b)** Oedematous phase. **(c)** Angulated phase. **(d)** Declining phase. **(e)** Inactive phase following the declining phase. (Courtesy of F.E.F. Lindsay)

A good correlation has been demonstrated between the onset of ovulation and phase 'A1', and there is a good relationship between the fertilization period and phases 'A1' to 'A3' (see Figure 5.2). Generally, the onset of the fertile period can be detected by observing the onset of mucosal shrinkage without excessive angulation, whilst gross shrinkage of entire mucosal folds with obvious angulation is characteristic of the fertilization period.

Breeding is best planned to occur approximately 4 days after the first detected mucosal shrinkage, or at the onset of the period of obvious angulation of the mucosal folds. The end of the fertilization period can be detected by observing sloughing of the vaginal epithelium and development of a variegated appearance to the colour of the mucosal surface.

Queen

In the queen modern breeding protocols may hamper reproductive performance because of a misunderstanding of the normal reproductive physiology. Although there are methods to improve the timing of mating in the queen, careful observation and an understanding of the induction of ovulation as well as the recurrence of oestrus are key. It is important to allow cats sufficient time and a number of matings over several days in order to induce ovulation. Unsuccessfully mated queens will return to oestrus within 2 weeks if ovulation has not occurred, or between 35 and 45 days if the mating itself was successful, but fertilization has not taken place. Thus, the timing of the return to oestrus is an important indicator of possible problems.

Reproductive physiology

Ovulation is induced by an adequate surge of plasma LH, which is released following coitus. Repeated mating on the same day may be necessary to ensure an LH surge of sufficient magnitude to induce ovulation, and interestingly there is a variable response depending on the day of oestrus on which mating occurs. In a recent study it was demonstrated that a single mating on day 1 of oestrus resulted in 60% of queens ovulating, whereas 70% ovulated when mated three times on day 1. Furthermore, 83% of queens ovulated when mated once on day 5 of oestrus, whilst 100% ovulated when mated three times on day 5 (Tsutsui *et al.*, 2009).

Therefore, it can be seen that restricted mating regimens, or attempts at mating that are too early, which are common breeding practices, may result in failure of ovulation in a high proportion of queens. It would appear to be important that not only are multiple matings attempted but also that normal courtship be allowed, to ensure that mating occurs on an appropriate day of the cycle.

Optimal breeding time

Clinical examination

Vulval swelling does not occur during oestrus in the queen; therefore, clinical assessments of the queen are of little value for determining the optimum breeding time. A small amount of white vulvar discharge may be noticed occasionally during oestrus.

Onset of oestrous behaviour

It is difficult to identify accurately the stages of pro-oestrus or oestrus in the queen; however, unlike in the bitch, the behavioural events are more reliable. During the 1–2 days of pro-oestrus the queen refuses copulation but is more active and may show interest in the male; this period can only truly be identified in the presence of a male. Oestrus may last between 3 and 20 days, with an average of 8 days. During this time the queen displays a crouching and lordotic stance that facilitates mounting by the male. This response can be elicited by firmly grasping the queen by the skin on the back of the neck.

Hormone concentrations

Progesterone concentrations do not rise before ovulation in the queen, but because ovulation is stimulated by coitus detection of impending ovulation is not necessary. A key clinical problem in the queen is ovulation failure due to inadequate mating (see above). Measurement of plasma progesterone concentration from 2 or 3 days after mating onwards can be used to demonstrate that ovulation has occurred. Concentrations >15 nmol/l (>5 ng/ml) indicate that ovulation has taken place. Progesterone concentrations may be measured by radioimmunoassay, quantitative or qualitative ELISA or immunochemiluminescence assay as previously discussed.

Exfoliative vaginal cytology

The examination of exfoliative vaginal cytology is useful in the queen for determining the stage of the oestrous cycle; however, the technique does not enable the prediction of the onset of oestrus. Up to one-third of queens may show signs of oestrus before the vaginal smear contains evidence of cornified cells. The technique is therefore most useful for verifying oestrus.

Collection and examination: The smear may be collected using either a moistened cotton swab or by irrigation with an eye-dropper containing sterile saline. Care should always be taken when collecting vaginal epithelial cells because the technique may induce ovulation. Staining of the epithelial cells can be achieved using a variety of stains, including a modified Wright–Giemsa stain ('Diff-Quick').

Interpretation: Erythrocytes are not found within the vaginal smear because uterine diapedesis is not a feature of oestrus in the queen. The changes in the vaginal smear are therefore limited to alterations in the morphology of epithelial cells, because polymorphonuclear leucocytes are also usually absent except during early metoestrus and pregnancy. The percentage of epithelial cells that are cornified in appearance changes in a similar manner to that seen in the bitch. During oestrus >50–80% of cells are cornified. If the queen does not ovulate, the exfoliative cells return to a state similar to that observed during anoestrus or early pro-oestrus. Early metoestrus is characterized by increasing numbers of parabasal and small intermediate epithelial cells, whilst debris, mucus and polymorphonuclear leucocytes also become evident.

References and further reading

Hotston Moore A and England G (2008) Rigid endoscopy: urethrocystoscopy and vaginoscopy. In: *BSAVA Manual of Canine and Feline Endoscopy and Endosurgery*, ed. P Lhermette and D Sobel, pp. 142–157. BSAVA Publications, Gloucester

Johnston SD, Root Kustritz MV and Olson PNS (2001) *Canine and Feline Theriogenology*. WB Saunders, Philadelphia

Tsutsui T, Higuchi C, Soeta M *et al.* (2009) Plasma LH, ovulation and conception rates in cats mated once or three times on different days on oestrus. *Reproduction in Domestic Animals* **44 (2),** 76–78

6

Clinical approach to the infertile bitch

Gary C.W. England

Introduction

Infertility is a term that is used to reflect a reduced ability to produce young. It does not denote the complete inability to reproduce (which is termed sterility), although often these terms are incorrectly used interchangeably. It is not uncommon for breeders of dogs to consult veterinary surgeons requesting investigation of both infertility and sterility. In some breeds this is associated with a perceived reduction in fertility for that breed.

It is important to have an understanding of the normal expectation of fertility, which can vary according to breed (especially in relation to litter size), to ensure an appropriate context for clinical investigation. Importantly, older bitches may cycle less frequently and have reduced pregnancy rates and smaller litter sizes compared with younger bitches; peak breeding potential generally occurs at approximately 3 years of age. Recent data from the author's laboratory (England et al., 2009) suggest that normal ovulation rates, conception rates and whelping rates for fertile bitches are 97–100%, 8–92% and 86–100%, respectively.

A further important consideration is that infertility may reflect a problem with the male, or be associated with a behavioural problem which results in ineffective mating. Other common causes of infertility include abnormal oestrous cycles, and anatomical or functional abnormalities of the reproductive tract, whilst a minority of causes are related to neoplasia or infectious agents, or are iatrogenic.

The clinical approach to examination of the allegedly infertile bitch needs to be logical, ordered and within the context of the likelihood of various common conditions and the fact that infertility may be difficult to recognize (because failure to conceive does not result in a rapid return to oestrus) and may therefore be long established before it has been recognized by the breeder.

This chapter details the clinical approach to infertility based on the way in which these cases present to the veterinary surgeon. Further details on conditions that affect the reproductive tract and the associated clinical signs are given in Chapter 18.

Case investigation

The basis of the case investigation is collection of a relevant reproductive history and performance of an examination of the reproductive tract. Importantly, defining the clinical problem should be within the context of the wide breed variations in normal fertility parameters (especially litter size), the breed-related specific fertility problems (e.g. in the Irish Wolfhound) and the normal variations in fertility that occur as a result of ageing.

Clinical history

In many cases infertility is the result of a misunderstanding of the normal physiology (breeders commonly assume that ovulation occurs on a specific day of the oestrous cycle, and therefore encourage mating at an incorrect time). Establishing the clinical history can therefore be very helpful in understanding the nature of alleged infertility. The general aims are to define the problem, understand the previous breeding history and circumstances of breeding, and to determine the general health and management of the bitch. It is also important to establish the current fertility of any males that have been used in the breeding attempts. This should include the number of oestrous cycles at which the bitch has been mated and the males that were used for mating. It may be important to examine these animals if suitable information about their true current fertility is not available.

In collecting a clinical history, important aspects include the number of oestrous periods observed, the interval between the onset of successive pro-oestrus periods, and the duration, clinical signs and sexual behaviour of the bitch displayed at various stages of the cycle. It is also important to establish whether appropriate pregnancy diagnoses were made, the cycles at which pregnancy was established, whether pregnancy loss was confirmed, the timing of parturition, the number of puppies born and the occurrence of any stillbirths or early neonatal losses.

After the collection of the history it should be possible to place the problem in one of the following categories:

- Misunderstanding of the normal biology
- Failure to show oestrus
- Abnormal oestrous cycle length
- Abnormal mating
- Apparently normal mating but infertility
- Pregnancy failure
- Male infertility.

Clinical examination

A general clinical examination should be undertaken to rule out systemic disease, as well as to establish the bodyweight and body condition of the bitch compared with expected averages for the breed. A specific examination of the reproductive tract should then be performed.

Assessment of reproductive behaviour

Observation of the demeanour of the bitch may demonstrate the presence of oestrous behaviour (especially if stroking the perineal region results in flagging of the tail and lifting of the vulva) or possibly nursing behaviour. This may be useful to establish the possible stage of the cycle.

Palpation of the mammary glands

Inspection and palpation of the mammary glands can help establish the current stage of the oestrous cycle because mammary gland enlargement is quite significant within the normal luteal phase, and at this time lactation may also be detected (see Chapter 1). Furthermore, careful palpation of the mammary glands may indicate the presence of nodules, neoplasia or mastitis.

Inspection of the perineum and vulva

Inspection of the vulva should be undertaken to establish the presence of any discharge and to ascertain its colour and nature; this may help to indicate the stage of the oestrous cycle or the presence of pathology. Furthermore, the relative size and degree of swelling of the perineum and vulva may be used to help establish the current stage of the oestrous cycle because there is normally significant enlargement during pro-oestrus and oestrus, and the external reproductive tract is at its smallest during anoestrus.

Examination of the vestibule and vagina

Digital examination of the vestibule and vagina should be performed to establish the normality of the caudal vault of the reproductive tract, and particularly to eliminate the presence of obstructive disease of the vulva, vestibule or vagina, as well as other possible pathologies of the vagina and urethra. These abnormalities might, for example, present an obstruction to normal intromission.

Collection of a vaginal smear

Examination of cells collected from the cranial vagina should be used to establish the current stage of the oestrous cycle (see Chapter 5), as well as to evaluate subjectively the number of white blood cells and bacteria. The latter may also be useful to document uterine or vaginal pathology.

Endoscopic examination of the vaginal cavity and cervix

An endoscopic examination of the vagina should always be performed to confirm normality, help establish the current stage of the oestrous cycle, and identify the presence of any normal or abnormal discharge from the cervix or external urethral orifice. Endoscopic examination is particularly useful for localization of pathological changes that result in a vulval discharge.

Palpation and imaging of the uterus

Palpation of the uterus can be difficult in large or obese bitches, but it may be possible to establish the shape, size and texture in other females. Such examination may help to localize pathological changes, or to confirm the presence of pregnancy.

Diagnostic real-time ultrasonography may be used to image the uterus accurately and confirm its presence, size and internal architecture. Ultrasonography may be useful to confirm the stage of the oestrus cycle, to document and diagnose uterine pathology, and to confirm pregnancy or pregnancy failure. Radiography can be useful to detect uterine enlargement, to confirm pregnancy after mineralization of the fetal skeletons has occurred, and to confirm the number of fetuses by counting the number of fetal heads.

Measurement of plasma hormones

In clinical practice it is not common to measure concentrations of the gonadotrophin hormones. However, it is possible to measure concentrations of luteinizing hormone (LH), oestrogen, progesterone and relaxin. It is not feasible to measure prolactin concentrations in practice owing to the unavailability of commercial assays.

Luteinizing hormone: Measurement of concentrations of LH may be used to indicate the absence of ovarian tissue (for example in the neutered bitch or a bitch with ovarian agenesis); in this case LH concentrations are significantly elevated because of the lack of negative feedback from ovarian steroids (Figure 6.1). Furthermore, stimulation tests using gonadotrophin-releasing hormone (GnRH) or human chorionic gonadotrophin (hCG) may be used to confirm the function of the pituitary gland and ovary, and the function of the ovary, respectively (Figures 6.2 and 6.3).

Oestrogen: Measurement of the basal oestrogen concentration has limited clinical value, especially because elevated oestrogen concentrations can be inferred from the detection of anuclear epithelial cells within a vaginal smear. Bitches with oestrogen-producing ovarian neoplasms will have elevated plasma oestrogen, but this can be identified by the clinical signs of persistent oestrus and the presence of increased numbers of large epithelial cells in the vaginal smear. A more useful diagnostic test would be ultrasound examination of the ovaries to determine their structural appearance, which may help to differentiate normal follicles from follicle cysts and ovarian neoplasia.

Progesterone: A hormone that has great value in establishing the stage of the oestrous cycle of the bitch is progesterone. Plasma progesterone concentrations are increased immediately prior to ovulation, and they remain elevated throughout the luteal phase of non-pregnancy and pregnancy. Measurement of plasma progesterone may therefore be used to predict impending ovulation, confirm ovulation and to document the active pregnant or non-pregnant luteal phase. Plasma progesterone concentrations are low throughout anoestrus (Figure 6.4).

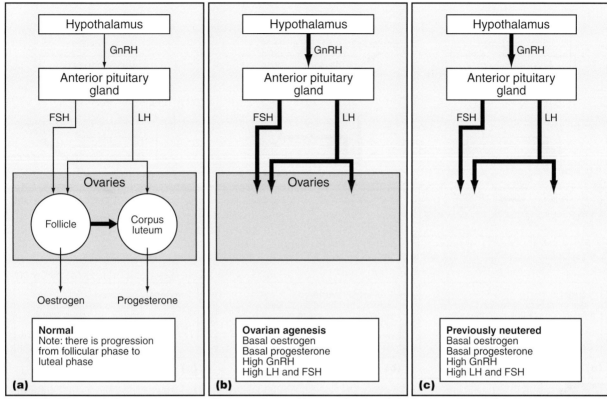

6.1 Changes in hormone concentration associated with normal and pathological conditions of the hypothalamic–pituitary–gonadal axis. **(a)** Normal ovarian function (i.e. sequential hormone production; first oestrogen during oestrus and then progesterone after ovulation). **(b)** Ovarian agenesis. **(c)** Absence of the ovaries due to previous neutering. FSH = follicle-stimulating hormone; GnRH = gonadotrophin-releasing hormone; LH = luteinizing hormone.

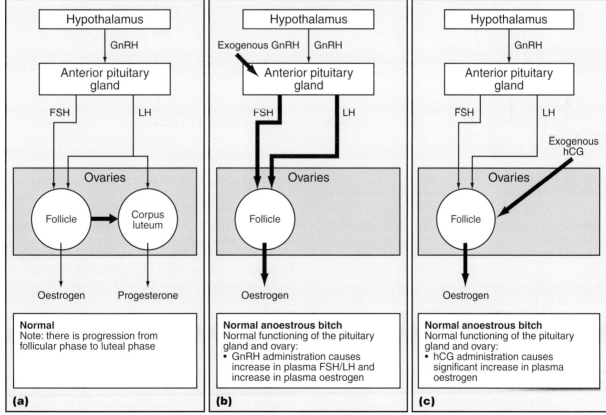

6.2 Tests to confirm the function of the pituitary gland and ovary. **(a)** Basal hormone secretion. **(b)** Changes in hormone concentration following administration of exogenous gonadotrophin-releasing hormone (GnRH) to a normal anoestrous bitch. **(c)** Changes in hormone concentration following administration of exogenous human chorionic gonadotrophin (hCG) to a normal anoestrous bitch. FSH = follicle-stimulating hormone; LH = luteinizing hormone.

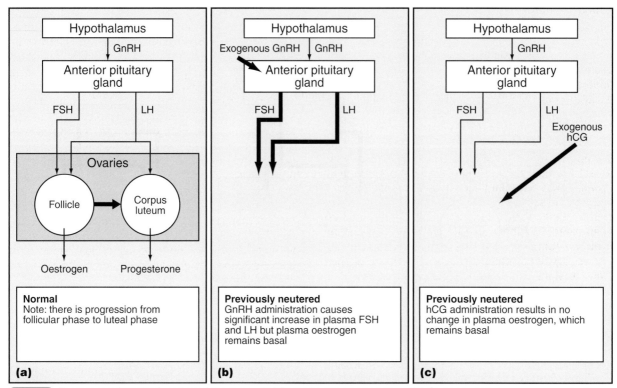

6.3   Tests to confirm the function of the ovary. **(a)** Basal hormone secretion. **(b)** Changes in hormone concentration following exogenous gonadotrophin-releasing hormone (GnRH) administration to a previously neutered bitch. **(c)** Changes in hormone concentration following exogenous human chorionic gonadotrophin (hCG) administration to a previously neutered bitch. FSH = follicle-stimulating hormone; LH = luteinizing hormone.

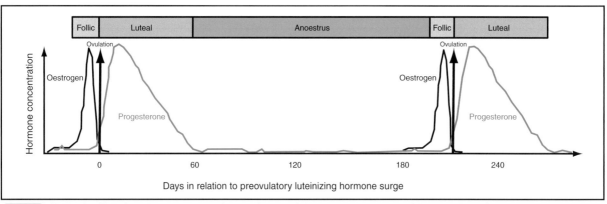

6.4   Follicular (follic) and non-pregnant luteal phases of the oestrous cycle and the duration of anoestrus in relation to plasma oestrogen and progesterone concentrations and the time of ovulation in normal bitches.

Relaxin: Measurement of plasma concentrations of relaxin may be used to confirm the presence of placental tissue, and this is commonly used as a method of pregnancy diagnosis. Values may remain elevated for a few days after pregnancy loss, so may be useful in confirming this condition if the bitch is examined shortly after the presumed pregnancy loss and there are other confirmatory clinical signs (such as uterine enlargement and luminal fluid detected upon ultrasound examination).

Microbiological culture
There is widespread confusion about the role of microorganisms in infertility in the bitch. However, it is clear that, with the exception of *Brucella canis* (which is not present in the UK), other bacteria are commensal organisms and are not primarily responsible for

infertility, abortion, anoestrus or weak puppies; neither are these organisms transmitted to the male to cause infertility. In this author's opinion the routine bacteriological screening of bitches with no clinically significant vulval discharge (and in countries where *Brucella canis* is not present) is pointless. In some circumstances examination for viral disease using culture, polymerase chain reaction (PCR) or serological means may be valuable, specifically when investigating causes of pregnancy failure.

Other techniques
Several other techniques may be useful for examination of the reproductive tract, including examination of the ovaries with ultrasonography, transcervical collection of uterine cells, and possibly inspection of the tract at laparotomy or laparoscopy.

Misunderstanding of the normal biology

The reproductive cycle of the bitch is complicated because there is significant variation around what is understood as a normal average cycle, and this variability occurs both between bitches and within individual bitches from one cycle to another. The greatest variability is in the day of the cycle on which ovulation occurs. It is generally thought that ovulation is a reliable event that occurs on day 12 of the cycle. In fact the day on which ovulation occurs is hugely variable, with some bitches ovulating as early as day 5 and some as late as day 25 after the onset of pro-oestrus (see Figure 6.6). Interestingly, it has also been clearly demonstrated that whilst some bitches show repeatability from one cycle to the next (i.e. always ovulating 'early' or 'late'), in many bitches there is significant variability between cycles. It is most important that the breeder appreciates this variability, and uses a sensible approach to determine the optimal time for breeding. The methods available for assessing breeding status are given in Chapter 5.

Failure to exhibit oestrus

Failure to cycle may be either a primary problem where there has been no oestrous activity before 24 months of age (i.e. the bitch shows delayed puberty) or a secondary problem where oestrous activity has not occurred within 12 months of a previous cycle (i.e. the bitch has prolonged anoestrus).

Delayed puberty

Although the phrase 'delayed puberty' appears to indicate a primary endocrinological failure, this condition may be associated with poor management or observation on behalf of the owner as well as aberrant physiology and pathology.

Normal physiological variation

Despite a common perception that puberty occurs reliably at 9 months of age, there is significant variation both within and between breeds. Generally, bitches of larger breeds are slower to reach puberty, and often this does not occur until several months after adult bodyweight is reached. In the author's opinion it is sensible not to embark upon clinical investigation until the bitch is at least 24 months of age, because this is still within the normal physiological range for the onset of puberty. In some breeds, particularly those undertaking significant work (such as Greyhounds), bitches may not reach puberty until 3 or even 4 years of age. Puberty may also be delayed if there is an inadequate diet or chronic disease resulting in the bitch being underweight, once these are corrected there is usually a rapid onset of puberty.

Inadequate observation by the owner or 'silent oestrus'

It is not uncommon for bitches to be very fastidious at their first oestrus, and if there is frequent cleaning of any discharge by licking, and no male dog in the vicinity, oestrus may be unobserved by the owner. Detection of elevated plasma progesterone (>2.0 ng/ml) will demonstrate that oestrus has occurred within the last 60 days but has not been noticed (see Figure 6.4). In bitches that have ovulated and are examined during the luteal phase, careful clinical examination may also reveal enlargement of the mammary glands, sometimes with residual secretion. In some cases there may also be a history of vague signs of pseudopregnancy. It is not uncommon for the owner to be adamant that oestrous behaviour has not occurred, and therefore it is frequently reported that 'silent oestrus' (where there are no external signs of oestrus) is a common clinical condition. In the author's opinion it is rare for bitches to have absolutely no external clinical signs, and some evidence of oestrus can usually be detected with regular and careful inspection.

Abnormal sexual differentiation

In a small proportion of bitches there may be chromosomal abnormalities that result in failure to reach puberty, or weak and variable signs of oestrus. Often these bitches have an approximate female phenotypic appearance, but careful examination of the external genitalia demonstrates that the vulva is small and it may be prepuce-like and positioned more cranially than in the normal female. There may also be enlargement of the clitoris noted at the time of puberty.

Establishing the complement of chromosomes present (the karyotype) can help to differentiate the normal bitch (78XX) from the abnormal; these are commonly 77XO, 79XXX, 79XXY or 78XX/78XY. Such chromosome complements reflect non-dysjunction during meiosis or mitosis in the process of embryonic development. These animals will never be fertile and it is best to remove the gonads and tubular genitalia surgically to prevent subsequent pathology.

Ovarian agenesis/aplasia

This rare condition, in which there is absence of the ovaries, results in an absence of oestrus. The condition may be difficult to diagnose without resorting to laparotomy, although there will be increased concentrations of follicle-stimulating hormone (FSH) and LH, and low concentrations of oestrogen. Administration of either GnRH or hCG will not cause any change in plasma oestrogen concentrations, which remain low.

Treatment

For bitches in which no underlying abnormality has been detected it may be possible to induce oestrus. The ethical considerations relating to this are not particularly significant as long as the bitch has reached a suitable age. Oestrus may be induced by the daily administration of 20 IU/kg equine chorionic gonadotrophin for 5 days and 500 IU/bitch hCG on day 5. Fertility at the induced oestrus is variable but is usually relatively low. Administration of a prolactin inhibitor such as cabergoline, although very effective in cases of prolonged anoestrus, has a variable effect in bitches that have not yet reached puberty.

Prolonged anoestrus

The normal interval between the beginning of one oestrus and the beginning of the next is approximately 7 months (not 6 as commonly assumed). However, there is significant individual variation around this value, and for the Basenji there appears to be only one cycle every 12 months. Because of this variability it is probably best to define prolonged anoestrus as an interval of more than 12 months. An interval of this length may be associated with oestrus being unnoticed by the owner, or be related to a pathological condition.

Inadequate observation by the owner

Inexperienced owners may not understand the signs of oestrus, or may not examine bitches frequently or carefully enough. These factors may be compounded in bitches that are kept in groups, where there is not the close proximity of a male, and in particular breeds in which clinical signs of oestrus are not marked (e.g. Greyhounds). Some bitches are especially fastidious and frequently lick away any vulval discharge. These presentations may be described as 'silent oestrus' (see above).

Detection of elevated plasma progesterone will demonstrate that oestrus has occurred within the previous 2 months (see Figure 6.4). To ensure that the subsequent oestrus is not missed, bitches should be examined carefully twice weekly taking note of their behaviour, the appearance of the vulva and any discharge; wiping the vulva with cotton wool can be useful for detecting any change in coloration of the discharge. Vaginal cytology, examined every other week, can be extremely useful for documenting the onset of an oestrogenic phase.

Poor body condition or systemic illness

Prolonged anoestrus can be associated with disease affecting other body systems where there is resultant poor body condition and debility. In most cases it is clear from the general clinical examination that there is an ongoing disease process, and it is rare for cyclicity to be influenced by anything other than significant concurrent disease. In most cases, when the underlying disease is corrected there is a rapid return to cyclical activity.

Prolonged luteal phase

Whilst this is not a true cause of prolonged anoestrus (because it is the luteal phase that is extended), these cases may present with a failure to return to cyclical activity. It is extremely rare for progesterone secretion to continue for longer than the normal 65 days unless there is significant ovarian disease such as luteal cysts or ovarian neoplasia. The diagnosis and treatment of these conditions are discussed below.

Endocrinopathies

There has been wide reporting in textbooks of specific endocrinopathies such as hypothyroidism and hyperadrenocorticism influencing the return to cyclical activity and causing prolongation of anoestrus. However, there is little within the scientific literature to substantiate these claims; indeed most recent reports suggest that endocrinopathies are very rarely associated with a delayed return of oestrus. In the author's experience it is rare to see an effect upon reproductive cyclicity in the absence of other signs of systemic disease such as poor coat, significant weight increase or decrease, and lethargy.

Premature ovarian failure

Although referred to in many textbooks, a condition akin to menopause in women where there is (here premature) cessation of ovarian function is very rare in the bitch. If present the principal problem is thought to be an inadequate ovarian response to pituitary gland gonadotrophin stimulation. In these cases concentrations of FSH and LH are elevated, but because the ovary is non-responsive, concentrations of oestrogen are low. In the investigation of these cases, a GnRH or hCG stimulation test would not result in any significant increase in plasma oestrogen concentration.

Treatment

In bitches with prolonged anoestrus the primary treatment is correction of the underlying abnormality. In those cases in which no abnormality can be detected it may be possible to attempt treatment using regimes for induction of oestrus. Clinically, the most successful regimes are those that involve the continual daily administration of prolactin inhibitors (e.g. cabergoline at 5 µg/kg) until the first day on which signs of vulval swelling and a serosanguineous discharge occur. Bitches that are going to respond usually do so within 30 days, and if treatment is stopped once pro-oestrus has begun there is usually a high pregnancy rate, similar to natural cycles. There are recent reports of using GnRH agonist preparations to treat prolonged anoestrus, but these products are not yet widely available in clinical practice.

Abnormal oestrous cycle length

Rapid return to oestrus

The normal expectation for the return to oestrus should be approximately 7 months after the start of the previous cycle. There are a number of situations in which oestrus recurs more rapidly than this; these are usually associated with an absence of ovulation and a rapid return to oestrus associated with the development of another follicular wave.

Puberty

At the first (pubertal) cycle it is common for there to be a failure of progression to oestrus and ovulation. These bitches have signs of pro-oestrus (and sometimes behavioural signs of oestrus) but these signs then wane and there is no ovulation. It is assumed that these 'problems going through puberty' are associated with either inadequate secretion of LH or insufficient LH receptors, resulting in lack of ovulation. The consequence of this is that there is a return to pro-oestrous behaviour within a few weeks. This new pro-oestrus may progress to oestrus and ovulation, although in some bitches the whole sequence of events is repeated several times before finally there is a normal progression to oestrus and ovulation. This condition, which has the same manifestation as 'split oestrus' (see below), requires monitoring but no treatment.

Split oestrus
Bitches with split oestrus have a very short interval (2–12 weeks) between clinical signs of pro-oestrus. The condition is common at the time of puberty (see above) and in bitches of <4 years of age. It is assumed that there is follicle growth and oestrogen production but that the follicles regress prior to ovulation. The condition may lead to confusion because it is often assumed that ovulation has occurred, and for some owners the sanguineous vulval discharge at the subsequent pro-oestrus is mistakenly thought to represent pregnancy loss.

Recognition of the condition is important to ensure that mating occurs at the correct time. Usually the condition requires no treatment (particularly in the bitch at puberty), but if it is a recurrent problem attempts may be made to supplement LH release by exogenous administration of hCG, usually at the peak of cornification of the vaginal epithelial cells as noted on a vaginal smear.

Shortened interval between cycles
When the interval between oestrous periods is short but this is not associated with split oestrus or pubertal oestrus (as noted above) it is most likely that there has been failure of ovulation (because ovulation is normally followed by an obligatory luteal phase), or shortening of the luteal phase or of anoestrus.

Ovulation failure
Ovulation failure is not normally recognized at the time of oestrus (unless serial progesterone sampling has been undertaken); rather it is recognized when pro-oestrus returns earlier than expected. In these cases, because there has been no ovulation, there will be no luteal phase, and so oestrus occurs 2 months earlier than expected (Figure 6.5). In fact, it is not uncommon for the interval between one oestrus and the next to be as short as 4 months. These cases are likely to have a similar aetiology to split oestrus (possibly caused by low concentrations of LH or LH receptors). Ovulation failure appears to occur in approximately 1% of oestrous cycles, although it appears to be far more common in particular lines of German Shepherd Dogs, and dogs of similar breeds.

When the clinical problem has been recognized at one oestrus, attempts may be made at the subsequent oestrus to ensure that ovulation occurs. This is

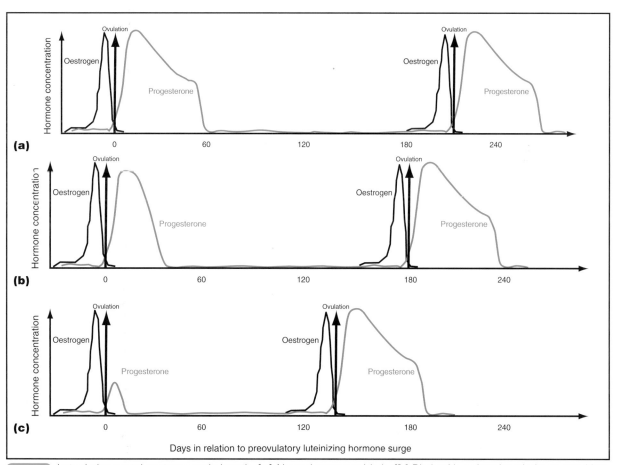

6.5 Luteal phase and oestrous cycle length. **(a)** Normal pregnant bitch. **(b)** Bitch with a short luteal phase, resulting in pregnancy failure followed by a normal ovulation and normal pregnant luteal phase. **(c)** Bitch with ovulation failure with a very short rise in progesterone (presumably as a result of follicle luteinization) followed by a normal ovulation and normal pregnant luteal phase. Note: it is not uncommon for bitches with this condition to have repeated cycles with an abnormal luteal phase rather than for it to be followed by a normal cycle as shown here.

best done by the administration of 500 IU of hCG administered every day for 3 days, commencing when the first rise of progesterone is detected (i.e. at the time of onset of luteinization of the follicles).

Short luteal phase

Bitches with a short luteal phase appear to have ovulated but have inadequate production of progesterone from the corpora lutea. The condition can only be detected with serial monitoring of plasma progesterone concentrations (which is not common practice), but it may be suspected when there is loss of a confirmed pregnancy and a shortened interval to the onset of the next pro-oestrus, especially when this presentation occurs in a German Shepherd Dog or similar breed, in which the condition is most common. The condition may occur at a single cycle or may be repeated in subsequent cycles. In affected cycles, serial monitoring of plasma progesterone will demonstrate that the values are lower than expected, and progesterone concentrations return to basal within 20–40 days after ovulation (rather than the normal 65-day luteal phase) (see Figure 6.5).

Reported treatments for a short luteal phase include the supplementation of progesterone throughout pregnancy, although this is not without the risk of inducing cryptorchidism in male puppies and masculinization in female puppies.

Short anoestrus

There is considerable variation in the length of the normal anoestrus period of bitches, which accounts for the variability in the time to return to pro-oestrus in normal animals (because the luteal phase is tightly regulated). It is believed that a variety of environmental cues may be responsible for controlling anoestrus. That most commonly observed is a shortening of anoestrus in bitches that come into contact with other bitches that are in oestrus. In large breeding establishments, housing of anoestrous bitches with bitches in oestrus may be used in an attempt to hasten oestrus in those bitches in anoestrus.

Lengthened interval between cycles

The interval between the onset of one cycle and the next may be delayed because of a prolongation of anoestrus (see above), as a normal physiological variant, or where there is pathological extension of the luteal phase.

Normal physiology

As noted previously, whilst the normal interval between periods of pro-oestrus is approximately 7 months, there is considerable variation around this figure and it is not uncommon or abnormal for the interval to be up to 11 months. Anecdotally, the interval between cycles increases as the bitch gets older.

Prolonged luteal phase

It is important to remember that the normal non-pregnant luteal phase of the bitch is as long as the pregnant luteal phase, and that there is not a rapid reduction in progesterone and return to oestrus in the absence of pregnancy as observed in other species. Thus, it is rare for there to be an extended luteal phase beyond the normal 65 days, unless there is significant ovarian pathology.

Luteal cysts: Rarely, luteal cysts may be identified in bitches. Most commonly they are found in cases with pyometra, where presumably the extended luteal phase has allowed the development and progression of this condition. Luteal cysts may be suspected when there is a failure to return to cyclical activity and serial blood samples show persistently elevated plasma progesterone concentrations. There are no published data on the response of luteal cysts to treatment, but repeated administration of a low dose of prostaglandin would be a prudent option, as would administration of a prolactin inhibitor (such as cabergoline), the latter in an attempt to remove the support of the luteal structures.

Care should be taken with the diagnosis of cystic ovarian disease when fluid-filled structures are noted close to the ovaries on ultrasonography because parabursal cysts are common and asymptomatic.

Progesterone-producing ovarian neoplasms: In some rare cases, progesterone-secreting granulosa cell tumours have been identified in bitches. Often the clinical presentation results from the mass effect produced by the tumour, or is a consequence of the continued secretion of progesterone, allowing the establishment of pyometra. Clinically, these bitches have an absence of oestrus. The treatment is ovariectomy or ovariohysterectomy.

Prolonged anoestrus

Conditions that are associated with prolonged anoestrus are described above.

Lengthened period of pro-oestrus or oestrus

Many textbooks suggest that pro-oestrus and oestrus last a total of approximately 20 days, and breeders may therefore expect bitches to have ovulated and for oestrus to have ended within 3 weeks of its onset. However, the length of pro-oestrus and oestrus may be longer than this, either as a normal physiological variant or as a result of delayed ovulation or ovarian neoplasia.

Normal physiology

The average bitch generally ovulates on approximately day 12 after the onset of pro-oestrus. However, the time of ovulation is highly variable, both between bitches and within the same animal. It is not uncommon for ovulation to occur as late as day 25 after the onset of pro-oestrus, or as early as day 5 (Figure 6.6). These 'late ovulations' are normal and part of the population variability. Unfortunately, many owners do not recognize that some bitches will be at the extremes of normality and will insist on

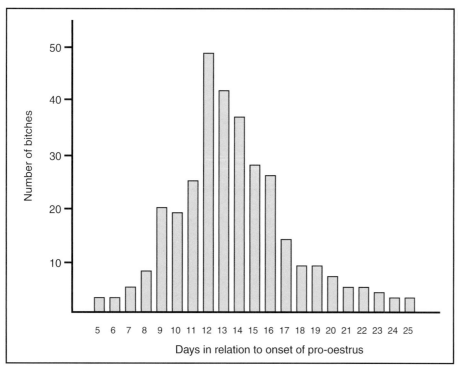

6.6

Day on which ovulation occurred in relation to the first signs of vulval swelling and serosangineous discharge in 220 normal bitches, which became pregnant when mated.

breeding earlier, resulting in conception failure. A key indicator that ovulation has occurred late is the duration of oestrus, because the signs do not normally subside until at least 7 or 8 days after ovulation. These animals do not need treatment, and they will become pregnant if they are mated at the correct time. It is best to monitor the time of ovulation using vaginal cytology and measurement of plasma progesterone concentration (see Chapter 5).

Prolonged pro-oestrus or oestrus
Most cases of prolonged pro-oestrus or oestrus are part of the normal physiological range (as noted above). However, in some cases ovulation has not occurred by as late as 30 days after the onset of pro-oestrus. This may be associated with failure of adequate LH production or insufficient LH receptors, and some of these bitches will respond to administration of hCG.

Follicular cysts
Bitches that have a period of pro-oestrus or oestrous behaviour which extends for more than 30 days should be subjected to imaging of the ovaries with ultrasonography to establish whether follicular cysts are present. Follicular cysts are usually large (8–12 mm in diameter), thin-walled and filled with anechoic fluid. They need to be differentiated carefully from normal follicles, cavitated corpora lutea and parabursal cysts. It must be remembered that follicular cysts are extremely rare and should be at the bottom of any list of differential diagnoses.

The consequence of not treating follicular cysts is unknown but in some cases bone marrow suppression occurs as a result of the persistently elevated oestrogen levels, and therefore in most cases attempts are made to induce ovulation using hCG

(500 IU/bitch) on three occasions. In cases that do not respond, progesterone administration may be used in an attempt to cause regression of the cysts (although progesterone will increase the risk of pyometra in the oestrogen-primed uterus), or an ovariectomy may be performed.

Oestrogen-producing ovarian neoplasms
Ovarian tumours are not common in the bitch but they may secrete oestrogen and result in clinical signs of persistent oestrus as well as bone marrow suppression. Although the mean age at occurrence is 8 years, oestrogen-secreting granulosa cell tumours have been identified in bitches as young as 7 months of age. Ultrasound examination of the ovaries is usually diagnostic because in these cases the ovary is often partly solid but contains multiple fluid-filled cavities. Ovariectomy is the treatment of choice and during surgery care should be taken not to traumatize the neoplasm because metastatic spread may occur via transcoelomic seeding.

Conditions that mimic oestrus
In some circumstances bitches may be thought to be in oestrus when they are not. In most cases this is because there is interest in the bitch from a male dog, and usually the bitch shows no behavioural signs typical of oestrus.

Vaginitis
Bitches with vaginitis may have a vulval discharge that is attractive to male dogs. Commonly vaginitis is caused by overgrowth of commensal bacteria due to an anatomical or structural abnormality or a foreign body within the vagina. Careful examination of the bitch, including vaginal cytology, will differentiate vaginitis from pro-oestrus and oestrus.

Abnormal mating

Whilst normal mating behaviour is exhibited by the majority of bitches, there are particular circumstances when mating does not occur. There are a variety of conditions that may be related to this, including those associated with abnormal behaviour at mating and those associated with physical obstruction of the mating procedure.

Behavioural problems

It is important to ensure that the bitch's first experience of mating is not traumatic so that normal mating behaviour is established.

Previous poor experience

Bitches that have a poor initial experience, such as being presented for mating at an inappropriate time and therefore being force-mated, or when the stud dog is too dominant, may be reluctant or refuse subsequent mating attempts. In these cases it is imperative that a calm and experienced male is chosen, and that the bitch is relaxed and has ample time to display normal courtship with the dog. It is always best to monitor oestrus using vaginal cytology and measurement of plasma progesterone to ensure that the bitch is presented to the dog at the correct time. It is often helpful for the dog and bitch to be introduced several days prior to the mating attempt. In some circumstances where registration authorities allow, artificial insemination may be contemplated.

Poor response to the male dog

In a small number of cases bitches may demonstrate poor behavioural responses to the male, which can include marked submission or aggression. This may be observed in those cases where the bitch lives principally as part of the human household and has poor canine social skills, or sometimes when the dog and bitch normally cohabit and are 'too familiar'. These cases can be difficult to manage and require careful introduction of the male and prolonged periods of play before attempted mating, but still may require artificial insemination if registration authorities allow.

Abnormalities of the reproductive tract

It is rare for the experienced breeder to overlook external abnormalities of the female reproductive tract that cause obstruction to normal mating. However, some abnormalities occur within the vagina and are only identifiable upon digital exploration. It is prudent to teach experienced breeders how to perform an aseptic digital examination of the vestibule and vagina, and to encourage such examination of every bitch immediately prior to mating.

Vestibular and vaginal malformations

During fetal development there is opportunity for error in the formation of the caudal female reproductive tract. These are observed most commonly at the junction of the vestibule and vagina, within the vagina, and more rarely at the junction of the vulva with the vestibule.

Vestibulovaginal bands: Dorsoventrally directed fibrous bands or circumferential fibrous constrictions at the junction between the vestibule and the vagina are not uncommon. These abnormalities, which are found just proximal to the external urethral orifice, tend to be detected when they cause pain and failure to achieve intromission. In some cases the bands result in vaginitis, which appears to develop as a result of trapping of uterine/vaginal discharge in which commensal bacteria overgrow.

Thin dorsoventral bands can be broken down manually per vaginum and small constrictions can be dilated using a similar approach. However, for larger dorsoventral bands, general anaesthesia and an episiotomy approach is usually necessary to enable visualization, ligation and transaction. In these cases, if treatment is undertaken in pro-oestrus the bitch can be mated during the same oestrus. However, it is important to establish the extent of the abnormality because in some cases the fibrous bands may divide the vagina over a significant length, and these are not amenable to treatment.

Vaginal hypoplasia: Vaginal hypoplasia is a rare condition that presents most commonly with pain at attempted intromission, or more rarely with vaginitis similar to that seen in cases of vestibulovaginal bands. Usually the majority of the vagina is affected and is small in diameter and non-dilatable from the junction with the external urethral orifice to the cervix. It may be possible to pass a small diameter endoscope along the length of the vagina, and this can be used to confirm the severity of the condition. Extensive lesions are not amenable to surgical correction.

Vulvovaginal bands: Vulvovaginal bands are less common than vestibulovaginal bands. Most commonly the bands are circumferential and produce a tight ring at the vulval opening that prevents normal intromission. There is usually an acute pain response at attempted mating, and the bitch will then refuse subsequent mating attempts. Diagnosis is obvious by clinical examination. In mild and moderate cases progressive manual dilation can be achieved under general anaesthesia.

Vaginal hyperplasia

Hyperplasia of the vaginal wall is a normal response to elevated oestrogen concentrations during oestrus. In bitches with clinical disease there appears to be an exaggerated response to normal concentrations of oestrogen, such that prolific thickening of the vaginal wall occurs especially in the regions just cranial to the external urethral orifice. In fact, most commonly, it is mainly the ventral vaginal wall that is affected; it is thrown into a thick tongue-shaped mass with a wide base of attachment that obstructs the vaginal lumen (Figure 6.7). The hyperplastic tissue develops during pro-oestrus and regresses when oestrogen concentrations decrease and progesterone concentrations increase at the beginning of the luteal phase.

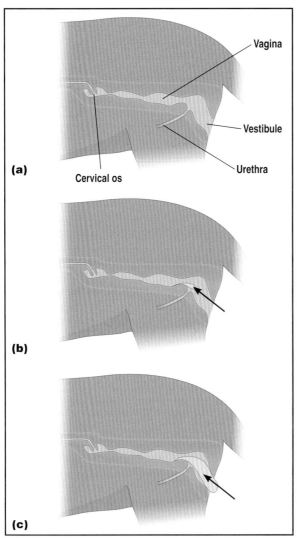

Vagina

Vestibule

Urethra

(a)   Cervical os

(b)

(c)

6.7   Anatomy of the vagina and vestibule.
(a) Normal bitch. **(b)** Bitch with mild vaginal hyperplasia (arrowed) protruding into the vaginal lumen. **(c)** Bitch with extensive vaginal hyperplasia (arrowed) protruding from the vulval lips.

In mild cases the mass may go unnoticed, except that it causes a pain response at attempted intromission. In more severe cases the hyperplastic tissue may be so extensive as to protrude from the vulval lips (Figure 6.7c). In these cases there may be some debate as to whether it is a vaginal polyp or neoplasm. Hyperplasia can be confirmed usually because of the relationship to oestrus, the origin of the tissue (just cranial to the urethral orifice) and the wide base of attachment. In some very severe cases there may be circumferential hyperplasia with protrusion of a ring of tissue from the vulval lips. In cases where there is protrusion of the hyperplastic mass there may be excessive licking and self-trauma, and the tissue may dry and crack and become significantly ulcerated and devitalized.

Given that the condition usually regresses fully at the onset of the luteal phase, most cases can be managed conservatively by preventing self-trauma and using protective creams. Bitches with protrusion of a small mass can be bred by holding the mass ventrally and allowing the dog's penis to penetrate

dorsal to the mass. In bitches with masses that obstruct the vagina or with moderate to large protrusions, the only option for breeding at that particular cycle is artificial insemination, if allowed by regulatory authorities. In bitches that become pregnant the hyperplasia will have regressed by the time of parturition and will not contribute to dystocia.

Although it has not been studied extensively, it appears as though there is some progression in the severity of the hyperplasia with each subsequent cycle, and surgical neutering may be advisable in these cases. When continued breeding is required, surgical resection of the hyperplastic tissue may be contemplated. This surgery is undertaken during pro-oestrus and is complicated by the oedema and increased vascularity of the vagina. Care must be taken to protect the integrity of the urethra during surgical resection.

Apparently normal mating but infertility

The two broad causes of infertility following an apparently normal mating are those where mating has occurred distant from the time of ovulation such that fertilization is unlikely, and those where the mating time is appropriate but some other factor has prevented conception.

Mating remote from ovulation
Mating remote from ovulation is probably the most common cause of infertility and is described above (see also Chapter 5).

Mating close to ovulation
When a bitch has been mated at an appropriate time during the fertile or fertilization period but has not conceived there are a number of conditions that should be considered. These include abnormalities of the female reproductive tract that would prevent sperm transport or interfere with development of the fertilized embryo, and male infertility. The latter should always be of prime concern despite an owner's contention that the male is fertile.

Uterine disease
It is not uncommon to identify cystic uterine disease in a bitch that fails to become pregnant despite mating with a fertile dog at an appropriate time. Although there are no published data on this relationship, it is the author's contention that the underlying endometrial disease contributes to infertility by allowing persistence of commensal vaginal bacteria within the uterus. These bacteria are introduced into the uterus during oestrus when the cervix is open and fail to be cleared because of the uterine pathology. Treatment regimes for this clinical presentation have not been evaluated appropriately.

Uterine tube disease
It is not common for aplasia of the uterine tube to be identified as a cause of infertility; indeed, most commonly the condition is unilateral and therefore does not apparently influence fertility, although litter size is reduced. Bilateral uterine tube aplasia is rare, but if

present is usually best diagnosed at laparotomy. Blockage of otherwise normal uterine tubes has not been reported, although inflammatory lesions that affect the uterine tube have been described in a small number of bitches.

Infertile male

Breeders often have very limited information about the fertility of the male that they have used at a specific breeding. This is commonly because the dog is used infrequently and it is believed that there will have been no change in his fertility since the last mating. However, it is clear that semen quality may deteriorate over a short period of time, and also in some males there may be variability in semen quality, so previous fertility may not indicate current status. Furthermore, there may be different presentations of male infertility (see also Chapter 8):

- The dog with azoospermia or such severe semen abnormalities that he is sterile
- The dog with small numbers of normal sperm (usually with large numbers of abnormal sperm), where frequent mating is required to achieve the number of sperm for success
- The dog with sperm abnormalities that result in reduced longevity of sperm in the female tract. In this case careful monitoring of the bitch and mating within the fertilization period are essential.

Pregnancy failure

Where pregnancy has been confirmed in a bitch but there has been subsequent pregnancy failure, it is important to follow a logical and ordered protocol to establish the underlying cause of the problem. This should include:

- Histopathological examination of the aborted material
- Virological and microbiological examination of the aborted material and vulval discharges or vulval swabs
- Serological monitoring of the bitch
- Examination of the uterus using diagnostic ultrasonography to document any pathology.

Although endocrinological evaluation is often suggested, this is usually pointless at the time of resorption or abortion because it cannot distinguish whether any abnormality has caused or been caused by the problem. Detailed descriptions of abnormalities of pregnancy are described in Chapter 13.

References and further reading

England GCW, Russo M and Freeman SL (2009) Follicular dynamics, ovulation and conception rates in bitches. *Reproduction in Domestic Animals* **44**, 53–58

7

Clinical approach to the infertile queen

Eva Axnér

Introduction

In order to evaluate an infertile queen and make a correct diagnosis it is important to have an understanding of the normal physiology. The expected conception rate in the cat is 70–80%. The previous breeding history of both the male and the female may indicate which of the animals should be evaluated. If a queen has failed to conceive after three matings with a male of proven fertility, a female reproductive problem should be suspected. Reproductive failure can be caused by failure to show normal oestrus, failure to mate, anovulation, failure of fertilization, early embryonic loss, resorption or abortion. The first step in the diagnostic plan is to identify at what stage the reproductive failure has occurred; a thorough history is therefore essential. This should include a complete breeding history of the queen including age at first oestrus, length of oestrus, normal oestrous interval, observations of mating, occurrence of postcoital reaction after breeding, oestrous interval after mating, and previous reproductive performance of both the male and female. In addition the history should include information about any medications given to the queen, a complete vaccination history, and any information about previous heath problems or diseases in the cattery. If the owner has other cats, signs of reproductive failure or infectious diseases in other animals in the same household may indicate a cattery problem.

Abnormalities of the oestrous cycle

Failure to exhibit oestrus

Failure to exhibit oestrus can be divided into primary anoestrus and secondary anoestrus. Primary and secondary anoestrus must be differentiated from silent oestrus. A previous ovariohysterectomy should be ruled out if the complete history of the cat is not known (see Chapter 19). A thorough clinical examination may reveal underlying diseases or developmental abnormalities. Causes of failure to exhibit oestrus are listed in Figure 7.1.

Primary anoestrus

A queen with primary anoestrus has never achieved puberty. The first oestrus is normally seen between 4 and 21 months of age, with an average of 9–12 months. Given that puberty can occur as late as 21 months of age, a diagnosis of primary anoestrus should not be made until the queen is at least 2 years old. The causes of primary anoestrus can be the same as for secondary anoestrus but in addition a developmental disorder cannot be ruled out. Chromosomal disorders are rare but have been reported in the cat. Primary anoestrus can be caused by X-chromosome monosomy (X0), which results in ovarian dysgenesis with lack of normal ovarian germ cells, or by testicular feminization syndrome. Testicular feminization syndrome is caused by defective androgen receptors. Affected animals have male gonads in the abdomen and produce testosterone, but androgen-dependent masculinization cannot occur because of a lack of functional androgen receptors. Given that developmental disorders cannot be ruled out in cases of primary anoestrus, establishing the karyotype may be warranted. Laparotomy to explore the internal genital organs and histology of the gonads may be necessary for a diagnosis in intersex animals.

Secondary anoestrus

In secondary anoestrus puberty has occurred, but the interval since the last oestrus has been

Condition	Clinical sign	Diagnostics	Treatment
Primary anoestrus	A queen that is at least 2 years old and has never shown oestrus	Rule out previous ovariohysterectomy, progestogen treatment, pregnancy or silent oestrus	Light stimuli; social stimuli; karyotyping; laparotomy
Secondary anoestrus	The interval since the last oestrus is abnormally long	Rule out previous ovariohysterectomy, progestogen treatment, pregnancy or silent oestrus	Light stimuli; social stimuli; hormonal induction of oestrus
'Silent oestrus'	A queen does not show oestrous behaviour although she is cycling. Cannot be differentiated from other causes of anoestrus on the basis of clinical signs	Vaginal cytology during the reproductive season and breeding when the smear is cornified	Stimulation of oestrous behaviour by change of environment for subordinate females or exposure to a tom cat

7.1 Causes of failure to show oestrus.

abnormally long. The length of the cycle in most queens is 2–3 weeks during the reproductive season. However, there are large individual variations in the length of the cycle, number of cycles per year and the duration of anoestrus. Spontaneous ovulation and resulting pseudopregnancy may also affect the length of the cycle. Given that the cat is a seasonal long-day breeder, the amount of light will affect cyclicity.

Although seasonality varies with the breed, the individual and the exposure to natural and artificial light, it is normal for many cats to be in anoestrus during seasons with shorter days. Some queens will cycle more or less throughout the year, whilst others only have one or two oestrous periods per year. Therefore, it may be difficult to determine whether the interval has been abnormally long. However, more than a year since the last oestrus must always be considered abnormal. The most common reason for a long period of anoestrus is the season or an inadequate amount of light.

An association between other endocrine disorders (e.g. hyperadrenocorticism or hypothyroidism) and anoestrus has, to the author's knowledge, not been described for the cat but is a possibility.

Iatrogenic secondary anoestrus: Iatrogenic secondary anoestrus can be caused by administration of progestogens or androgens. Progestogen compounds are marketed to be used for contraception and suppression of oestrus, and breeders are usually aware of this effect. After injection of a long-acting depot formulation the time to the next oestrus may be unpredictable and it may sometimes take more than a year before the queen comes into heat again. If the cat has changed owner during this time the new owner may not be aware of previous treatments. High doses of corticosteroids can potentially suppress cyclicity.

'Silent oestrus'

A queen may have normal oestrous cycles and follicular waves without displaying overt oestrus. This is more likely to occur in timid cats that are low in the social scale. Regular vaginal cytological samples (three times per week) can be collected to diagnose oestrus. The presence of >80% cornified epithelial cells indicates that the queen is in oestrus. Repeated blood samples for analysis of oestradiol could also be used, but this is less cost-effective than vaginal cytology. Serum progesterone is, in contrast to the

bitch, usually of no diagnostic value to evaluate cyclicity. If progesterone concentrations are elevated the queen has ovulated and is cycling, but because ovulation typically is induced by mating (although spontaneous ovulation may also occur) progesterone concentrations are usually basal after oestrus.

Induction of oestrus

Artificial induction of oestrus with exogenous hormones is rarely indicated in normal cat breeding. Natural methods such as light and social stimuli are preferable before hormone treatment. However, induction of oestrus can also be used for diagnostic purposes in anoestrous cats.

Treatment with follicle-stimulating hormone (FSH) or equine chorionic gonadotrophin (eCG) (which is FSH-like in action) usually causes some ovarian hyperstimulation, and a delayed secondary wave of follicle development with formation of corpora lutea even at recommended doses. Although pregnancies and birth of kittens have been achieved with different protocols using exogenous hormones, concerns have been raised that this may induce an abnormal endocrine environment which may inhibit the development of normal pregnancy.

There are no compounds authorized for the induction of oestrus in the cat, but different protocols have been described. Treatment should be given in interoestrus or anoestrus to avoid hyperstimulation. Before induction of oestrus is attempted serum progesterone concentrations should be analysed to rule out pregnancy or pseudopregnancy. Prepubertal queens should not be treated with exogenous hormones because they may develop high numbers of cystic follicles that do not ovulate. Protocols that involve exogenous hormones often include induction of ovulation. This is because the protocols are usually developed for assisted reproduction technologies where there are no mating stimuli. Artificial induction of ovulation is not necessary in cats if there is a natural mating at the time when the follicles are mature. Protocols for artificial induction of oestrus are listed in Figure 7.2.

Light and social stimuli: The most natural method to induce oestrus in the queen is to increase the amount of daylight. Cats kept in a minimum of 12–14 hours of light per 24 hours will cycle year-round. A light flash for 1 hour during the dark period shortens the time to oestrus. It may be necessary to have full control over the amount of light to be successful with

Oestrus induction	Ovulation induction	Reference
Porcine FSH s.c. q24h for 4 days in daily decreasing doses to a total of dose 3–5 IU/cat	3 IU porcine LH on day 5 of treatment	Pope *et al.*, 2006
FSH 2.0 mg/cat i.m. q24h for 5 days or until oestrus	Mating 3 times per day from day 1 in oestrus for 3 days, or mating once daily throughout oestrus	Wildt *et al.*, 1978; Goodrowe *et al.*, 1988a
FSH 2 mg/cat i.m. on day 1 followed by 1 mg/cat i.m. on subsequent days until oestrus. Total dose restricted to 9 mg/cat	Mating 1–3 times on day 1 or 2 of oestrus	Tsutsui *et al.*, 1989
eCG 100–150 IU/cat i.m. once	hCG 75–100 IU/cat i.m. once 80–88 hours after injection with eCG, or mating 80 hours after eCG injection	Goodrowe *et al.*, 1988b; Donoghue *et al.*, 1992; Roth *et al.*, 1997

7.2 Example protocols used to induce oestrus in the queen. eCG = equine chorionic gonadotrophin; FSH = follicle-stimulating hormone; hCG = human chorionic gonadotrophin; LH = luteinizing hormone.

light stimuli (i.e. no windows in the room). Contact with other cycling females or with a tom cat may also have a stimulatory effect.

Follicle-stimulating hormone: Owing to the short half-life of FSH, repeat injections are necessary to achieve oestrus induction. Different experimental regimens have been described for oocyte collection and embryo transfer rather than for natural breeding. Doses of 2.0 mg/day until oestrus resulted in a pregnancy rate of 71% after natural mating, with a mean litter size of 5.8 kittens (Wildt *et al.*, 1978). In another study, 2.0 mg FSH was injected on day 1 followed by administration of 1.0 mg FSH i.m. on subsequent days until oestrus. Oestrus occurred a mean of 4.6 days after the first injection. Ovulation was induced by natural mating. It was observed that cats in induced oestrus had lower plasma progesterone concentrations than cats bred during natural oestrus. The conception rate was low in the queens in induced oestrus and embryonic death occurred in the pregnant females (Tsutsui *et al.*, 1989).

Equine chorionic gonadotrophin: Due to the availability and longer half-life of eCG compared with FSH, eCG is probably the first choice for hormonal induction of oestrus in clinical practice. Although some older protocols describe repeat injections, single doses of 100–150 IU eCG are sufficient to stimulate oestrus; high doses compromise fertility. Optimal oocyte maturation seems to occur 80–88 hours after injection with eCG; therefore, breeding should be attempted at this time. In the absence of mating stimuli (e.g. when using artificial insemination), ovulation can be induced with 75–100 IU human chorionic gonadotrophin (hCG) i.m. 80–88 hours after injection of eCG. Repeat treatments with eCG at short intervals can be inefficient owing to formation of anti-gonadotrophin antibodies. An interval of at least 4 months between treatments is therefore recommended to avoid ovarian refractoriness.

Induction of ovulation

For successful induction of ovulation there must be mature follicles in the ovaries. Successful induction of ovulation is therefore most likely in mid-oestrus. Ovulation can be induced with luteinizing hormone (LH), gonadotrophin-releasing hormone (GnRH) or hCG; however, LH is not readily available in clinical practice and is used mostly in experimental protocols. Administration of hCG in doses of >100 IU/cat i.m. once may be detrimental for oocyte quality, while doses of 75–100 IU/cat i.m. once are successful in inducing ovulation. Use of GnRH at a dose of 25 µg/cat i.m. once has also been successful in inducing ovulation in oestrous females.

In addition to hormonal induction, ovulation can also be induced with mechanical stimulation of the vagina to mimic the mating stimuli. This stimulation can be performed with a cotton-tipped swab.

Prolonged oestrus

Normal ovaries

The normal length of oestrus in the queen varies between 2 and 19 days with a mean of 5–8 days. An apparent prolongation of oestrous behaviour can be caused by a short interoestrus interval. An owner may believe that the cat is in permanent oestrus, although oestrous behaviour has in fact been interrupted by a few days of interoestrus. Differentiation between prolonged oestrus and short interoestrus can often be made with a thorough history and by study of vaginal cytology. Some cats will continue to show oestrous behaviour in periods of interoestrus and in the absence of active follicles. These queens are normal and may become pregnant if bred at the correct time.

Although queens normally have distinct follicular waves, sometimes these waves overlap so that the queen shows oestrous behaviour before the next wave of follicles starts to produce oestradiol. Prolonged oestrous behaviour in a queen with normal ovaries can be interrupted by breeding, induction of ovulation or treatment with progestogens. However, treatment with progestogens during oestrus may increase the risk of uterine pathology. Continuous or frequent oestrus is, on the other hand, also a risk factor for uterine disease and may in addition lead to decreased appetite and loss of bodyweight.

Cystic follicles

Oestrogen-producing follicular cysts may be a reason for prolonged oestrus. There are few reports of the occurrence of follicular cysts in the queen. Prolonged secretion of oestradiol is likely to increase the risk of development of cystic endometrial hyperplasia. Treatment options are bilateral or unilateral ovario(hyster)ectomy, an attempt to induce ovulation with 500 IU eCG or 25 µg GnRH per cat i.m. once or surgical resection of the cysts. Another alternative is to induce regression of the cysts with oral progestogens. However, treatment with progestogens after a prolonged period of oestradiol stimulation is likely to increase the risk of development of pyometra.

Other types of cysts are encountered more frequently. Cysts in the rete ovarii seem to be fairly common in the cat. Large rete ovarii cysts can cause compression of the ovary. Cystic remnants of the mesonephric duct may also be confused with follicular cysts. Findings of cystic structures around the ovary on ultrasonography or during abdominal surgery should be considered as incidental findings unless accompanied by clinical signs.

Ovarian neoplasia

Ovarian tumours are not common in the cat but can be a cause of prolonged oestrus. The most common ovarian neoplasm in the cat is the granulosa cell tumour, which often produces oestradiol and causes prolonged oestrus. Approximately 50% of granulosa cell tumours are malignant and have metastasized when diagnosed. Continuous oestrous behaviour can also be caused by ovarian dysgerminomas or luteomas. In addition to abnormal oestrous cyclicity, other signs of hyperoestrogenism that can be observed in cats with oestradiol-secreting tumours are anorexia, hair loss and cystic endometrial hyperplasia. The recommended treatment is ovariohysterectomy unless the tumour has already metastasized.

Failure to mate

Failure to mate is usually related to the mating regimen rather than to pathology of the male or female. Cats are territorial animals and this is why mating is often more successful if the female is brought to the male, so that the tom cat is in familiar surroundings. However, some very nervous or shy females may be more likely to accept mating if they are in their own territory. Young and inexperienced queens may need more time to accept the male than more experienced females. The peak of oestrus is often seen around day 3 or 4, although this may vary between individuals. Failure to accept mating may be caused by attempts to breed the queen too early, when she is only in pro-oestrus. Partner preference sometimes causes mating failure. In one study, queens would not allow the male to breed in 15/38 attempts despite confirmed oestrus at the time of attempted mating (Root *et al.*, 1995).

Congenital vaginal abnormalities are potential causes of failure to mate, but to the author's knowledge have not been confirmed to cause these problems in the cat. This could be attributed to the fact that the anatomy of the queen makes it difficult to diagnose such defects. Acquired strictures after difficult parturition may also prevent mating.

Mating problems related to the male are described in Chapter 8.

Mating but no pregnancy

Infertility in a queen with normal cyclicity and mating behaviour can be caused by failure of sperm transport to the fertilization site, failure of ovulation, early embryonic loss or resorptions/abortions. Failure to achieve a pregnancy can also be caused by an infertile male. A pregnancy diagnosis with ultrasonography or abdominal palpation may help to differentiate between fertilization failure and embryonic/fetal loss.

Developmental disorders

Failure of sperm transport to the site of fertilization can be caused by failure in the development of the Müllerian duct. Although rare, aplasia of the cranial vagina and absence of the cervical os have been described in the cat. Diagnosis is usually made after surgery and inspection of the female tubular organs.

Anovulation and timing of mating

In order to diagnose failure of ovulation, information about the interoestrus interval is valuable. If the queen comes into oestrus within 3–4 weeks after mating she has not ovulated, since normal dioestrus is 25–55 days, giving an interoestrus interval of 6–10 weeks. If the interoestrus interval is >6 weeks the queen may have ovulated, especially if her interoestrus interval is usually shorter when not bred. A blood sample for analysis of progesterone concentration, collected 1–3 weeks after mating, is diagnostic. A high progesterone concentration indicates that the female has ovulated, whilst a basal value indicates that she has not ovulated and cannot be pregnant.

Failure to ovulate is usually related to management and timing of mating rather than to pathology. A common cause is that the cats have not mated at all. Furthermore, mating too early or too late in oestrus, or too few matings, can also lead to anovulation. A single mating is often insufficient to release enough LH to stimulate ovulation, and some queens may not reach the threshold of LH required for ovulation if they are mated at the beginning of oestrus (see Chapter 1). At least three matings per day on day 3 or 4 of oestrus will usually stimulate ovulation. Ovulation can be induced using hCG or GnRH, but there is little need for artificial induction of ovulation in normal cat breeding.

Spontaneous ovulation may interfere with the timing of breeding. If spontaneous ovulation occurs in early oestrus, 48 hours or more before breeding, the oocytes may be incapable of being fertilized and this can be a cause of infertility.

Pregnancy failure

Resorption or abortion of fetuses can be caused by non-infectious or infectious causes. Early pregnancy failure may cause only a pronged interoestrus interval, while post-implantation loss may be accompanied by clinical signs such as a haemorrhagic vulval discharge, abdominal discomfort, signs of general illness and/or expulsion of fetuses.

Early embryonic loss

Not all oocytes that are ovulated after mating will result in implanted fetuses; some oocytes fail to fertilize whilst others die after fertilization. Early embryonic loss is difficult or impossible to differentiate from fertilization failure in clinical practice. The queen with early embryonic loss will have a prolonged interoestrus interval caused by the luteal phase. Little is known about the causes of early embryonic loss in the cat.

Non-infectious causes

Cystic endometrial hyperplasia

Cystic endometrial hyperplasia is a common condition in queens above 3–5 years of age. The condition is characterized by a hyperplastic endometrium with cystic dilatation of the uterine glands. It is believed to be progressive and caused by repeated stimulation of the endometrium by oestradiol and progesterone. The degenerative endometrial changes are likely to interfere with the normal process of implantation and normal placental function, and are associated with infertility. Cystic endometrial hyperplasia can be diagnosed by ultrasonography or hysterography. However, in the absence of large cysts, degenerative changes of the endometrium may be difficult to diagnose. No efficient treatment of cystic endometrial hyperplasia in the cat has been described. Repeated oestrous cycles and treatment with progestogens are likely to be predisposing factors (see also Chapter 19).

Metritis, endometritis and pyometra

Metritis/endometritis/pyometra syndrome refers to an inflamed endometrium that is contaminated with normal vaginal bacteria. Low-grade endometritis may not result in obvious clinical signs but can cause a small amount of vaginal discharge. Low-grade endometritis is therefore extremely difficult to diagnose in the queen and its incidence and importance for infertility are unknown. Uterine biopsy could be diagnostic but is a very invasive procedure. The condition may progress to accumulation of pus in the uterus (referred to as pyometra), a condition that usually is easier to diagnose than low-grade endometritis (see Chapter 19). Bacterial infection of the uterus during pregnancy may cause resorption/abortion and neonatal death. Bacterial culture from aborted fetuses and placentas is a valuable diagnostic aid.

Low-grade endometritis can be treated with antibiotics after bacterial culture and sensitivity testing during oestrus (when the cervix is open). Vaginal cultures cannot be used to diagnose metritis because healthy cats have the same vaginal flora as infertile females, but can be used to choose an appropriate antibiotic once a presumptive diagnosis has been made. Pyometra should not usually be treated with antibiotics alone (see Chapter 19). Because of the risk of selection of resistant bacterial strains it is not appropriate to treat infertile cats with antibiotics without a thorough clinical examination to rule out other causes for the infertility. The first choice of antibiotic in the non-pregnant queen is enrofloxacin. If the queen is pregnant enroflaxacin should be avoided and amoxicillin or clavuanate are suitable first choices.

Nutritional disorders

Nutrition is of vital importance for normal reproduction. Nutritional disorders are probably unlikely to cause pregnancy failure in cats that are fed balanced controlled commercial diets. However, with increasing interest in feeding cats homemade diets, the risk of nutritional imbalances may also have increased because these homemade diets often have an inadequate nutrient content (Remillard, 2008). The bones and raw food diet (BARF) is, for example, sometimes fed to pregnant queens. Cats fed diets with insufficient taurine, arachidonate or copper have lower conception rates. Taurine deficiency will also result in an increased frequency of fetal resorption. Diets with inadequate essential fatty acids and low energy content can also have a negative effect on reproductive performance.

Iatrogenic causes

Medication of pregnant queens should be avoided if possible (lists of drugs that can cause fetal damage have been published elsewhere). Before a drug is given to a pregnant queen the safety of the drug during pregnancy should be checked.

Genetic aberrations

Genetic abnormalities as a cause of embryonic or fetal loss are difficult to confirm but have been described in a few cases.

Stress

Although it has not been proven that stress can cause pregnancy failure in the cat it is likely to be beneficial to avoid stress in pregnant queens. The adrenal glands can produce both corticosteroids and progesterone in response to stress. Abnormal hormonal profiles in early pregnancy are known to interfere with embryo transport in the oviducts in other species.

Infectious causes

Pregnancy loss can be secondary to general illness in the queen. However, some infectious agents may infect and/or cross the placenta resulting in embryonic or fetal death, congenital abnormalities or birth of infected kittens without causing clinical illness in the queen.

Viral infections

Feline parvovirus: Feline parvovirus (FPV) is transmitted by contact with saliva, urine, faeces or vomit from infected cats. The virus is extremely stable and can persist for a long time in the environment, which is why direct contact between cats is not required for transmission of the virus. Infection with parvovirus can cause reproductive problems without any other obvious signs in the queen. Reproductive problems can occur at all stages of pregnancy and can be manifested as early embryonic death, resorptions, abortions or congenital defects, depending on the stage of gestation at the time of infection. The queen may remain healthy. Sometimes only some kittens in the litter are affected. Fetuses infected in late gestation may be born with cerebellar hypoplasia resulting in ataxia, or with hydrocephalus. Vaccination with live vaccines during pregnancy can cause the same problems as the infection. Diagnosis is made by virus isolation from aborted fetuses. Vaccination before breeding is recommended to avoid fertility problems caused by FPV.

Feline leukaemia virus: Infection with feline leukaemia virus (FeLV) can affect pregnancy at any stage and has been associated with unsuccessful matings, infertility and abortions. The queen usually has no other clinical signs at the time of reproductive failure. Pregnancy loss is often diagnosed by palpation of uterine enlargement followed by a sudden disappearance of the fetal swellings. Because FeLV is unstable and only survives in the environment for a short time, transmission usually requires direct contact between cats. FeLV can therefore be controlled by only introducing animals to a cattery that have tested negative for the disease. FeLV was previously considered to be a common cause of reproductive failure but it is probably not very common any more because many breeders routinely test their animals.

Feline immunodeficiency virus: Infection with feline immunodeficiency virus (FIV) results in an immunodeficiency syndrome. Bite wounds are the major route of transmission. However, after infection it may take months to years before clinical signs develop. The virus can cross the placenta and cause diverse manifestations of reproductive failure, such as arrested fetal development, mummification of fetuses, abortion, stillbirth, low birth weight and birth of congenitally infected kittens. FIV is most prevalent in free-roaming tom cats and is not very common in pedigree cats that are kept under control. Therefore, FIV is not likely to be a common cause of reproductive failure in pedigree cats. Serological testing and elimination of seropositive animals is effective for prevention of the disease in catteries.

Feline herpesvirus and calicivirus: A tendency for an association between feline herpesvirus (FHV) and reproductive problems has been observed. However, no clear association with fetal resorption/abortion has been demonstrated after natural infection, although abortion secondary to severe debilitating upper respiratory disease of the queen is a possibility. Infection *in utero* has only been demonstrated after experimental intravenous inoculation of the virus, whilst the virus failed to cross the placenta after intranasal infection. In contrast, feline calicivirus has been demonstrated to cross the placenta and probably cause fetal death. However, transplacental infections with calicivirus seem to be a rare event. Vaccination is recommended to protect cats from clinical signs of disease.

Bacterial infections

Bacteria that cause infection *in utero* and reproductive failure are usually derived from the normal vaginal flora. For example, *Escherichia coli* can cause placentitis that leads to abortion. Pregnancy loss caused by bacteria from the normal vaginal flora cannot be diagnosed by culture because healthy cats will usually also have positive results. However, post-mortem examination of expelled fetuses and their placentas may be diagnostic.

Bacteria can also be ingested by the queen. *Salmonella typhimurium* has been reported as a cause of fetal death in a clinically ill queen that was fed raw chicken. To avoid infection by ingestion of pathogenic bacteria, the feeding of raw meat to pregnant queens should be avoided.

Toxoplasmosis

Cats are both definitive and intermediate hosts for *Toxoplasma gondii*. Cats are mainly infected by ingestion of tissue cysts from prey or raw meat, but they can also be infected by sporulated *T. gondii* oocysts in the faeces of infected cats. Infected cats usually remain clinically healthy. If queens are infected during pregnancy, transplacental infection can result in abortion of mummified fetuses, premature birth and birth of weak kittens. It is possible that the strain and number of *T. gondii* organisms affects the outcome. Antibody titres often remain high several years after infection, which is why serology of the queen is of little value unless a rising titre demonstrating a recent infection can be found. If the queen is seronegative, toxoplasmosis is an unlikely cause of pregnancy failure. Reproductive failure caused by toxoplasmosis can be avoided by preventing pregnant queens from hunting and not feeding them raw meat.

Diagnostic plan

A diagnostic plan for the infertile queen with a normal cycle and mating behaviour includes the following:

1. Check the medical history of the queen, including possible medications given during pregnancy.
2. Check the feeding regimen, including possible supplements, and determine whether the queen has been given raw food (a risk factor for toxoplasmosis and some bacterial infections).
3. Check the vaccination history and correct the vaccination routines if indicated.
4. Confirm that the female has ovulated after mating. A short interoestrus interval indicates that there has been no ovulation. If anovulation is the likely cause for infertility no further testing is warranted but the owner should instead be advised to change the mating regimen and/or to use a different tom cat.
5. Confirm that the queen is negative for FeLV and possibly also FIV.
6. Perform a thorough clinical investigation with particular emphasis on palpation of the uterus and examination for the presence of a vulval discharge.
7. Perform an ultrasound examination of the uterus and ovaries.
8. In the presence of signs of metritis (vaginal discharge) and in the absence of severe cystic endometrial hyperplasia or other severe diseases, collect a bacterial culture from the vagina during oestrus for selective antibiotic treatment.
9. If there are signs of general disease in the cattery, further testing based on clinical signs may be warranted.
10. Correction of the routine in the cattery may be warranted (e.g. keeping cats in smaller groups to minimize stress and the spread of infectious agents).

In addition, any aborted fetuses should be sent together with their fetal membranes for post-mortem examination as soon as possible. The submitted material should be accompanied by a complete history of the case.

References and further reading

Axnér E, Ågren E, Båverud V *et al.* (2008) Infertility in the cycling queen: seven cases. *Journal of Feline Medicine and Surgery* **10**, 566–576

Donoghue AM, Johnson LA, Munson L, Brown JL and Wildt DE (1992) Influence of gonadotrophin treatment interval on follicular maturation, *in vitro* fertilization, circulating steroid concentrations, and subsequent luteal function in the domestic cat. *Biology of Reproduction* **46**, 972–980

Goodrowe KL, Howard JG and Wildt DE (1988a) Comparison of embryo recovery, embryo quality, oestradiol-17β and progesterone profiles in domestic cats (*Felis catus*) at natural or induced oestrus. *Journal of Reproduction and Fertility* **82**, 553–561

Goodrowe KL, Wall RL, O'Brien SJ, Schmidt PM and Wildt DE (1988b) Development competence of domestic cat follicular oocytes after fertilization *in vitro*. *Biology of Reproduction* **39**, 355–372

Johnston SD, Root Kustritz MV and Olson PNS (2001) Clinical approach to the complaint of infertility in the queen. In: *Canine and Feline Theriogenology*, ed. R Kersey, pp. 486–495. WB Saunders, Philadelphia

Meyers-Wallen VN, Wilson JD, Fisher S *et al.* (1989) Testicular feminization in a cat. *Journal of the American Veterinary Medical Association* **195**, 631–634

Pope CE, Gomez MC and Dresser BL (2006) *In vitro* embryo production and embryo transfer in domestic and non-domestic cats. *Theriogenology* **66**, 1518–1524

Remillard RL (2008) Homemade diets: attributes, pitfalls and a call for action. *Topics in Companion Animal Medicine* **23**, 137–142

Root MV, Johnston SD and Olson PN (1995) Estrous length, pregnancy rate, gestation and parturition lengths, litter size and juvenile mortality in the domestic cat. *Journal of the American Animal Hospital Association* **31**, 429–433

Roth TL, Wolfe BA, Long JA, Howard JG and Wildt DE (1997) Effects of equine chorionic gonadotrophin, and laparoscopic artificial insemination on embryo, endocrine and luteal characteristics in the domestic cat. *Biology of Reproduction* **57**, 65–71

Tsutsui T, Sakai Y, Matsui Y *et al.* (1989) Induced ovulation in cats using porcine pituitary gland preparation during the non-breeding season. *Japanese Journal of Veterinary Science* **51**, 677–683

Wildt DE, Kinney GM and Seager SWJ (1978) Gonadotrophin induced reproductive cyclicity in the domestic cat. *Laboratory Animal Science* **28**, 301–307

8

Clinical approach to the infertile male

Daniele Zambelli and Xavier Levy

History and clinical examination

Testing fertility is one of the most frequent consultation requests in male reproduction. When a mating has been unsuccessful it is very often male 'infertility' that is suspected first. However, before any prognosis can be made or treatments suggested it is important that a thorough clinical history is taken and a semen sample analysed. The causes of infertility are divided into several categories:

- Absence of production or emission of an ejaculate
- Normal coitus but absence of pregnancy
- Subfertility when the pregnancy rate is <75% (Meyers-Wallen, 1991).

Clinical history

The clinical history should always include age and breed. In young animals puberty is defined as the time when the first fertile spermatozoa are produced; however, the quality of the ejaculate at this time is often poor. As animals get older, the volume and concentration of the ejaculate increase and the percentage of abnormal spermatozoa reduces. Therefore, young males presented for infertility should be reassessed after a few months. In contrast, ageing males may show a reduction in libido and quality of semen (due to physiological testicular degeneration).

Any systemic illness that may have occurred within the last 6 months can influence semen quality. Treatments (e.g. corticosteroids, antimycotics, steroids, chemotherapy) administered during the last 6 months (spermatogenesis process (>60 days) plus duration of pharmacological effect of the medications) should be recorded because they can affect spermatogenesis. Any trauma or disease localized in the genitourinary system (sheath, penis, scrotum, testis, prostate gland and bladder) should also be noted.

Breeding history

It is important to record the following details from the breeding history:

- Dates of previous inseminations and/or matings, normal coitus (intromission, tie, ejaculation), pregnancy result, number of puppies/kittens and any semen evaluations that have been performed in the past

- Evaluation of the females mated (or inseminated) should include fecundity, fertility and prolificacy
- Evaluation of breeding management, including the method used for monitoring of optimum mating time, and the handling and methods used for mating or artificial insemination.

Genetics

Many Kennel Clubs now recommend breed-specific health screening (e.g. hip scoring, eye testing) to prevent the most common inherited diseases for each breed. In some countries the father and grandfather must have a veterinary certificate to prove that both testes were descended in order to register puppies. Many breed societies, especially for breeds with small numbers of puppies registered per year, offer analyses of inbreeding coefficients and recommendations for suitable mating partners.

Physical examination

All genital organs should be assessed by inspection and palpation:

- Scrotum (symmetry, mobility of the different layers, presence of hair, pigmentation)
- Testes (size, shape, symmetry, position, consistency, mobility, pain)
- Epididymis: head, body and tail (size, position, consistency, pain)
- Penis sheath (hair, shape, pigmentation, injuries)
- Penis (size, shape of different parts, curvature and ability to protrude)
- Appearance and colour of the penile mucosa (inflammation)
- Presence of adhesions or urethral prolapse.

Dogs

The onset of puberty in the dog is determined by age, weight and breed. It normally occurs between 7 and 10 months, but may be as early as 5 or as late as 12 months. Given that mating at the wrong time is still the most common cause of infertility, it is important to also take into consideration breeding management and the fertility of the female. Older dogs often present with testicular atrophy, associated with a reduction in the prostatic parenchyma. However, any reduction in semen quality is most often linked to disease (e.g. prostate hyperplasia or prostatitis). The prostate gland has a crucial role in the constitution of the ejaculate. Most dogs develop some degree of benign prostate hyperplasia from the age of 5 years old, which may influence semen quality without any clinical manifestations. Dogs with high inbreeding coefficients (0.125–0.558) have lower reproductive performance (reduced number of puppies, fewer live puppies) compared with dogs with less inbreeding (Wildt *et al.*, 1982).

Physical examination allows assessment of any systemic diseases and evaluation of the external genital tract. The testes should be measured (measurement using ultrasonography seems to be the most accurate technique when compared with a calibrator) and slight asymmetry is common. The prepuce should be retracted past the bulbus glandis part of the penis to assess any adhesions, pain reactions and possible presence of lymphoid follicles. The penile bone should be palpated for the presence of pain, callus or a disjunction. The prostate gland is assessed by digital rectal palpation, noting the position, size, symmetry, consistency and any pain.

Tom cats

It can be difficult to obtain a breeding history in the tom cat, but as much information as possible should be acquired. This should include vaccination records, diet, any signs of clinical disease and knowledge of recent medication. Infectious diseases in the tom cat or in the cattery can also be the cause of infertility. Viral diseases are most common, but poor hygiene or large numbers of animals closely confined can also lead to bacterial infections. Any injury or trauma should be recorded. Pedigrees should be evaluated and questions asked about possible inbreeding.

A complete breeding history including libido, mounting, intromission and pelvic thrust should be obtained. In the cat, successful copulation can also be evaluated indirectly through the queen's post-coital reaction, which is induced by stimulation of the vaginal mucosa by the penile spines. It is important to record the number of queens bred to the tom, the number of matings per queen, pregnancy rates and litter sizes.

Physical and reproductive tract examinations allow the assessment of general health and the external genitalia. The prepuce should be retracted to establish normal mobility of the penis. The presence of a 'hair ring' around the base of the glans is usually associated with unsuccessful mating. The penis is covered by cornified spines and a thickened dermis, indicating normal androgenic activity. Normal tom cat testes are freely movable within the scrotum, round and symmetrical and have a firm consistency. Chromosomal disorders are frequently associated with testicular hypoplasia and azoospermia. The prostate and bulbourethral glands may be evaluated by digital palpation per rectum, but sedation is usually required (Johnston *et al.*, 2001).

Semen collection and evaluation

Semen evaluation is a crucial part of male fertility testing, although it will not give conclusive results (Rodriguez–Martinez, 2003). Only mating or artificial insemination can truly evaluate male fertility and only a few sperm parameters show significant correlation with the prediction of fertility *in vivo*. Therefore, males should not be excluded from breeding on the basis of a single semen evaluation (Root Kustritz, 2007). The dog should be re-examined not less than 2 months after the first examination because recovery of spermatogenesis may take at least this long from an initial trauma. No guidelines for the timing of semen collection in relation to sexual activity have been published for dogs or cats. In humans it is recommended to perform semen collection after sexual rest of at least 2 but no more than 7 days, and to evaluate two ejaculates collected 7–21 days apart (Root Kustritz, 2007).

Collection

The most common techniques for semen collection in the dog are digital manipulation and use of an artificial vagina. Electroejaculation is the method most commonly used in the tom cat, although collection by artificial vagina is possible in previously trained males. If semen collection is not possible, aspiration from the vagina of a recently mated queen, or urinary catheterization after ejaculation by the male, can be used to assess the presence of sperm. The sperm collected via these techniques cannot be used for a complete semen evaluation, but will differentiate azoospermic males from males that can produce sperm.

Semen collection should be performed in a quiet room on a non-slip surface and in the presence of an oestrous female. In some males collection is possible without a teaser female, or by using swabs taken of the discharge of oestrous females, which can be frozen and stored.

After the collection, the ejaculate should be assessed immediately because some parameters change over time. The ejaculate can be kept between 20 and 30°C (at room temperature or in a water bath) and out of direct sunlight. Rapid changes in temperature during handling should be avoided because they damage the sperm.

Dogs

Whenever possible, semen should be collected in the presence of a bitch in oestrus. Male libido can be assessed either by observing copulation or during manual collection. Some breeds and lines are known to have a low libido, despite normal endocrinology and physiology. Other males refuse collection by masturbation (with or without the presence of a bitch in oestrus).

The success of copulation is judged by the speed of mounting the female, obtaining a full erection, intromission, ejaculation and tie. During mounting and thrusting, conditions unrelated to reproduction such as musculoskeletal lesions and arthritis may cause pain and prevent intromission and ejaculation. Prostatic disease and trauma to the os penis can also result in pain reactions during this phase of copulation. This may also be observed during manual collection.

Semen is ejaculated in three fractions: the first pre-sperm fraction, which probably originates from the prostate gland, the semen-rich fraction, and then prostatic fluid. For semen evaluation only the first and second fractions should be collected.

Tom cats

The most common methods used to collect semen from the tom cat are electroejaculation and use of an artificial vagina (Zambelli and Cunto, 2006). Both techniques have advantages and disadvantages, but they both enable the collection of whole ejaculate of good quality that can be used for different purposes. Semen collection may also be performed using urethral catheterization after administration of medetomidine (130–140 mg/kg of bodyweight) (Zambelli *et al.*, 2008). This technique can be used for all tom cats and is easier than the other two techniques. The ejaculates obtained in this way are high in concentration and small in volume and can be used both for cryopreservation and insemination. Owing to the extremely small volumes (µl) of the feline ejaculate it is frequently necessary to use the entire ejaculate for a complete semen evaluation.

Evaluation

- *Macroscopic* parameters that are determined are volume and appearance.
- *Microscopic* parameters that are determined are motility, morphology, concentration and total sperm output (TSO).
- It is important to assess all the parameters to be able to make any predictions or recommendations regarding the fertility of the animal.
- *Further investigations* may include membrane integrity, pH, hypo-osmotic swelling test, sperm penetration assays, identification of other cell types, sperm culture and seminal chemistry, which will give further information for diagnosis (Zambelli and Cunto, 2006).

Macroscopic parameters

Volume: The volume is measured in a calibrated test tube or using a pipette.

Appearance: The normal colour of the sperm-rich fraction of the ejaculate is white or creamy. There may be urinary contamination (yellow), inflammatory cells or purulent material present (yellow to green) or blood (red to brown). Clear or cloudy fluid without any change in colour, indicates a low density or absence of spermatozoa.

Microscopic parameters

Motility: Sperm motility is assessed by placing a drop of undiluted semen on a microscope slide with a cover slip and examining it at X100–200 magnification. Sperm motility is highly sensitive to temperature and the slide should be warmed to 37°C. Motility is evaluated subjectively to give the percentage of total motility (from 0–100%) and progressive motility. Progressive motility is divided into five categories from 0 (non-motile spermatozoa) to 5 (rapidly progressive) (Zambelli and Cunto, 2006). Good quality semen should have at least 70% of sperm in category 5. Semen with a high sperm concentration should be diluted 1:1 with physiological saline before assessment.

Cold or dilution shock, contact of the spermatozoa with lubricants, latex or detergent, and any reproductive tract infection may cause a decline in motility. Progressive motility can increase after the addition of an extender and with repeated collections (Meyers-Wallen, 1991). The speed of progression and percentage of progressively motile spermatozoa are not necessarily correlated with the fertilizing ability of the male (Johnston *et al.*, 2001).

Morphology: Sperm morphology can be assessed using wet preparations or by fixation and staining. An unfixed and undiluted wet smear will give a first impression with few artefacts.

Semen samples can be preserved for later evaluation in Hancock and Gledhill solution. Different stains will show up a variety of morphological abnormalities.

- Fast-green FCF–rose Bengal staining gives the best differentiation of sperm structures (especially acrosome integrity) and results in fewer stain-induced artefacts than other stains.
- Wright–Giemsa may be used to identify contamination of the ejaculate with blood, prostatic cells or spermatogonia.
- Nigrosin–eosin, so called vital staining, will show up live (white) and dead (pink) sperm by penetrating the membranes of dead spermatozoa.

At least 100–200 spermatozoa should be evaluated at X400–1000 magnification. The numbers of live and dead, normal and abnormal spermatozoa should be recorded. In a fertile ejaculate 70–80% of the spermatozoa should be morphologically normal. Sperm defects may be classified as primary, secondary or tertiary, which occur during spermatogenesis, the epididymal phase, or during collection and processing, respectively. Sperm defects may also be classified in relation to the sperm structures (head, acrosome, midpiece, tail) they affect. Other causes of sperm abnormalities are prolonged sexual rest, testicular trauma, reproductive tract infections and pyrexia (Johnston *et al.*, 2001). In the dog, abnormalities of the midpiece are correlated with infertility (Johnston *et al.*, 2001).

Concentration: The concentration may vary greatly depending on the amount of prostatic fluid collected (dog) or the technique of sperm collection (cat). The number of spermatozoa per millilitre can be determined using an electronic counting chamber that is calibrated for dog semen, or manually with a haemocytometer counting chamber.

TSO: The TSO is defined as the ejaculatory volume multiplied by the concentration and it is a more meaningful measure of actual sperm output. It is affected by sexual arousal, frequency of ejaculation and age (it is decreased in young or old animals), and is positively correlated with testicular weight.

Computer-based automated semen analysis (CASA): CASA systems may be used for the assessment of motility, the percentage of morphologically normal sperm, and sperm concentration in dogs (Root Kustritz *et al.*, 2007). This system offers the possibility to collect data more objectively, but owing to its cost it is rarely available to practitioners.

Dogs

Macroscopic parameters: The volume of the ejaculate depends on the size of the dog and testes and is normally:

- First fraction: 0.5–2.0 ml
- Second fraction: 0.5–3.0 ml
- Third fraction: 5–20 ml.

Prostatic fraction: Small volumes of prostatic fluid may be caused by lack of libido (general or due to collection technique), retrograde ejaculation or low output (especially in older dogs with prostatic disease). The prostatic fraction (first and third fractions of the ejaculate) should be clear. The presence of blood (haematospermia) during collection of the third fraction is often a sign of benign prostatic hyperplasia. Dogs with haematospermia are not necessarily infertile, but the underlying cause should be investigated. The presence of purulent contents in the ejaculate can be seen when prostatic abscesses are present and is not always accompanied by systemic disease. Ejaculates contaminated in this way have poor semen quality and should not be used for artificial insemination.

Microscopic parameters: There may be significant variations in semen quality over time. The quality of semen is highest in spring and lowest in summer.

Motility: In fertile dogs, sperm show typically >80% progressive forward movement and anything above 70% is considered normal (Threlfall, 2003). Motility is not affected by sampling frequency (England, 1999).

Morphology: There is no direct correlation between the percentage of abnormal sperm and fertility. However, semen is considered normal when ≥80% of the spermatozoa are normal, and ejaculates showing <60% normal live spermatozoa are classed as subfertile. Repeated sampling does not alter the rate of abnormality (England, 1999). In practice it is sometimes possible to see an improvement in the quality of the semen of giant-breed dogs when a second collection is taken a week after the first one. This observation was not confirmed in a study looking at variations in semen over time (Taha *et al.*, 1983).

Other cells: Epithelial cells (urethral and prostate gland), spermatogonia, erythrocytes and inflammatory cells can often be found in an ejaculate. In healthy dogs concentrations of up to 2000 white blood cells (WBC)/µl in the first two fractions are considered normal. In contrast, in one study 44% of dogs without inflammatory cells had aerobic bacteria isolated from the ejaculate (Amann, 1986).

Tom cats

The normal values for volume (µl) and sperm concentration (x10^6/ejaculate) in feline semen vary with the collection technique and are as follows:

- Artificial vagina: volume = 40 x (10–120); sperm concentration = 57 x (3–143)
- Electroejaculation: volume = 233 x (140–738); sperm concentration = 28 x (9–153)
- Urethral catheterization: volume = 10.5 ± 5.3; sperm concentration = 21 ± 18.1.

There is considerable variation, both within and between cats, in the volume and concentration of spermatozoa produced in response to electrical stimulation. Similar variations are reported among different techniques, and care must be taken when comparing data from different sources or techniques. The higher percentage of normal spermatozoa in ejaculates collected from cats during the breeding season than during the non-breeding season may indicate a possible seasonal effect on male fertility (Axner and Linde Forsberg, 2007). However, the percentage of abnormal sperm in an ejaculate is not always correlated with infertility, and in one study a tom cat with fewer than 40% normal spermatozoa sired kittens. It is probably more important to consider the proportion of different defects than the percentage of normal spermatozoa, because different defects may have different effects on fertility. Recording all defects, even if a spermatozoon has more than one, seems to result in more reliable information relating to fertility (Axner, 2008).

Sperm morphology is likely to be associated with fertility, although results must be interpreted with caution, and both morphology and fertility may change over time (Axner and Linde Forsberg, 2007). The most common sperm abnormalities in the cat are: pyriform sperm; micro- and macrocephaly; abnormal acrosome; bicephalic and biflagellate; abaxial flagellar attachment; coiled and abnormal midpiece formation; immature sperm; proximal and distal retained cytoplasmic droplet; detached heads; bent flagella or neck (Howard, 1992).

Hormone measurement

Steroid hormones (testosterone, oestradiol) are frequently investigated by practitioners without prior knowledge of their dynamics or significance. Some males may have normal fertility with a normal libido and low testosterone concentrations, whilst others have serious functional impairments despite normal or elevated hormone concentrations. This is due to the pulsatility of secretion of testosterone from the Leydig cells. In order to evaluate testicular function a stimulation test with comparative blood results should be performed (Johnston *et al.*, 2001). There are several protocols for Leydig cell stimulation tests.

Blood samples may be taken before and 24 hours after intramuscular injection of human chorionic gonadotrophin (hCG; 30–50 IU/kg i.m.) and analysed for testosterone. When using intravenous hCG the second sample must be timed carefully 1.5 hours after stimulation, otherwise testosterone concentrations may return to basal concentrations. Gonadotrophin-releasing hormone (GnRH) may also be used as a stimulating agent (0.1–1 µg/kg i.v. or i.m.), with the second blood sample taken 1–2 hours after the injection. The expected values post-stimulation are at least two to three times higher than the basal values: a post-stimulation concentration of <2 ng/ml is considered diagnostic for hypogonadism in both dogs and cats. There are no maximum levels.

Dysfunction of the hypothalamus or the pituitary gland can affect the hypothalamic–pituitary axis, inhibiting the secretion of follicle-stimulating hormone (FSH) and luteinizing hormone (LH) and leading to secondary impairment of spermatogenesis. Quantitative assays of FSH and LH are limited to research laboratories. However, there is a semi-quantitative test which can be used with some success.

Dogs

Fertile dogs with a normal libido rarely show concentrations of testosterone <0.4 ng/ml. Dogs with testosterone concentrations of >30 nmol/l (9 ng/l) post-hCG stimulation are within the normal range, whereas values <20 nmol/l (5.71 ng/ml) are considered to be low and may indicate hypogonadism.

Any disorder of the adrenal glands (Addison's disease or Cushing's syndrome) and hypothyroidism can have a marked effect on gonadal function. The best way to evaluate thyroid function is to determine free thyroxine (T4) levels in combination with an assay for thyrotrophin-releasing hormone (TRH).

Testicular tumours often secrete hormones and will therefore have an effect on both testes. Serum oestrogen concentrations of >10 pg/ml indicate active Sertoli cell or Leydig cell tumours. Hyperoestrogenism (>1 ng/ml) has also been described in Sertoli cell tumours, accompanied by feminization of male dogs. Feminization may also be caused by a rise in inhibin and low concentrations of LH and FSH.

Tom cats

Testicular tumours in the cat are uncommon, and feminization associated with Sertoli cell tumours has not been reported. Androgen activity may be evaluated in the cat by the presence or absence of cornified spines on the penis (see Chapter 4). Normal testosterone concentrations range from 0.05–3 ng/ml. Owing to the pulsatility of testosterone release in the cat, assessments should be made after stimulation with hCG (250 IU i.m.) or GnRH (25 µg i.m.). The testosterone concentration should be 3.1–9 ng/ml 1 hour post-hCG and 5–12 ng/ml 4 hours post-GnRH administration. An increase in serum testosterone concentration following hCG or GnRH stimulation can be used to confirm retained teste(s). Seasonal variations in epididymal sperm quality and testosterone production between spring and autumn have been found (Blottner and Jewgenow, 2007).

Diagnostic imaging

Radiography was more commonly used in the past, but has been replaced in some areas by ultrasonography.

Radiography

Radiography is still used widely to investigate any fractures of the os penis, or to perform contrast studies of partial/total urethral obstruction or colonic compression by an enlarged prostate gland. It can also be used to check for bone or thoracic metastasis in cases of primary prostatic tumours. Serial contrast radiographs can highlight connections between the prostatic urethra and intraparenchymatous cavities, revealing urethral rupture or fine irregularities of its surface. This can also be useful to 'map' large prostatic lesions before surgery.

Ultrasonography

Ultrasonography is particularly useful for the testes, prostate gland and any other soft tissue within the abdominal cavity associated with the reproductive tract. A 5 or 7.5 (preferred) MHz transducer is sufficient for most routine examinations, but higher frequency transducers may be necessary to assess superficial or small structures (e.g. testicular vascular cones, feline prostate/bulbourethral glands). Ultrasound-guided fine needle aspiration and biopsy are useful techniques to provide specimens for cytology, bacteriology and histology.

Dogs

Radiography allows assessment of the penile bone and prostate gland. The size of the prostate gland in comparison to the size of the dog can be evaluated by measuring the distance between the sacrum (S) and the tip of the pubis (P) in a straight line. Prostatic hyperplasia is diagnosed when the diameter of the prostate gland is >70% of the S–P distance on lateral views. If the prostate gland occupies >50% of the width of the internal pelvic diameter on ventrodorsal views it is also considered enlarged. Radiography as a technique tends to overestimate the size of the prostate gland, and standards do not take into account the age and breed of the dog. A study of all available literature was used to develop a guide for normal prostate gland size with regard to age and weight (Ruel *et al.*, 1998). The problem was that significant deviations of up to 20% occurred depending on whether the prostate gland was measured in transversal or longitudinal section (Atalan *et al.*, 1999).

Tom cats

Ultrasonographic imaging of the accessory glands is also possible in the tom cat, but no studies have been published about its specific use to diagnose reproductive disorders. In an unpublished study (Zambelli, 2009) the ultrasonographic appearance of the normal prostate and bulbourethral glands has been evaluated. The prostate gland surrounds the pelvic urethra, close to the pelvic brim, and appears as a globoid gland (0.6–0.9 cm). The bulbourethral glands are spherical in shape (0.3–0.5 cm) and located lateral to the urethra, close to the perineum. Both glands show a medium to fine echoic texture and a homogenous and moderately echogenic parenchyma.

Poor semen quality

Many conditions can reduce the quality of semen; these may be systemic or specific to the reproductive tract. Poor semen quality may be associated with prostate gland, testicular or epididymal diseases, or urinary conditions. Other conditions related to semen quality are hormonal disorders, infectious diseases of the reproductive tract (*Brucella canis*, aerobic bacteria, ureaplasma), genetic abnormalities (chromosomal abnormalities, inbreeding) ongoing medication (steroids, antifungals, certain antibiotics) and miscellaneous causes such as scrotal hyperthermia, age, pollution (dioxins) and stress. The ability to define different anomalies during semen analysis will help to identify the cause and suggest specific therapies.

Oligozoospermia

This term describes a low total number of spermatozoa per ejaculate. The TSO for a normal healthy dog should be >300 million sperm (20 million TSO/kg) (Amann, 1986), and for a fertile mating at least 150–200 million spermatozoa must be produced. Some very small toy breeds may produce less sperm and still be considered normal. Oligozoospermia is not uncommon and such dogs are not necessarily infertile. A low concentration of semen in the ejaculate may be caused by frequent use of the stud dog (ejaculations more than every 48 hours), resulting in depletion of the epididymal reserve (Taha *et al.*, 1983). Seasonal oligozoospermia may be a problem in hot countries.

A decrease in sperm production may be secondary to hormonal suppression associated with unilateral Sertoli cell tumours, hypothyroidism and hyperadrenocorticism. Prostatic disease can affect the quality and quantity of semen. Infectious diseases that cause local inflammation and hyperthermia and therefore increase intrascrotal temperature can also cause oligozoospermia. Immune-mediated orchitis with decreased testicular size has also been described.

Treatment of oligozoospermia depends on the aetiology, and improvements will take at least 60 days, i.e. until a new spermatogenic cycle has been completed. Removal of tumours and treatment of any infection may bring improvement. If extensive inflammatory and degenerative changes have taken place a return to normal fertility is unlikely.

Teratozoospermia

Although teratozoospermia (>60% of spermatozoa with morphologically abnormal forms) is associated with infertility in dogs, there are few descriptions of specific abnormalities associated with reduced fertility. The anomalies most often cited to be responsible for infertility are anomalies of the midpiece, head and proximal cytoplasmatic droplets. An increase in abnormal spermatozoa may be caused by testicular tumours, orchitis, prostatitis or high fever, or may be

of idiopathic aetiology. The treatment depends on the underlying cause, and the response will take at least 60 days.

Haematospermia

Haematospermia is defined as an ejaculate that contains macroscopically noticeable amounts of blood. It is most common in older dogs with benign prostatic hyperplasia and does not necessarily lead to infertility. If the semen sample has been collected for freezing it should be centrifuged and resuspended in extender to avoid any long-term damage. Malignant neoplasia of the prostate gland or testes can also cause haematospermia and such dogs should undergo ultrasound examination of both organs.

Asthenozoospermia

Asthenozoospermia is defined as progressive motility of <70% and is often associated with teratozoospermia. Causes include testicular tumours, infections of the reproductive tract or contaminated collection equipment, and idiopathic cases also occur. In some cases collection of the ejaculate straight into extender solution may be helpful.

Azoospermia

The absence of spermatozoa and the collection of a clear sample is a reason to investigate possible azoospermia. In dogs, this situation may arise because the male either does not have any sperm or does not ejaculate the sperm-rich fraction. Dogs can ejaculate the first fraction, bypass the sperm-rich fraction, and then release the prostatic fluid. This may be associated with stress, lack of libido, immaturity or an experienced stud dog uncomfortable with manual collection.

In order to differentiate between true azoospermia and the failure to release the sperm-rich fraction during ejaculation, the complete sample can be analysed for alkaline phosphatase (ALP). The level of ALP is much higher in epididymal fluid than in prostatic fluid. Samples from dogs that have failed to provide a complete ejaculate normally contain <5000 IU/l ALP. Dogs with true azoospermia have ALP concentrations of 5000–20,000 IU/l. The diagnostic investigation of azoospermia (Figure 8.1) involves determination of the level of ALP in the fluid collected.

True azoospermia may be the end stage of many of the conditions described for oligozoospermia. It may also occur in intersex animals and with germinal cell aplasia, bilateral cryptorchidism, testicular injury and trauma, and testicular neoplasia (producing high ALP concentrations). In cases with genetic abnormalities karyotyping will be useful. Post-testicular causes of azoospermia include obstructions, resulting from aplasia of parts of the epididymis, spermatoceles and sperm granulomas (low ALP concentration). In all such cases the prognosis for future fertility is guarded or poor.

Retrograde ejaculation can occur during collection, and spermatozoa will be found in urine samples following ejaculation. The underlying cause, possibly inflammation of the urethra, bladder or prostate

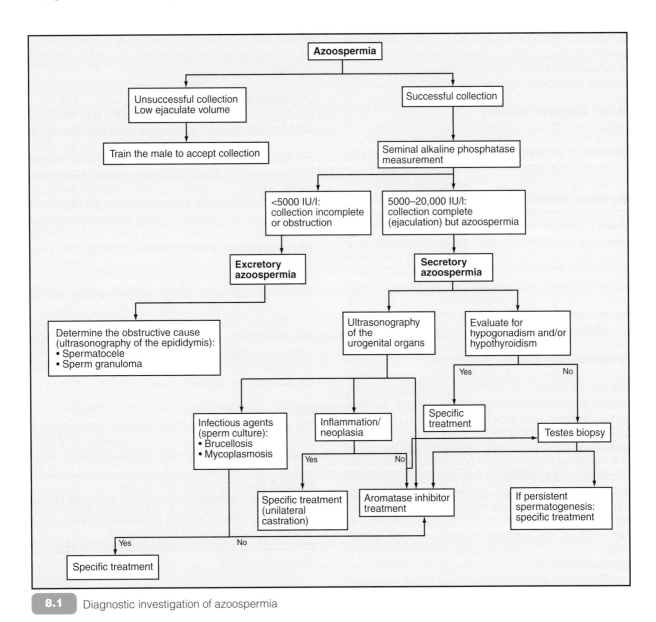

8.1 Diagnostic investigation of azoospermia

gland, cannot always be found, but treatment with phenylpropanolamine (3–4 mg/kg q8–12h) can be attempted in dogs (Beaufays *et al.*, 2008).

Urospermia
The presence of urine is common in the first fraction of canine sperm. Urine is toxic to sperm and often causes asthenozoospermia. Urospermia is common in dogs that have been excited for prolonged periods (e.g. a dog living with an oestrous female before collection). Other causes include urethritis and cystitis. Collection techniques used during electroejaculation (urinary catheter inside the bladder, excessive voltage during electrical stimulation) may also lead to contamination of the ejaculate during collection. Determination of creatinine concentrations in the ejaculate (they should be zero) will detect even small amounts of urine.

Dogs
Older dogs frequently show haematospermia attributable to prostatic disorders. This does not affect fertility unless it is associated with asthenozoospermia

or teratozoospermia. Aspermia in dogs is uncommon and retrograde ejaculation should be ruled out.

Tom cats
The cat is frequently affected by teratozoospermia (Pukazhenthi *et al.*, 2001), and this has been associated with decreased genetic variation and low circulating testosterone concentrations. Causes of azoospermia include administration of long-term and/or high doses of glucocorticoids, other steroids, fever and systemic illness. More common testicular causes are intersexuality, cryptorchidism, trauma, orchitis, degeneration, hypoplasia and neoplasia. Conditions that affect the duct system include epididymal cysts, epididymitis, spermatoceles and sperm granulomas. Congenital or acquired obstruction of the epididymis and ductus deferens may induce azoospermia or oligozoospermia.

Localization and analysis of the possible causes of the defect is necessary before attempting therapy. The clinical approach includes physical examination, ultrasonography, semen collection and evaluation, evaluation of urinary ALP (after ejaculation) to evaluate retrograde ejaculation and sperm in the urine (indicating partial or total duct impatency), together with testicular cytology (or preferably histology). The prognosis is often poor. Treatment is only possible in very few cases (i.e. those with fever, systemic illness, unilateral testicular trauma, orchitis and neoplasia, or unilateral obstruction of the duct system) when the testes have not been damaged permanently and semen quality may improve after treatment.

Failure to achieve an erection

Sexual arousal normally leads to erection in the male. Failure to achieve or to maintain an erection is a common cause of mating failure. Erection may never be achieved in the first place or may subside before a successful mating has been completed. The reasons for this may be psychological, anatomical, endocrinological or congenital.

Psychological problems are very common in mating situations. The male and female may have never met before the oestrous female is introduced to the stud dog. Aggressive or dominant bitches may refuse mating and make their dislike known to the dog. Usually mating takes place in the home environment of the male to give him some advantage. Inadequate or stressful environments (such as slippery floors), small areas that prevent the animals from getting away from each other, and inexperienced or cautious owners can make mating impossible. Males that have had previous unpleasant experiences may not want to mate with a female again. In some cases sexual behaviour has been discouraged by the owner over a long period because it is inappropriate in an everyday context, and so the male cannot understand why this should be different in the new situation. Finally, the two animals chosen by the breeders may simply not like each other.

A successful mating that achieves erection, intromission and a tie requires a certain amount of agility and mobility on the part of the male. If at any stage the animal experiences pain, this may make him abort the attempt. Pain may be caused by congenital or acquired phimosis, penile traumatic lesions (e.g. to the os penis), testicular conditions such as orchitis or torsion, and acute prostatitis. Musculoskeletal problems such as degenerative joint disease, lumbosacral disease and other conditions may also prevent the male from maintaining the correct posture during mating. Endocrine problems, congenital conditions such as testicular hypoplasia, degeneration/atrophy or hypotrophy, and intersexuality will also prevent the male from mating the female.

The diagnostic approach to erection failure requires a clinical examination to exclude congenital or acquired defects and/or to localize painful conditions. Observation of breeding behaviour permits the clinician to collect useful information about psychological problems or painful diseases. Assay of the serum testosterone concentration following stimulation and karyotyping may be necessary in the investigation of non-painful conditions.

Treatment should be aimed at resolving the specific problem and, in normal males, libido may be improved before sperm collection or mating by administration of GnRH. Care must be taken when using GnRH because frequent use may induce a decrease in endogenous testosterone concentrations caused by negative feedback from the pituitary gland.

Dogs
Pelvic nerve (parasympathomimetic) stimulation generates an increase in arterial flow and occlusion of venous return, which leads to a rapid increase in the volume of the spongious and cavernous bodies, and thus a rapid increase in penis length and volume. The dog protrudes the penis from the prepuce before the erectile bulbs swell. Bad experiences caused by incorrect handling by the owner, bitch aggression, pain at the time of an earlier mating, musculoskeletal problems (os penis, joint diseases), injuries and malformations (phimosis) are the most common causes of failure to achieve an erection. Lower urinary tract disease (prostate gland or bladder disease) is rarely the cause of erection failure but must be ruled out. Dogs with poor libido can be trained to ejaculate in response to manual stimulation, which should be repeated at regular intervals. Injection of GnRH before sampling may enhance libido in the short term. In the case of phimosis, surgical correction will be necessary.

Tom cats
Erection is mediated through the parasympathetic nervous system, especially portions of the second sacral nerve roots. Assessment of full erection in the tom cat during copulation is not easy owing to the size of the penis and breeding behaviour. Evaluation of libido on the basis of the behaviour of the male in the presence of an oestrous queen may provide some information. Aggressive queens, unfavourable environments and mate preferences are the most common causes of psychological problems. Penile hair rings (more frequent in longhaired cats), congenital phimosis, post-traumatic phimosis and orchitis (e.g. secondary to bite wounds from other cats) are painful conditions that may prevent erection. Bites may also induce painful muscular abscesses. In each case specific therapies may correct the problem and restore normal libido. Administration of GnRH (25 µg i.m.) followed by sperm collection or mating 1 hour later should also improve performance. If necessary, selection of another queen may be tried.

Failure to achieve normal copulation

Failure to achieve normal copulation may be caused by male or female problems. Age should be considered; immature males may be physiologically unable to copulate, and aged males can fail because of the normal decrease in serum testosterone concentration with age. Endogenous testosterone concentrations can decrease following testicular conditions such as hypoplasia, degeneration/atrophy, hypotrophy and neoplasia, or secondary to drug administration (exogenous testosterone, anti-androgens, progestogens, glucocorticoids, ketoconazole, cimetidine, anabolic steroids, oestrogens, chemotherapy). Painful (musculoskeletal problems of the neck, spinal column and joints), neurological (protruding intervertebral discs and degenerative myelopathy), abdominal (including prostatic enlargement), urinary and genital (orchitis or persistent frenulum) conditions may cause problems with copulation.

Many of the conditions mentioned above can be diagnosed with a physical examination and other diagnostic tests, and treated with specific therapies. If the condition preventing a successful mating cannot be treated, semen may be collected and used for artificial insemination, although Kennel Club regulations on the use of artificial insemination should be considered if relevant.

In the male, if no pathological condition can be diagnosed, physiological causes should be considered. Good socialization for puppies, positive sexual experiences at puberty and optimal breeding management can be used by owners to prevent many psychological problems in the stud dog later in life. Libido and successful copulation can also be influenced by genetic selection.

Dogs

Inexperienced dogs that get excited without mounting may experience premature swelling of the erectile tissue, which may prevent extension from the prepuce and/or intromission. Older dogs may also have difficulties in mounting, decreased libido (Taha *et al.*, 1983) or musculoskeletal problems. The constant presence of oestrous females can also reduce libido, which will be restored after 2 days of rest. A dog may need some time to approach the bitch. It can take several minutes or hours before a mating can be achieved. Repeated negative reinforcement when showing sexual interest, stressful environments or uncooperative dominant females may prevent the male from mating. Some dogs have an idiopathic lack of libido. Stimulation with GnRH can be attempted 2–3 hours before mating.

Tom cats

Normal copulation in the cat involves the tom approaching the queen, investigating her perineal region and after a few minutes immobilizing the female by biting her neck. Pelvic thrusting will achieve intromission and ejaculation in a short time. Multiple matings over the first 2–3 days of oestrus usually induce ovulation in the female. Observation of the behaviour of the male during mating will define normal copulation. Prolonged pelvic movements or several attempts to mount without biting the neck of the queen are considered abnormal. Copulation with complete intromission induces the characteristic post-coital reaction in the queen: ovulation and a rise in serum progesterone concentration about 2–3 days later.

Despite normal erection and copulation, ovulation may fail. Male causes for this problem are too few matings or inadequate vaginal stimulation. Young

males, vaginal anomalies in the queen, and some of the reasons described above for failure to achieve an erection may also cause failure to achieve normal copulation. Gingivitis, stomatitis and other painful conditions of the mouth are common in the cat and may prevent female immobilization during coitus. Some psychological problems may be resolved if another female (more receptive, experienced and tolerant) is introduced. With inexperienced tom cats the queen should be introduced into the male environment. Multiple copulations should be permitted and encouraged to ensure normal ovulation rates. If ovulation cannot be induced by mating, administration of hCG (250 IU i.m.) to the female may resolve the problem.

Failure to ejaculate

Ejaculation requires the emission of sperm from the testis and epididymis into the prostatic urethra (through the vas deferens), the closure of the bladder sphincter, and propulsion of the ejaculate through the penile urethra. The ejaculate is composed largely of secretions from the prostate (in the dog and cat) and the bulbourethral glands (cat only). The orthosympathetic system controls the closure of the urethral sphincter. The propulsion of semen through the penile urethra is controlled by the somatic system, which induces contraction of the bulbospongiosus and ischiocavernosus muscles. Ejaculation follows intromission of the penis in both cats and dogs in less than 1 minute.

The absence of ejaculation in the presence of an erection and ejaculatory behaviour is defined as aspermia. Aspermia may be caused by sexual immaturity, pain, psychological factors, or iatrogenic and neuropathic conditions (commonly of idiopathic origin, or secondary to diabetes mellitus or spinal cord injury). It is possible to swab or aspirate spermatozoa after mating in 65% of cases if the procedure is performed within 24 hours (Bowen *et al.*, 1985). During mating itself, it is important to note the behaviour of the male (overlap, nibbling in the cat). If mating cannot be achieved or no sperm can be recovered from the vagina post-mating, a semen sample should be taken using either masturbation or electroejaculation.

Dogs

The dog produces the first fraction of the ejaculate within 20 seconds of thrusting movements, and the spermatic fraction is released once movement has ceased, usually in less than a minute. During semen collection it is important to note the behaviour of the male during ejaculation, the emission of the different fractions without interruption, and the presence of penile muscle contractions. Reluctant or immature dogs may only release the first and third fractions without the spermatic fraction (there will be a pause during collection). In cases of normal collection behaviour and persistent aspermia, urine analysis to investigate retrograde ejaculation and ALP measurement to rule out azoospermia should be performed. In cases of behavioural aspermia, GnRH stimulation (1–2 µg/kg, 2–3 hours before sampling) is attempted. In cases of retrograde ejaculation, an attempt may be made to treat the underlying cause, or semen collection can be performed on a dog with a full bladder, followed by treatment with phenyl-propanolamine (3 mg/kg orally q8–12h) combined with a non-steroidal anti-inflammatory drug (NSAID) for 5 days.

Tom cats

Intromission and ejaculation occur in 1–27 seconds in the cat, after which the male dismounts rapidly. Aspermia in the cat and failure to ejaculate are not well described in the literature. Possible causes are sexual immaturity and pain at the moment of ejaculation. It is possible to obtain spermatozoa from a vaginal swab after copulation. Different degrees of spermatic retrograde flow are considered a normal component of the ejaculatory process in the cat; in fact 15–90% of the ejaculate is retrograde and flows into the urinary bladder (Dooley *et al.*, 1991). The amount of retrograde ejaculation is not influenced by the voltage used during electroejaculation.

References and further reading

Amann RP (1986) Reproductive physiology and endocrinology of the dog. In: *Current Therapy in Theriogenology*, ed. DA Morrow, pp. 532–538. WB Saunders, Philadelphia

Atalan G, Holt PE and Barr FJ (1999) Ultrasonographic estimation of prostate size in normal dogs and relationship to bodyweight and age. *Journal of Small Animal Practice* **40**, 119–122

Axner E (2008) Updates on reproductive physiology, genital diseases and AI in the domestic cat. *Reproduction in Domestic Animals* **43**(Suppl. 2), 144–149

Axner E and Linde Forsberg C (2007) Sperm morphology in the domestic cat, and its relation with fertility: a retrospective study. *Reproduction in Domestic Animals* **42**, 282–291

Beaufays F, Onclin K and Verstegen J (2008) Retrograde ejaculation occurs in the dog, but can be prevented by pre-treatment with phenylpropanolamine: a urodynamic study. *Theriogenology* **70**, 1057–1064

Blottner S and Jewgenow K (2007) Moderate seasonality in testis function of the domestic cat. *Reproduction in Domestic Animals* **42**, 536–540

Bowen RA, Olson PN, Behrendt MD *et al.* (1985) Efficacy and toxicity of estrogens commonly used to terminate canine pregnancy. *Journal of the American Veterinary Medical Association* 186, 783–788

Dooley MP, Pineda MH, Hopper JG and Hsu WH (1991) Retrograde flow of spermatozoa into urinary bladder of cat during electroejaculation, collection of semen with an artificial vagina, and mating. *American Journal of Veterinary Research* 52, 687–691

England GCW (1999) Semen quality in dogs and the influence of a short-interval second ejaculation. *Theriogenology* 52, 981–986

Howard JG (1992) Feline semen analysis and AI. In: *Kirk's Current Veterinary Therapy XI: Small Animal Practice*, ed. Kirk RW and Bonagura J, pp. 929–938. WB Saunders, Philadelphia

Johnston SD, Root Kustritz MV and Olson PNS (2001) *Canine and Feline Theriogenology*. WB Saunders, Philadelphia

Meyers-Wallen VN (1991) Clinical approach to infertile male dogs with sperm in the ejaculate. *Veterinary Clinics of North America: Small Animal Practice* 21, 609–633

Pukazhenthi B, Wildt DE and Howard JG (2001) The phenomenon and significance of teratospermia in felids. *Journal of Reproduction and Fertility* 57(Suppl.), 423–433

Rodriguez–Martinez H (2003) Laboratory semen assessment and prediction of fertility: still utopia? *Reproduction in Domestic Animals* **38**, 312–318

Root Kustritz MV (2007) The value of canine semen evaluation for practitioners. *Theriogenology* 68, 329–337

Ruel Y, Barthez PY, Mailles A and Begon D (1998) Ultrasonographic evaluation of the prostate in healthy intact dogs. *Veterinary Radiology and Ultrasound* 39, 212–216

Taha MS, Noakes DE and Allen WE (1983) The effect of frequency of ejaculation on seminal characteristics and libido in the Beagle dog. *Journal of Small Animal Practice* 24, 309–315

Threlfall WR (2003) Semen collection and evaluation. In: *The Practical Veterinarian: Small Animal Theriogenology*, ed. MV Root Kustritz, pp. 97–123. Butterworth-Heinemann, St. Louis

Wildt DE, Baas EJ, Chakraborty PK, Wolfle TL and Stewart AP (1982) Influence of inbreeding on reproductive performance, ejaculate quality and testicular volume in the dog. *Theriogenology* 17, 445–452

Zambelli D and Cunto M (2006) Semen collection in cats: techniques and analysis. *Theriogenology* 66, 159–165

Zambelli D, Prati F, Cunto M, Iacono E and Merlo B (2008) Quality and in vitro fertilizing ability of cryopreserved cat spermatozoa obtained by urethral catheterization after medetomidine administration. *Theriogenology* 69, 485–490

9

Artificial insemination in dogs

Wenche K. Farstad

Introduction

Artificial insemination (AI) is the process by which spermatozoa are placed into the reproductive tract of a female for the purpose of impregnation using means other than natural mating. The main indications for the use of AI in dogs are:

- The geographical distance between dam and sire – for use of chilled or frozen semen
- Physical difficulties in mating or unwillingness to mate on the part of the bitch or dog – for use of freshly ejaculated semen.

At times when natural mating may not be possible or desirable, AI is an invaluable tool. The advantages of using frozen semen include the ability to store genetic material well beyond the lifetime of the donor dog, the possibility of transporting semen to bitches worldwide, and the convenient storage of semen until the optimum breeding time.

Today, nearly 40 years after its introduction, the interest in exchanging frozen canine semen among dog breeders is increasing rapidly. This is due to: a greater risk of disease transmission associated with international travel; variable national rules concerning physical modifications of live animals (e.g. tail docking), which may prevent importation of an animal to be used for show or competition purposes; and the economic incentive to develop elite breeding stock, e.g. guide dogs for the blind, sniffer dogs or police dogs. In addition, the improved fertility results after AI and the availability of fresh, chilled and frozen semen contribute to the increasing interest.

Despite considerable progress in, and prevalence of, the use of AI in dogs, there is still marked variation in the success rate between different clinics and laboratories. The success in achieving canine pregnancies with AI still rests on semen quality, accurate timing of insemination and the site of deposition of the semen.

Semen collection

Semen collection from a stud dog is usually simple if the dog is sexually excited. The presence of an oestrous female is useful (although not always necessary) and makes collection easier, particularly in dogs that are not trained for semen collection or

dogs that are inexperienced or timid. When the dog attempts to mount the bitch, the penis is redirected manually into a collection funnel or an artificial vagina, and digitally stimulated to cause ejaculation. Collection of semen should not be attempted until full erection has occurred and the thrusting movements have ceased (if a collecting funnel is used), so as not to damage the mucosa (Figure 9.1).

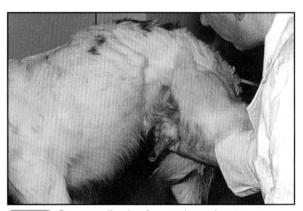

9.1 Semen collection from a dog using a teaser bitch. The prepuce is manually deflected caudally and the penis gripped firmly behind the penile bulb, whilst the collector's fingers constrict the base of the penis. Semen may be collected directly into a disposable plastic funnel.

The ejaculate consists of three fractions: the first and third fractions consisting of prostatic fluid, and the second being rich in spermatozoa. It should be noted that prostatic fluid accounts for >95% of the volume of the canine ejaculate. Prostatic fluid (2–20 ml) is believed to facilitate intrauterine deposition of the semen by flushing the female reproductive tract during the coital tie. There have been some conflicting reports concerning the effect of prostatic fluid on sperm function in ejaculated and processed semen. Some authors have demonstrated that prostatic fluid compromises sperm function following semen preservation, whereas other studies have shown that prostatic fluid enhances the fertility of frozen–thawed semen, as well as has a positive effect on the post-thaw quality of epididymal sperm.

Hence, in the author's laboratory the following procedure has been adopted. When the ejaculate is well fractioned, only the second (sperm-rich) fraction should be collected if the semen is to be diluted and frozen, or transported in the fresh diluted state. In cases where the semen will be used immediately for AI, 1–2 ml of prostatic fluid can be allowed into the funnel, and the semen may be used directly with no other extender. Semen may be collected twice at 30-minute intervals from most dogs, and if the second sample appears to be dilute, centrifugation at 700–1000 g for 5 minutes can be undertaken. After removal of the supernatant, the resulting sperm pellet may be added to the first sample.

Semen examination

After collecting the semen and before further processing, a sample from the whole ejaculate or, preferably, the sperm-rich fraction should be checked to ensure that it is sufficiently concentrated and that the spermatozoa are progressively motile and appear normal morphologically. The semen should also be examined for colour. The normal colour is that of skimmed milk to slightly creamy depending on density; it should never be yellow (indicates contamination with urine or inflammatory cells) or red (blood-tinged). If the colour is abnormal, the cause should be established by microscopic evaluation. Semen with leucocytes, erythrocytes or urine may be washed by centrifugation at 700–1000 g for 5 minutes in TRIS extender (see Figure 9.3) or saline with antibiotics, the supernatant discarded, and the pellet re-suspended in fresh extender with antibiotics. The semen may then be used for AI.

Cryopreservation
Full details on semen evaluation methods are given in Chapter 8, but are described briefly below with particular reference to selection of a sample for cryopreservation.

Sperm concentration
The total number of spermatozoa in canine semen varies from 200 to 1200 million, with a concentration of 100–700 million spermatozoa per ml. Assessment of semen density (i.e. the cell concentration in spermatozoa per ml) can be made using microscopy and a counting chamber (e.g. bright line haemocytometer), or using photometers calibrated for dog semen (e.g. SpermaCue™) which provide a digital readout of the concentration in million spermatozoa per ml. From the volume and concentration the total number of spermatozoa can be calculated. Another benchtop piece of equipment that allows precise sperm counts as well as measurement of viability is the Nucleocounter SP-100™, which also facilitates accurate dilution of canine semen.

Progressive motility
Progressive motility is usually estimated by visual inspection using a light contrast microscope at 100X magnification. Progressive motility (the percentage of spermatozoa with forward movement) is estimated to the nearest 5%, and values normally range from 75–90% in undiluted freshly collected semen. If the semen has been frozen, progressive motility post-thaw should be a minimum of 50% if it is to be used for AI. The speed of forward progression (velocity) is also important. This can be scored subjectively on a scale of 0 to 5: where 0 denotes no forward progression (necrospermia); and 5 denotes the most rapid forward movement. Normal fertile canine semen is usually graded between 3 and 5, and semen to be used for cryopreservation should show mostly fast forward progression (i.e. preferably grade 5) prior to freezing.

Low temperatures reduce sperm motility, and the use of a warming plate for the microscope, or use of pre-warmed slides, is recommended, particularly when frozen–thawed semen is examined. Extenders that contain glycerol or chelating agents also depress motility temporarily; however, motility is usually restored after 2–3 minutes on the warming slide. Normal canine ejaculates should have a minimum of 75–90% progressively motile and rapidly moving spermatozoa, and <25% should show morphological defects (see Chapter 8). Only the higher motility ejaculates with rapid speed of progression and good morphology are recommended for cryopreservation.

Morphology
Morphology is given as the percentage of spermatozoa with a normal appearance and is evaluated using a phase contrast microscope at a minimum of 200X magnification. The percentage of sperm showing normal morphology in canine semen varies from 75–90%. The minimum percentage of morphologically normal sperm in a canine ejaculate that is required to maintain normal fertility is not established, but values below 60% usually indicate some disturbance in testicular or epididymal function. For semen intended for cryopreservation at least 75% of spermatozoa should be morphologically normal prior to freezing.

Semen dilution and artificial insemination dose

Normally, freshly ejaculated sperm is not diluted (except for the small volume of prostatic fluid) if it is to be used for AI immediately after collection. However, if the semen is going to be transported with a delay before insemination of 2–3 hours or longer, it should be diluted with either a semen extender containing a cooling protectant (e.g. egg yolk) and transported at 5°C, or an extender containing compounds that ensure chemical preservation (in which case egg yolk is not necessary). Unless such chemicals are used, the longevity of canine semen is considerably longer at 4°C than at 22°C. The author has successfully used the two different extenders for the transport of fresh semen (Figures 9.2 and 9.3). Commercial extenders are now available that claim to maintain a minimum of 70% relative motility over a 10-day period at 5°C.

Component	Concentration
Pasteurized cream (12% fat)	8 ml
Egg yolk (20% v/v)	2 ml
Potassium penicillin	1000 IU/ml

9.2 Composition of the egg yolk cream extender for dilution and transport of fresh chilled semen.

Component	Concentration
Trishydroxy methyl aminomethane	6.056 g
Citric acid	3.400 g
Fructose	2.500 g
Double distilled water	200 ml
Note: this solution is boiled and cooled before the addition of antibiotics and egg yolk (and glycerol for freezing medium)	
Crystalline penicillin (may be substituted with sodium benzylpenicillin or potassium penicillin)	200,000 IU (1000 IU/ml)
Dihydrostreptomycin	0.2 g
Note: the egg yolk is added just prior to dilution (20% v/v) – 2 ml egg yolk to 8 ml TRIS buffer base and antibiotics	
Cryopreservation	
This buffer can be used with the following modifications for cryopreservation:	
The volume of glycerol added to the solution is either 12 ml (6% v/v) or 16 ml (8% v/v) depending on the freezing regime: 6% glycerol for automatic freezing; 8% glycerol for manual freezing. The distilled water content in the solution is adjusted accordingly to either 188 ml (6% glycerol) or 184 ml (8% glycerol), instead of the original 200 ml.	

9.3   Composition of the TRIS extender (without glycerol) and buffer base with antibiotics for transport of fresh chilled semen or with the addition of glycerol, for cryopreservation.

Another important factor that affects sperm viability and fertility during cooled transport is the extender-to-semen dilution ratio. Semen should be diluted to the appropriate concentration for transport or freezing by careful dropwise addition of extender at 35°C, after evaluation of sperm morphology and concentration. If chilled, the dilution should be at least 2:1 of extender-to-semen rich fraction of the ejaculate, to ensure that enough cooling protectant is added.

For canine ejaculates with normal sperm concentrations the semen is extended by 1:3–6, i.e. to a semen concentration of 100–200 million spermatozoa per ml. The dilution rate depends on the original sperm concentration and number of AI doses required. There is a limit to the insemination volume, particularly with intrauterine AI. It is usually not beneficial to dilute the semen to a higher volume than is needed for one or two AI procedures if the semen is going to be used for one bitch only, because large volumes may overload the uterus and elicit backflow. Sperm doses for vaginal or non-surgical intrauterine AI typically contain a minimum of 100 million sperm per ml, and frequently 200 million per ml. In non-surgical AI the semen is deposited into the body of the uterus, whereas in surgical AI the semen is often deposited into the uterine horns or even occasionally into the oviduct/uterine junction, hence a lower number of sperm may be needed for surgical AI. If frozen semen is used, a larger total number of sperm is required than with fresh semen, owing to the loss of viability by 20–50% as a result of damage to the sperm during freezing and thawing. Therefore, even at the time of dilution it is necessary to keep in mind how the semen will be used.

Freezing of semen

A variety of freezing regimes, extenders and thawing protocols have been published in the literature. The author has worked mainly with the TRIS–fructose extender with egg yolk as a protectant for cooling, and glycerol for freezing. The TRIS-EYG extender (see Figure 9.3) has given good results for dog and fox semen over a number of years. Most commercial extenders for canine semen are based on modifications of the TRIS buffer, with either fructose, lactose or glucose sugars, and sometimes with membrane protective agents, such as sodium dodecyl sulphate (Orvus ES-Paste™ or Equex STM Paste™). Extenders based on skimmed milk/glucose have also been used, and extenders free of egg yolk may have to be used in the future owing to the zoosanitary risks.

The following procedure for the processing and freezing of canine semen is used in the author's laboratory:

1. Preparation and addition of extender.
2. Cooling and equilibrium.
3. Packaging.
4. Freezing.
5. Thawing.

Preparation and addition of extender

The TRIS base solution with antibiotics and glycerol (cryoprotectant) is used as the freezing solution (see Figure 9.3). The TRIS base solution with glycerol is mixed with 20% egg yolk just prior to use. The egg yolk is obtained from specific pathogen-free hens for export of semen; for domestic AI eggs sold for human consumption are used. The yolk is separated from the egg white and then rolled over a clean piece of paper tissue, gently broken and allowed to run into a funnel. The yolk is broken down by vigorous beating with a glass rod to minimize the size of the egg yolk globules. This reduces the frequency of attachment of sperm heads to the egg globules, which makes semen evaluation more difficult. The egg yolk is then mixed with the TRIS buffer which has been pre-warmed to 30°C. The mixture is stirred well and warmed to 35°C. A premixed TRIS–egg yolk extender can be kept frozen for 2 months prior to use. Gentle thawing in a water bath at 35°C is recommended.

After microscopic evaluation of the semen, the sperm-rich fraction of the ejaculate is diluted by dropwise addition of the 35°C extender until the desired concentration is reached. In the author's laboratory a total concentration of 100 million spermatozoa per ml is now used routinely for freezing.

Cooling and equilibration
After dilution the sample of sperm is examined microscopically, and the extended semen is then poured into plastic centrifugation tubes which are placed into a beaker holding water at 35°C. The beaker is then put in a refrigerator at 5°C and left for 2 hours. During this time the semen is cooled to 5°C, and the cryoprotectant glycerol penetrates through the spermatozoal membrane. The cryovials that will be used to package the semen should be placed in the refrigerator at the same time as cooling and equilibration of the semen commences, to avoid re-warming of the semen when it is put into the vials. Loading of the semen into the cryovials is done after cooling and equilibration.

Packaging
The semen is stirred gently to remix the sperm and extender after the 2-hour period of equilibration and cooling. Semen may be frozen in pellets or straws. Pelleted semen is produced by dropping a small volume (normally 100 µl) of the diluted and chilled semen directly on to blocks of frozen carbon dioxide (dry ice). These drops are then collected to give proper breeding doses in screw-top 1–2 ml Eppendorf or cell culture-safe plastic cryotubes; each vial usually contains one dose. The cryotubes are then stored in special racks in tanks of liquid nitrogen.

In Europe the most common cryovials for dog semen are PVC straws. The standard straw volume used for dog semen is 0.5 ml, although some laboratories may use 0.25 ml straws. Each straw has one open end and one filter tip end with two cotton filters, and can be filled manually using a suction device, or automatically with a small machine. The straw must be filled in such a way that the powder between the two filter tips at the filter end of the straw is filled with liquid and solidifies. The straw is first filled halfway, a small air bubble is then allowed to enter, before the rest of the straw is filled to approximately 1 cm from the top. The air bubble prevents semen from being expelled owing to changes of pressure within the straw during thawing. The straw is sealed either by ultrasound waves, or by using commercially available sealing balls or powders.

Cryotubes and straws may be labelled using a waterproof ink pen or an automated marker. Usually the cryovials are marked prior to cooling with the dog's breed, pedigree or pet name, Kennel Club registration number or chip/ear tattoo identification number, date of freezing and the name of the laboratory/clinic that collected and processed the semen (required by some countries at import).

Freezing
There are two principal alternatives for freezing: a manual, static protocol and an automatic, dynamic protocol. The amount of cryoprotectant added to the extender varies between the two freezing methods (see above for adjustment of water content in buffer base).

Manual
The manual protocol involves the use of a styrofoam box (30 cm x 40 cm x 30 cm), with a removable metal rack placed 10 cm below the edge of the box, which is filled with liquid nitrogen up to a level 4 cm below the rack.

The pre-filled straws (at most eight to ten at a time) are placed horizontally on top of the rack using forceps, and are then left on the rack in the nitrogen vapour for 10 minutes. The forceps are then cooled in liquid nitrogen, used to grasp the straws one by one, and the straws plunged into the liquid nitrogen. The straws are left for some time in the liquid nitrogen to stabilize, after which they are moved to the goblets inside the canister of the liquid nitrogen container for storage or shipment. This freezing protocol is called static because there is a constant non-regulated flow of vaporized nitrogen during cooling and freezing.

Automatic
Automatic freezing involves the use of a freezing machine. It is called dynamic because vaporized nitrogen is let into the freezing chamber at a variable speed by the preset freezing programme, which allows the rates of cooling and freezing to be regulated. One such automated freezing programme was developed for fox semen and is used for dog semen in the author's laboratory when large quantities are frozen. Straws are frozen horizontally on a rack with a removable lid, and several racks can be put into the freezing chamber. The freezing programme follows the regime:

- 2°C/minute from +4°C to −7°C
- 50°C/minute from −7°C to −100°C
- 25°C/minute from −100°C to −180°C.

After the programme is completed the whole rack is removed and placed directly into liquid nitrogen.

Thawing
Semen frozen in pellets can be thawed by direct immersion in a pre-warmed solution, which ensures rapid thawing and reduction of the cryoprotectant solution. Pelleted semen is removed from the cryotube using forceps under liquid nitrogen, and the pellets are dropped directly into the thawing solution (e.g. 0.9% saline solution, 3% sodium citrate solution or other more complex buffers) in an appropriate volume contained in an Eppendorf tube held in a water bath at 37°C. The Eppendorf tube is then swirled vigorously for 30 seconds.

The semen frozen according to the procedure used in the author's laboratory, using 0.5 ml straws, should be thawed in a water bath at 70°C for 8–9 seconds. The water should be dried off the straw with a clean, dry piece of paper tissue, and the sealed end of the straw should be cut with sharp scissors and the contents expelled into a dry and clean cell safe plastic tube without further dilution. For procedures that use thawing solutions, the straw should be opened and the semen emptied into a tube containing warmed thawing medium.

When the frozen semen is thawed, its motility will initially be slow owing to the lower temperature, but it will usually pick up in speed rapidly when examined using a microscope with a warming plate set at 35–38°C. Post-thaw progressive motility varies, but good frozen semen may typically show a motility of 60–70% or even higher. Motility may vary also with the extender used, but the loss of motility should ideally not be more than 20% after thawing when compared with the motility of the original fresh semen.

Quality

The quality of the processed semen may sometimes be a problem at the receiving end because not all clinics or commercial practices have the knowledge needed to handle semen properly. There are a small number of commercial agencies that process the semen themselves and have semen banks, and a few university/private clinics in Europe and Australia. The methods used by these laboratories to freeze semen, the extenders they use, and the type of vial (pellet or straw) are variable and often subject to proprietary restrictions.

Quantity

The number of semen vials to request is a common question. One factor to consider when requesting frozen semen is that many semen banks in the USA freeze semen at relatively low concentrations, because the use of surgical AI is much more common in that country. When the semen is to be inseminated using either the intravaginal or the transcervical intrauterine route, a larger number of sperm may have to be used. Hence, if surgical AI is not an option, the insemination doses should contain a higher number of spermatozoa per ml, usually a total of 100–200 million motile sperm per ml. In many countries, such as Norway (NKK requirement), keeping one vial for later identification of the sire is mandatory for frozen semen, so this extra vial should be allowed for when freezing the semen. Correct and meticulous identification of the vials is important, because mistakes may lead to refusal of registration or the birth of crossbred puppies.

International movement of canine semen

The international movement of canine semen is possible using either fresh diluted and chilled semen or frozen semen. Fresh chilled semen is increasingly used as an alternative to natural mating or the use of AI with frozen semen when time-limited (e.g. the bitch is already in oestrus) and the transport distances are not too far. Some of the new commercial diluents for fresh semen enable storage of chilled

semen for several days, which then abolishes transport distance as an obstacle to AI with the semen. Most countries have import restrictions that are very rigid for semen from domestic livestock, but less so for canine semen. However, there is a distinct difference between those countries that are rabies-free and those with recent outbreaks of rabies, or those which have not obtained rabies-free status according to the Regulations of the OIE (World Organisation for Animal Health). Additionally, the use of egg yolk is becoming increasingly troublesome owing to the possibility of infection with avian influenza of various types and subtypes, or other avian diseases, unless the egg yolk used is from specific pathogen-free flocks. Hence, there is a need to develop egg yolk-free extenders, which is a challenge, owing to the many beneficial effects of this natural compound (Farstad, 2009).

Fresh diluted semen

Fresh diluted semen may be sent over longer distances or kept for some time before insemination, or the semen may be cryopreserved at the facility where it is collected. Alternatively, after being collected at a veterinary practice and shipped in a fresh diluted state, the semen may be frozen in another laboratory. For cooled transport, dog semen is diluted with an appropriate extender and packaged in a container that facilitates slow cooling to 4–6°C during shipment. Fresh chilled semen is usually transported in containers with frozen cooling elements or crushed ice. Transport of fresh semen can be undertaken easily using commercial transport kits (Figure 9.4) or a styrofoam box or a stainless steel flask with crushed ice at the bottom. Water-tight plastic tubes with screw corks, such as cell culture proof centrifugation tubes, can be used. Tubes with rubber tops should be avoided because some rubbers and latex have spermicidal properties. The tubes should be secured in a holder during transport.

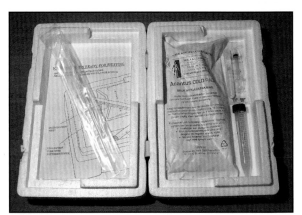

9.4 Commercial kits used for the transport of fresh semen.

Frozen semen

Transport of frozen semen requires the use of liquid nitrogen dewars. These vary, and some countries require the use of previously unused dewars. This has led recently to the development of commercially available lightweight disposable containers that can hold liquid nitrogen for 4 days. Small dry shippers are commonly used. These may be taken on board an aeroplane in the cabin or sent in the non-pressurized baggage compartment. Conventional storage containers that are filled with liquid nitrogen, as opposed to the dry shippers in which the foam inner wall of the container is saturated with liquid nitrogen, are not recommended for international transport owing to the increased chance of spillage of the liquid nitrogen. Containers of liquid nitrogen are considered to be dangerous goods because liquid nitrogen can displace oxygen when in a gaseous state at room temperature, and there is a danger of serious frost burns when the compound is spilled in its liquid state. All handling of liquid nitrogen (−196°C) requires the use of gloves and protective glasses. It is important to fill the tank some time in advance to check that it is patent, and several refills with liquid nitrogen may be needed to saturate a dry shipper.

Paperwork

Most clinics, units and laboratories that collect semen for AI use international freight companies. The paperwork is filled in by the person sending the semen, but an international account must usually be established by the owner of the bitch or stud dog, who is then able to follow the movement of the container to its destination. It is advisable that all legal aspects of ownership and responsibility for the transport of semen are allocated between the two parties and regulated in writing. The veterinary surgeon who collects the semen is responsible for the health of the dog, the quality of the semen at the time of dispatch, the use of a transport container appropriate for the type of preserved semen, and for providing a good user's manual for the receiving inseminating veterinary surgeon.

Timing of the insemination

The principles of oestrus detection and determination of ovulation in the bitch are covered elsewhere in this Manual (see Chapter 5). Measurement of plasma or serum progesterone provides a good indicator of the time of ovulation, and progesterone concentration is considered the most reliable tool to determine the optimum time for AI. Whether to inseminate once or twice depends on how accurately ovulation is estimated and on the quality of the semen. One AI may yield a conception rate and litter size as high as two procedures; however, some studies show that two AI procedures may increase litter size when frozen semen is used.

When fresh or fresh chilled semen is used, the semen should be inseminated on the day of ovulation, with a second insemination 2 days later. When frozen semen is used, its longevity is reduced, and the capacitation time is shorter because the freeze–thaw process influences the stability of the sperm acrosome and membranes. Owing to the necessity for canine oocytes to mature in the oviduct, insemination should be delayed until 2 days after ovulation, with a second insemination 24 hours after the first.

Rapid enzyme-linked immunosorbent assay (ELISA) techniques are available for the assessment of plasma progesterone in the practice laboratory. These may be used to give qualitative or quantitative assessment of the progesterone concentration and are more reliable than the rapid test kits, which by means of a colour change indicate the progesterone concentration to be below or above 5 ng/ml (15 nmol/l).

There is some discrepancy among studies concerning the range of progesterone concentration at ovulation and AI. The progesterone concentration should be approximately 35–54 nmol/l (12–18 ng/ml) on the first day and between 55 and 75 nmol/l (19–25 ng/ml) on the second day of insemination. The following regimen is used as a guide for timing AI in the author's clinic, based on retrospective analysis of results:

* Progesterone concentration of 15–24 nmol/l: first AI 2 days later, repeat after 24 hours
* Progesterone concentration of 25–34 nmol/l; AI 1 and 2 days later
* Progesterone concentration of 35–60 nmol/l; AI on the same day and 1 day later
* Progesterone concentration of >60 nmol/l; AI on the same day only.

Insemination techniques

There are essentially two different sites for the deposition of semen, vaginal and intrauterine, and there are two ways to reach the uterus, either via non-surgical transcervical passage (the Norwegian intrauterine insemination technique or the endoscope-assisted intrauterine technique), or by surgical (laparoscopic) insertion of the AI device.

Vaginal insemination

Vaginal AI may be performed using a simple plastic catheter (such as a bovine uterine flushing catheter cut to an appropriate length) to which a plastic disposable syringe (the veterinary surgeon should be aware that the rubber plunger inside the syringe may be spermicidal; those with plastic plungers should be used) that contains the semen is attached. Alternatively, readymade catheters may be obtained from a commercial supplier. The insertion of the pipette is done with the bitch in a standing position on an examination table. The pipette is inserted to the base of the vagina close to the false cervix (care should be taken to avoid the entrance of the urethra).

During the AI procedure and for up to 10 minutes afterwards the bitch is commonly held in a position with the hindquarters elevated and head down to ensure that the semen is not expelled through backflow, although reducing the time period with elevated hindquarters to 1 minute does not affect fertility. Special AI devices such as the Osiris gun (which is based on the inflatable balloon principle of the Foley catheter) have been used. This insemination gun has a flexible tube lining that has an inflatable part at the tip, which when fully inflated forms a ball that prevents semen backflow. This device is constructed to imitate the dog's erected penile bulb and is meant to increase the probability of intrauterine transport of the semen.

Intrauterine insemination

The main indications for intrauterine insemination are:

- Poor density of fresh semen
- The semen has been frozen and then thawed (which results in reduced longevity and motility of the sperm, so that it is unlikely that a sufficient number of spermatozoa of good motility will be at the fertilization site after vaginal AI).

Non-surgical techniques

Norwegian: The insemination equipment consists of a stainless steel catheter and a plastic guiding tube (Figure 9.5). The plastic tube is used to protect and steady the catheter, to stretch the vagina, and to protect the vaginal mucosa from damage during insertion of the catheter (Figure 9.6). The cervix is fixed through abdominal palpation, and the semen is deposited in the uterus by insertion of the catheter through the cervix. This is a non-surgical method, which requires training, but once learned allows AI in non-sedated standing bitches in a matter of a few minutes.

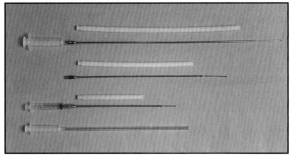

9.5 The Norwegian transcervical intrauterine catheter is available in three different sizes. At the bottom is a simple plastic catheter for intrauterine treatment in cows cut to an appropriate length for intravaginal AI in the bitch.

The Norwegian method is simple and inexpensive and may be performed by the veterinary surgeon in an owner's home or at a dog show. However, the owner cannot assess the entrance of the catheter into the uterus, because there is no visualization of the AI procedure on a screen, such as with endoscopic equipment. Failure to traverse the cervix occurs more often in giant breeds and obese animals than in medium and small breeds. The method is used by some veterinary surgeons in the Nordic countries, other European countries (France, Germany, UK) and the USA.

Endoscope-assisted: The endoscopic method involves the use of flexible, hollow plastic tubing within a rigid endoscope to enter the cervix through the vagina, and to pass through the cervical canal. The endoscopic AI technique was first described by Wilson (1993). The Davidson technique (Davidson, 2007) uses an elongated cystourethroscope for the procedure. The cystourethroscope comprises a 3.5 mm forward oblique telescope with 30° viewing angle, a 22 French protective sheath with two Luer lock adaptors and obturator, a telescope bridge with one 10 French instrument channel, and a cold light source. The working length of the endoscope is 29 cm. A rigid polypropylene catheter is passed through the endoscope into the caudal (vaginal) cervical orifice.

The endoscopic method requires expensive equipment and the client must be referred to a veterinary clinic. The owner may observe the process on a screen. Furthermore, this equipment enables studies of the reproductive tract as well as offering an opportunity for obtaining biopsy samples.

Sedation: Both methods need skill and practice, but the AI can be performed several times usually without any sedation or with only light sedation of the bitch. The author sometimes uses intravenous administration of xylazine (20 mg/ml). A low dose of 4–10 mg (0.2–0.5 ml) per bitch i.v. can be used to achieve abdominal relaxation, but minimize the emetic effect and reduce the duration of the sedation.

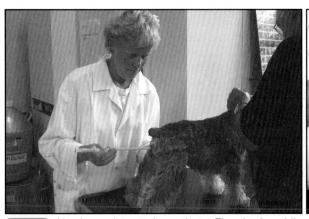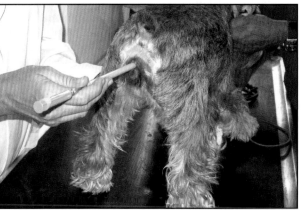

9.6 AI using an intrauterine catheter. The plastic guiding tube is placed over the metal catheter to protect the catheter from contamination and the vaginal mucosa from the stainless steel pointed tip of the catheter. The tube also aids in stretching the vagina sufficiently to locate the cervix through the abdominal wall. The guide tube is inserted to the entrance of the false cervix, whilst the cervix is fixed in a horizontal position between the thumb and index fingers, and the tip of the catheter is inserted subsequently through the cervical canal using gentle pressure.

Insemination volume: With intrauterine AI it is important that the insemination volume is kept low to avoid uterine overload and hence increased reflux of semen into the vagina. A maximum of 1 ml for small breeds, 2 ml for medium, and 3–4 ml for large and giant breeds, respectively, may be used as a guide. No studies have shown the effect of different volumes and sperm numbers used in breeds of different sizes on fertility, but this may imply that small or miniature breeds should receive more concentrated semen and smaller volumes in the insemination dose than medium, large and giant breeds if a definite total number of sperm should be used for AI.

Surgical technique

Surgery may be needed if there are anatomical obstructions in the vagina or cervix that may prevent insertion of the endoscope or catheter, or if the veterinary surgeon is not familiar with either of these techniques. In the former case, a Caesarean operation may be necessary to deliver the puppies at term. There are relatively few reports on the procedures used for surgical AI in bitches, but all require anaesthesia and abdominal surgery (laparascopy). This technique requires surgical skills and good control over anaesthesia so that the duration of anaesthesia is minimized. The semen can be delivered either by an injection needle connected to a disposable syringe, or a by a tom cat catheter after the uterine wall has been incised with a scalpel. The semen should flow easily into the lumen. After deposition of the semen, the fingers are clamped around the uterine wall just cranial to the cervix to prevent reflux. The surgical method (standard ovariohysterectomy approach or explorative laparoscopy procedure) has been used to inseminate Greyhounds in the UK and is the most common method for AI in the USA. In the UK surgical AI is restricted by the RCVS as well as the Kennel Club. The RCVS states that generally surgical AI does not benefit the dog and is not in the best interest of the dog, but this is to be subject to rare circumstances when transcervical insemination is not a realistic option. The Kennel Club has to approve the registration of puppies conceived by means of surgical insemination through a special committee meeting.

Results

Canine semen generally produces good fertility results even when frozen and thawed. The results are in the range of a 70–95% whelping rate with the intrauterine procedures that are currently available. However, AI does not always reach the success rate seen after natural mating, depending on the technique and ability of those performing it. Part of this disparity is probably due to factors in either the male or female, such as semen quality, or the timing of ovulation, delayed ovulation or even failure to ovulate, which may cause breeding problems owing to the underlying endocrine pathology. This is relevant for all types of processing, and even for fresh unprocessed semen. However, for frozen semen there may also be a failure of the spermatozoa to recover fully from the damage caused by freezing and thawing.

From the few reports of the endoscopic method in the literature, it matches the fertility results of the Norwegian catheter method. This assumption is based on two publications claiming an 80% whelping rate, but very few publications are available to document the results over time. The results from surgical AI are claimed anecdotally to be excellent, but scientific publications presenting the results are remarkably scarce. One example is AI of 157 Greyhounds in a private clinic in the UK. The bitches had a whelping rate of 92% and the mean litter size was seven puppies with one AI, compared with a mean average whelping rate recorded for natural service of 81% and five puppies per litter (Boland, 2004).

Vaginal insemination yields acceptable results for fresh semen (>50% conception rate), whereas for frozen semen only one group has documented results comparable to intrauterine AI but using five or six AI procedures per bitch. Improvement of the vaginal technique and development of new extenders for frozen semen may make vaginal deposition of frozen–thawed semen more successful in the future, but so far the fertility results have generally been well below those of fresh semen.

Regulations, ethics and practice

AI is sometimes requested even though both dogs are present. This commonly occurs when either the dog or the bitch is a problem breeder. In these cases, either the male will not mount or show interest in the bitch or the female will not allow the dog to mount her. Often, these are simply problems associated with inexperienced dogs. However, there are instances of behavioural problems where dogs will not breed. The female may be in the correct stage of oestrus, but one or the other of the dogs lacks the natural desire to mate or is aggressive. This may sometimes be due to inadequate social adaptation to other dogs, or it may be a hereditary or an acquired behavioural disorder.

In any instance, failure to mate owing to disease or behavioural problems should give rise to ethical concerns. Unless either of the pair has mated naturally on a previous occasion, AI should not be used uncritically. Article 12 in the Fédération Cynologique International Breeding Rules states that 'artificial insemination is not to be used on animals, which have not previously mated by natural service. In the event the bitch is to be artificially inseminated, the veterinary surgeon collecting the stud dog's semen must provide a certificate to the organisation which keeps the stud dog book with which the litter is to be registered, stating that the fresh or frozen sperm was produced by the agreed stud dog. In addition, the stud dog agent has to give, free of charge, the documents listed in Article 7 (a-g) to the owner of the bitch.' When a veterinary surgeon performs the AI, they must confirm to the organization that keeps the stud book (usually the breed club or the national kennel club) that the bitch has been artificially inseminated with the semen of the stud dog originally nominated. The certificate should also include the place and date of the insemination, the name and stud book registration number of the bitch, and the name and address of the owner of the bitch.

Commercial AI kits are advertised through the internet and sold directly to dog owners for immediate use or for use with diluted and chilled semen. A certificate of collection, including the dog's identity data, should always accompany the semen. The use of freshly collected undiluted semen is the most common practice; in most cases when sperm is collected from the male, the female is present and the semen is inseminated within the shortest time possible. If the AI is carried out by lay personnel semen samples are not always evaluated, and hence whelping results may be highly variable. With diluted chilled and shipped semen the veterinary surgeon or other person (a veterinary nurse/technician or sometimes the owner of the stud dog) who sent the sample may have evaluated the motility of the spermatozoa prior to dispatch, and hence, there is some quality control. With correct handling and re-warming of the insemination dose the results may be good. However, the use of frozen semen for AI requires training in proper thawing of the sample, post-thaw quality control is mandatory, and intrauterine AI is usually required for success, at least when the number of AI procedures is reduced to one or two per bitch. Hence, the use of frozen dog semen for AI by lay personnel is unusual. Surgical AI requires a veterinary authorization and a licence to practice.

In Norway, Sweden and the UK public ethical concerns for animal welfare discourage the use of surgical AI because there is a well documented non-surgical alternative. In Europe many countries refer to welfare considerations if invasive procedures are used when non-invasive procedures are available. In the UK, the RCVS and the Kennel Club provide guidance on the use of AI, and in 2008 the Kennel Club published a policy statement concerning the use of surgical AI. The Kennel Club will not normally register an AI litter if the donor male is alive and domicile in the UK, with one exception – Irish Wolfhounds ≥8.5 years old and domicile in the UK can be used as donors in AI. The donor dog must have produced at least one registered litter naturally. This means that in the UK the semen used for AI is either imported from stud dogs that live abroad, or has been saved in a semen bank until after the death of the donor.

References and further reading

Boland P (2004) Surgical insemination and other ways. *The 4th Congress of the European Veterinary Society for Small Animal Reproduction.* Barcelona, Spain

Davidson AP (2007) Endoscopy as a tool in assessing the reproductive tract in bitches and queens. *Compendium of the Norwegian School of Veterinary Science Postgraduate Continuing Education Course in Canine and Feline Reproduction, Obstetrics and Neonatology.* Oslo, Norway

England GCW and Millar KM (2008) The ethics and role of AI with fresh and frozen semen in the dog. *Reproduction in Domestic Animals* **43**, 165–171

Farstad W (2009) Cryopreservation of canine semen – new challenges. *Reproduction in Domestic Animals* **44**, 336–341

Linde Forsberg C (2001) Regulations and recommendations for international shipment of chilled and frozen canine semen. In: *Recent Advances in Small Animal Reproduction*, ed. PW Concannon *et al.* International Veterinary Information Service (www.ivis.org), Ithaca, New York

Thomassen R, Sanson G, Krogenæs A *et al.* (2006) Artificial insemination with frozen semen in dogs: a retrospective study of 10 years using a non-surgical approach. *Theriogenology* **66**(Suppl), 1645–1650.

Wilson M (1993) Non-surgical artificial insemination in bitches using frozen semen. *Journal of Reproduction and Fertility* **47**(Suppl), 307–311

Useful websites

The Kennel Club
www.the-kennel-club.org.uk

The Royal College of Veterinary Surgeons (RCVS)
www.rcvs.org.uk

10

Pregnancy diagnosis, normal pregnancy and parturition in the bitch

Catharina Linde Forsberg

Introduction

It is important that the veterinary surgeon understands the normal physiology of pregnancy and parturition and is able to make an accurate diagnosis of pregnancy and non-pregnancy to provide a relevant clinical service to their clients. This will also provide a foundation for the investigation of clinical disease when there are abnormalities of pregnancy and parturition.

Pregnancy diagnosis

Early pregnancy diagnosis is commonly requested by breeders to enable them to schedule their holidays to coincide with the arrival of the litter of puppies, to decide whether they should mate additional bitches in the kennel or simply out of curiosity. Some early signs of pregnancy may be recognized easily by the owner, such as a persistent swelling of the vulva after oestrus and a slight enlargement of the nipples, which also become pinker in colour from the third week of pregnancy. These signs are easiest to recognize in the primiparous bitch. Enlargement of the mammary glands is usually observable from the fifth week (but this may also be a sign of pseudopregnancy). Finally, signs of malaise may occur during the third week coinciding with implantation of the embryos and/or during the fifth week as a result of pressure from the distended uterine horns on the stomach and liver. The malaise normally only lasts for a day or two and it is important to inform breeders that a longer duration may be an indication that there is a problem and the bitch should be examined.

At the time of implantation a slight mucoid or haemorrhagic discharge may occasionally be observed from the vulva, but this should be considered normal if it is a small volume and persists for just a day or two (Jones and Joshua, 1982). A scant amount of a clear, viscid and odourless discharge from the vulva, like raw egg white, which originates from the glands of the cervix, is often noted slightly later and may persist from weeks 5–7 of pregnancy. Distension of the abdomen is often obvious from the fifth week, and fetal movements can be seen and palpated from week 7. Occasionally, dogs may be as attracted to the pregnant bitch as to a bitch in oestrus.

Manual palpation

Diagnosis of pregnancy by palpation of the fetal swellings of the uterine horns can be performed between the third and fifth weeks of pregnancy. At 3 weeks the fetal swellings are approximately 15 mm in diameter. They are round in shape, firm and well separated from each other, like a string of hazelnuts along the uterine horns. At this early stage they are carried high in the abdomen. At 4 weeks they are approximately 25 mm in diameter and somewhat more oval in shape. They are still well separated from each other in the uterine horns, which by now as a result of the increase in weight have attained a more central position in the abdomen. At 5 weeks the uterine swellings are 30–35 mm in size, oval in shape and becoming soft to the touch owing to the increase in fetal fluids. They can no longer be palpated as separate structures, and are more ventral in position. After 5 weeks of pregnancy the increase in fetal fluids and size of the uterine horns makes it difficult to determine pregnancy by palpation. The exact number of fetuses cannot be determined using this method.

Ultrasonography

Using a simple Doppler ultrasound instrument the fetal heartbeats can be detected from approximately day 24–28 of pregnancy. With a B-mode real time scanner and using a 5 MHz, 7.5 MHz or 10 MHz transducer, the fetal and surrounding structures can be identified from approximately day 17 of pregnancy. However, because of the variation in the timing of mating during oestrus it is recommended to postpone the examination until around day 24–28, or to repeat the examination at that time point in case of an early initially negative diagnosis. As with manual palpation, ultrasonography cannot be used to assess the exact number of fetuses present, although early examinations are generally more accurate than later examinations.

Radiography

Although the outline of the pregnant uterus may be observed as early as day 21, pregnancy cannot be differentiated at this time from other causes of uterine enlargement. A positive diagnosis requires detection of fetal calcification, which is generally not observed until after day 42. Radiography is the most accurate method for determining the number and position of fetuses in late pregnancy (see below).

Relaxin assay

Relaxin is a pregnancy-specific hormone that is produced in the placenta and/or in the ovary depending on the species. In the bitch it reaches a detectable concentration in the peripheral plasma at approximately day 25 of pregnancy and peaks between days 40 and 50. In-house enzyme-linked immunosorbent assay (ELISA) kits can be used from around 25 days after ovulation to measure plasma relaxin concentration. These kits are generally very accurate and are used especially at clinics that do not have ultrasound equipment. In the case of an early negative result, the bitch should be retested 5–6 days later.

Normal pregnancy

Physiological changes and clinical monitoring

The increased metabolic demands during pregnancy cause maternal physiological alterations. Blood volume increases by 40%, which provides an adequate reserve to compensate for the large quantities of blood and fluids lost at parturition. The increase in volume is primarily formed of plasma, with a resulting haemodilution (the haematocrit is 30–35% at term). An increase in cardiac output occurs, caused by enhanced heart rate and stroke volume. The functional residual capacity of the lungs is decreased by cranial displacement of the diaphragm by the gravid uterus, and oxygen consumption during pregnancy increases by 20%. Pregnant animals also have delayed gastric emptying attributable to decreased gastric motility and displacement of the stomach.

It is recommended to perform a clinical examination of the bitch at about 45 days of gestation. This allows the veterinary surgeon to assess the health of the bitch, confirm pregnancy by radiography if necessary, count the number of fetuses and answer any questions about the expected parturition that the breeder may ask. At this examination ultrasonography is less accurate in determining the number of fetuses but is an excellent tool for assessing fetal viability.

The physiological increase in the progesterone concentration, which is typical of pregnancy and metoestrus (dioestrus), stimulates growth hormone (GH) secretion. In some individuals, typically middle-aged and older bitches that are pregnant or have recently been in oestrus, this may cause downregulation of insulin receptors and inhibition of post-receptor pathways. In addition to insulin resistance, pregnant bitches have a reduced ability to produce glucose via gluconeogenesis, glycogenolysis and lipolysis. The resulting type 2 diabetes is usually transient. Once the blood progesterone concentration returns to the level seen in anoestrus, and thus the stimulus for GH secretion declines, insulin resistance resolves.

In contrast to the hyperglycaemia expected with insulin antagonism, some bitches exhibit a proparturient hypoglycaemia. This condition is rare, but affected bitches have been reported to be in late gestation and have a short history of muscle weakness, convulsions or collapse. Low glucose concentrations can be detected readily in the blood. The condition improves dramatically after treatment with intravenous glucose solutions and resolves after parturition.

During the immediate prepartum period the mineralization of the fetal skeletons, initiation of lactation and increasing myometrial activity lead to an increased demand for calcium by the bitch. The influx of free ionized calcium into the myometrial cells is a prerequisite for adequate uterine contractility. The availability of free calcium may be restricted by inappetence in the late pregnant bitch, and by respiratory alkalosis from panting. This may lead to deficient secretion of parathyroid hormone (PTH) during this period of increased calcium demand. A rising pH (alkalosis) and declining $PCO2$ at the onset of whelping may decrease responsiveness to PTH, thus resulting in a transient decrease in blood calcium concentrations (Hollingshead *et al.*, 2010). Induced acidosis has been shown to increase PTH secretion in response to hypocalcaemia. Dietary formulas for pregnant bitches that regulate the cationic/anionic difference and reduce the pH of arterial blood may be beneficial, especially in bitches with a previous history of primary uterine inertia.

Other physiological changes include an increased level of 15-keto-dihydroprostaglandin F2α (PGF2α) 24 hours before parturition and again at the onset of whelping. Serum cortisol concentrations also increase at the time of delivery and generally remain high for 12 hours, declining to basal levels again after 36 hours.

Duration of pregnancy

The apparent duration of pregnancy in the bitch averages 63 days, with a range of 56–72 days if calculated from the first mating. This surprisingly large apparent variation is mainly due to the long and variable behavioural oestrous period of the bitch, and the long lifespan of the oocytes and spermatozoa. The length of gestation is quite predictable when calculated either from the preovulatory surge of luteinizing hormone (LH) (65 ± 1 days), from the day of ovulation (63 ± 1 days) or from the time of fertilization (60 ± 1 days), rather than from the time of mating. The length of gestation is also influenced by litter size, being longer for small litters and shorter with an increasing number of puppies. It has been proposed that each additional puppy above the average for the breed results in a shortening of pregnancy length by 0.25 days, and for each puppy less than the breed average a corresponding lengthening of pregnancy occurs (Bobic Gavrilovic *et al.*, 2008). Gestation length has also been shown to vary with breed, irrespective of body or litter size.

Prediction of the day of parturition

The ability to determine gestational age and to predict the day of parturition in the bitch is of considerable clinical importance, and is especially valuable in cases of threatened abortion, prolonged gestation, or when the bitch is scheduled for an elective Caesarean operation or has previously suffered from dystocia. A number of clinical indicators of impending parturition may be used, including:

- Measurement of progesterone and LH during oestrus
- Behavioural changes close to parturition
- Clinical signs close to parturition
- Decline in body temperature
- Measurement of progesterone in late pregnancy
- Diagnostic imaging.

Measurement of progesterone and luteinizing hormone during oestrus

On average, parturition occurs 65 ± 1 days from the LH peak, which coincides with the initial sharp rise in serum progesterone concentrations to ≥4.5 nmol/l (≥1.5 ng/ml). There are in-house LH kits available. The accuracy of prediction of the day of parturition within an interval of ± 1, ± 2, and ± 3 days using pre-breeding serum progesterone concentrations was 67%, 90%, and 100%, respectively, and was not influenced by bodyweight or litter size.

Behavioural changes close to parturition

Several days before parturition the bitch may become restless, seek seclusion or become excessively attentive, and she may refuse all food. She may exhibit nesting behaviour 12–24 hours before parturition, concomitant with the increasing frequency and force of uterine contractions. Shivering may be an attempt to increase body temperature. In primiparous bitches, lactation may be established <24 hours before parturition, whereas after several pregnancies colostrum can be detected as early as 1 week prepartum.

Clinical signs close to parturition

The relaxation of the pelvic and abdominal musculature and of the perineal region that occurs prior to parturition as a result of elevated concentrations of relaxin is a consistent but subtle indicator of impending parturition.

Decline in body temperature

The most consistent change is the decline in rectal temperature that is caused by the final abrupt decrease in progesterone concentration (Figure 10.1). During the last week before parturition the rectal temperature of the bitch fluctuates and ultimately declines sharply 8–24 hours before parturition. Approximately 10–14 hours after this decrease, the concentration of progesterone in the peripheral plasma is as low as 6 nmol/l (2 ng/ml). Progesterone is thought to be thermogenic, and when values decline there is a subsequent decrease in rectal temperature. Subsequently, the rectal temperature rises again and may become higher than the temperature recorded at the end of pregnancy.

To assess the prepartum decrease in body temperature properly, measurements should be made every 1–2 hours whilst the temperature decreases and then less frequently when the temperature is seen to increase again. The rectal temperature in bitches of miniature breeds can fall to

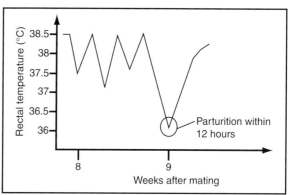

10.1 The best sign of impending parturition in the bitch is the marked drop in rectal temperature. During the last week of pregnancy, the temperature will fluctuate because release of prostaglandins causes a transient fall in peripheral plasma progesterone concentration, and progesterone is thermoregulatory. During the first stage of labour the temperature drop is more pronounced, and the bitch should be in the second stage of labour within 12 hours after reaching the lowest temperature and before the temperature has returned to normal.

35°C (95°F), and that in medium-sized bitches to around 36°C, whereas it seldom falls below 37°C (96.8°F) in bitches of the giant breeds. This difference is probably an effect of the ratio between surface area and body volume; the hair coat may also have an influence.

Measurement of progesterone in late pregnancy

As noted above there is a rapid decline in progesterone immediately prior to parturition, which results in a change in rectal temperature. The fact that serum progesterone concentrations decrease significantly from 12–15 nmol/l (4–5 ng/ml) to below 6 nmol/l (2 ng/ml) starting 24 hours before the onset of whelping can also be used to predict parturition if progesterone is measured using an ELISA test kit.

Diagnostic imaging

Ultrasonography: B-mode and four-dimensional (4D) colour Doppler ultrasonography have been used to assess the diameter of pregnancy structures and to estimate fetal size during pregnancy. The diameter of the inner chorionic cavity on day 18–37 following ovulation, and the fetal head diameter on day 38 to parturition show the best correlation with gestational age and day of parturition. Other fetal structures used to time pregnancies with ultrasonography are fetal limb buds, first detectable on day 33–35; eyes, kidney and liver on day 39–47; and intestine on day 57–63.

Radiography: Using radiography the fetal skeleton is rarely visible before day 42; the skull on day 45–49; pelvic bones on day 53–57; and teeth on day 58–63. Establishment using good quality radiographs of whether these structures are present may provide some clinical value in estimating the likely time of parturition.

Parturition

Physiology of parturition

An understanding of the course and control of normal parturition (eutocia) is necessary for the correct diagnosis and treatment of abnormal parturition (dystocia).

It is generally believed that stress produced by reduction of the nutritional supply from the placenta to the fetus stimulates the fetal hypothalamic–pituitary–adrenal axis, resulting in the release of adrenocorticosteroid hormone, thought to be the trigger for parturition. An increase in fetal and maternal cortisol is believed to stimulate the release of PGF2α, which is luteolytic, from the fetoplacental tissue. This results in a decline in plasma progesterone concentration. Withdrawal of the progesterone blockade of pregnancy is a prerequisite for the normal course of canine parturition; bitches given long-acting progesterone during pregnancy fail to enter normal parturition. Concurrent with the gradual decrease in plasma progesterone concentration during the last 7 days before whelping, a progressive qualitative change occurs in uterine electrical activity. In addition, a significant increase in uterine activity takes place during the last 24 hours before parturition with the final decrease in plasma progesterone concentration to below 6 nmol/l (2 ng/ml).

The change in the ratio of oestrogen to progesterone concentrations is probably a major cause of placental separation and cervical dilatation, although oestrogen has not been shown to increase before parturition in the bitch as it does in many other species. Sensory receptors within the cervix and vagina are stimulated by the distension created by the fetus and the fluid-filled fetal membranes (Figures 10.2 and 10.3). This afferent stimulation is conveyed to the hypothalamus and results in the release of oxytocin during the second stage of parturition. Afferent stimulation also participates in a spinal reflex arc; efferent stimulation of the abdominal musculature

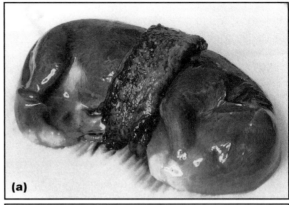

10.3 **(a)** A puppy in intact fetal membranes, following delivery by Caesarean operation. **(b)** The fetal membranes have been opened and the puppy is gasping for air.

produces straining. Relaxin causes the pelvic soft tissues and genital tract to relax, which facilitates passage of the fetus. In the pregnant bitch, this hormone is produced primarily by the placenta, although it has also been detected in the ovaries and uterus. The concentration of relaxin increases gradually over the last two-thirds of the pregnancy. It usually declines abruptly immediately after fetal death, abortion or parturition, but may remain at a detectable level for up to 9 weeks postpartum owing to invasion of trophoblast cells into the endometrium, and it can in those cases be used as an indicator of a previous pregnancy. Prolactin, the hormone responsible for lactation, starts to rise 3–4 weeks after ovulation and surges dramatically with the abrupt decline in serum progesterone just before parturition.

Premature onset of parturition

Premature onset of parturition may have infectious causes, but can also occur in the clinically healthy bitch. In some species low progesterone concentrations may be responsible for the premature onset of uterine contractions; however, this condition (hypoluteoidism) has been suggested by some to be important, but its role is not clear. Despite this, short-acting progestational compounds may be given by some clinicians in an attempt to maintain pregnancy. To inhibit premature uterine contractions adequately, tocolytic compounds such as terbutaline (a beta-adrenergic receptor antagonist) should also be

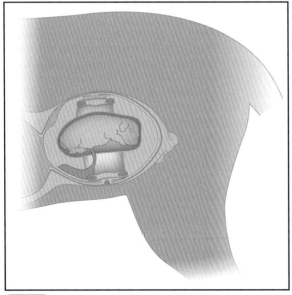

10.2 Fetus and the fetal membranes in the bitch.

administered orally or subcutaneously to effect, and should be discontinued 48 hours before the anticipated day of delivery.

Stages of parturition

Parturition is divided into three stages, with the last two stages being repeated for each puppy delivered.

First stage

The duration of the first stage of parturition is usually 6–12 hours, but it may last as long as 36 hours, especially in a nervous primiparous animal. Vaginal relaxation and dilatation of the cervix occur during this stage. Intermittent uterine contractions, with no signs of abdominal straining, are present. The bitch may appear uncomfortable, and the restless behaviour may become more intense. Panting, tearing up and rearranging of bedding, shivering and occasionally vomiting may be seen. Some bitches show no behavioural evidence of first-stage parturition. The unapparent uterine contractions increase both in frequency and intensity towards the end of the first stage. The fluid-filled fetal membranes are pushed ahead of the fetus by the propulsive efforts of the uterus and serve to dilate the cervix. During pregnancy the orientation of the fetuses within the uterus is 50% caudally and 50% cranially, but this changes during the first stage of parturition, when the fetus rotates on its long axis, extending its head, neck and limbs, resulting in 60–70% of puppies being born in anterior and 30–40% in posterior presentation (Figure 10.4).

Second stage

The duration of the second stage of parturition is usually 3–12 hours (in rare cases 24 hours). At the onset of the second stage the rectal temperature rises to normal or slightly above normal. The first fetus engages in the pelvic inlet, and the subsequent intense, expulsive uterine contractions are accompanied by abdominal straining. When the fetus enters the birth canal, the allantochorionic membrane may rupture and a discharge of some clear fluid may be noted. Covered by the amniotic membrane, the first fetus is usually delivered within 4 hours following the onset of the second stage of parturition. If the bitch

does not develop regular abdominal contractions, the owner should be instructed to try to stimulate straining initially by exercising the bitch, for instance by running the dog around the house or up some stairs. A considerable number of puppies are born in the car on the way to the veterinary clinic. Most of these would probably have been delivered in the home had the owners tried to induce straining, thereby giving the puppies a better start in life and possibly also resulting in the whole litter being born without further intervention.

Another means of stimulating straining in the bitch showing inefficient parturition is 'feathering' of the dorsal vaginal wall. Feathering is accomplished by inserting two fingers into the vagina and pushing them against the dorsal vaginal wall, thus inducing an episode of straining (the Ferguson reflex) (Figure 10.5). Feathering can also be effective in initiating parturition after correction of the position or posture of a fetus.

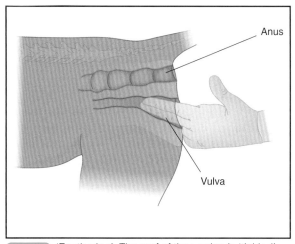

10.5   'Feathering'. The roof of the vagina is 'tickled' with the fingers to stimulate uterine contractions.

Voluntary inhibition of parturition may occur as a result of psychological stress, usually in a nervous primiparous animal. Reassurance by the owner or administration of a low-dose tranquillizer may remove the inhibition. Once the first fetus is born, parturition will usually proceed. Normally the bitch

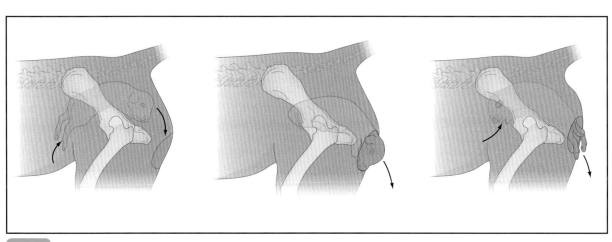

10.4   Normal birth of puppies in anterior and posterior presentation.

will break the amniotic membrane, sever the umbilical cord and lick the neonate. At times the bitch will need some assistance to open the fetal membranes to allow the puppy to breathe, and sometimes the airways will need to be emptied of fetal fluids. The umbilicus can be clamped with a pair of haemostats and cut with blunt scissors to minimize haemorrhage from the fetal vessels, leaving about 1 cm of the umbilical cord. In cases of continuing haemorrhage the umbilicus should be ligated.

Diagnosis: It is crucial that the veterinary surgeon is able to determine whether the bitch is in the second stage or still in the first stage of parturition. Inexperienced breeders tend to get nervous during the first stage of labour in the bitch, not fully understanding the function of this preparatory stage of parturition during which uterine contractions, the softening of the birth canal and the opening of the cervix take place.

Three signs indicate that the bitch has entered into the second stage of parturition:

- The rectal temperature returns to a normal level
- The first water bag (allantois) bursts and the fetal fluids are passed
- Visible abdominal straining occurs.

If one or more of these signs has been observed, the bitch is in the second stage of parturition. In normal parturition the bitch may show weak and infrequent straining for up to 2 hours, and at the most 4 hours, before giving birth to the first fetus. If the bitch is showing strong, frequent straining without producing a puppy, this indicates the presence of an obstruction, and she should not be left for more than 20–30 minutes before seeking veterinary advice.

The bitch should be presented for veterinary examination if one of the following occurs:

- Weak, irregular straining for more than 2–4 hours
- Strong, regular straining for more than 20–30 minutes
- Fetal fluid was passed more than 2–3 hours previously, but nothing more has happened
- Greenish discharge is seen but no puppy is born within 2–4 hours
- More than 2–4 hours have passed since the birth of the last puppy and more remain
- The bitch has been in the second stage of parturition for more than 12 hours.

Third stage
The third stage of parturition, which involves expulsion of the placenta and shortening of the uterine horns, usually follows within 15 minutes of the delivery of each fetus. However, two or three fetuses may be born before the passage of their placentas occurs. The bitch should be discouraged from eating more than one or two of the placentas because she may develop diarrhoea and vomiting, with risk of aspiration pneumonia. Lochia (the greenish postpartum discharge of fetal fluids and placental remains) may be seen for up to 3 weeks, or even more, but is most profuse during the first week. Uterine involution is normally complete after 12–15 weeks.

The bitch should be presented for veterinary examination if one of the following occurs (see Chapter 14 for further details):

- All the placentas have not been passed within 4–6 hours (although placental numbers may be difficult to determine if the bitch eats them)
- The rectal temperature is higher than 39.5°C (101.3°F)
- There is continuing severe genital haemorrhage
- The lochia are putrid or foul smelling
- The general condition of the bitch is abnormal
- The general condition of any of the puppies is abnormal.

Interval between births
The interval between births in normal uncomplicated parturition is 5–120 minutes. Expulsion of the first fetus usually takes the longest. In almost 80% of cases the fetuses are delivered alternately from the two uterine horns. When giving birth to a large litter, a bitch may stop straining and rest for more than 2 hours between the delivery of two consecutive fetuses. The second-stage straining will then resume, followed again by the third stage, until all the fetuses are born.

Completion
Parturition is usually complete within 6 hours following the onset of the second stage of labour, but it may last up to 12 hours. It should never be allowed to last for more than 24 hours because of the risks involved for both the bitch and the fetuses.

Litter size
Litter size in dogs ranges from just one puppy for some of the miniature breeds to the record number of 22 seen in a giant breed. Litter size is smaller in bitches of 1–2 years of age, increases up to 3–4 years, and decreases sharply after 5–6 years. A litter of only one or two puppies predisposes to dystocia, and thus fetal death, because of insufficient uterine stimulation and the large size of the puppies ('single-pup syndrome'). This can be seen in dog breeds of all sizes. Breeders of miniature dogs tend to accept small litters but should be encouraged to breed for litters of at least three to four puppies to avoid this complication.

Complications
While the majority of bitches have an uncomplicated parturition it is important to recognize that when there is dystocia, manipulation and medical treatment is successful in only 25–30% of cases. Approximately 65% of bitches brought to the clinic because of dystocia thus end up having a Caesarean operation. Furthermore, it is essential that breeders recognize that the incidence of fetal death increases from approximately 6% when bitches are presented to the veterinary surgeon within 1–4.5 hours after the beginning of second-stage parturition, to approximately 14% in the period between 5 and 24 hours. Early diagnosis and prompt treatment is therefore crucial in reducing the

neonatal death rate in cases of dystocia, and this will only occur when there is a clear understanding of the normal physiology such that difficult birth can be recognized quickly and accurately.

Clinical assessment in the peripartum period

When a pregnant bitch is examined at around the time of parturition, an accurate history and a thorough physical examination are important prerequisites for proper management. The presence of the three criteria of the second stage of parturition should be assessed: (a) rectal temperature has returned to normal; (b) passage of fetal fluids; (c) visible abdominal straining. An evaluation of the bitch's general health status should also be made and signs of any adverse effects of parturition noted.

Observation should be made of the bitch's behaviour and the character and frequency of straining. The vulva and perineum should be examined and the colour and amount of vaginal discharge noted. Mammary gland development, including congestion, distension, size and presence of milk, should be evaluated. Palpation of the abdomen, roughly estimating the number of fetuses and degree of distension of the uterus, should be carried out. Digital examination of the vagina using an aseptic technique should be undertaken to detect obstructions and determine the presence and presentation of any fetus in the pelvic canal. In most bitches it is not possible to reach the cervix during the first stage of parturition, but an assessment of the degree of dilatation and tone of the vagina may give some indication of the status of the cervix and the tone of the uterus. Pronounced tone of the anterior vagina may indicate satisfactory muscular activity in the uterus, whereas flaccidity may indicate uterine inertia.

The character of the vaginal fluids will also indicate whether the cervix is closed, with the production of scant fluid volume that is sticky and creates a slight resistance to the introduction of a finger. The cervix is likely to be open when fetal fluids lubricate the vagina, making exploration easy. When the cervix is closed the vaginal walls also fit quite tightly around the exploring finger, whereas with an open cervix the cranial vagina appears more open.

Radiographic examination is valuable to assess gross abnormalities of the maternal pelvis and the number and location of fetuses, to estimate fetal size and to detect congenital defects or signs of fetal death. In the dead fetus, intrafetal gas will appear 6 hours after death and can be detected radiographically, whereas overlapping of cranial bones and collapse of the spinal column will not be seen until 48 hours after the death of the fetus.

Ultrasonography can be used to identify fetal viability or distress. The normal fetal heart rate is 180–240 beats per minute, but it decreases at the time of parturition and in the compromised fetus. A fetal heart rate of ≤150 beats per minute at full term is an indication for intervention. Some bitches presented to veterinary surgeons for dystocia have already delivered all their fetuses or are showing pseudopregnancy. Pseudopregnancy is most commonly diagnosed as lactation without pregnancy, but it may include nesting behaviour and changes in personality that convince an owner that the bitch is pregnant.

A system for monitoring labour and parturition in the bitch is commercially available. These types of system are intended for use by veterinary surgeons or by breeders at home with veterinary guidance. They consist of a uterine tocodynamometer and a fetal Doppler probe (see Chapter 14 for further details).

Fetal disposition

Fetal disposition is the term used to describe the spatial arrangement of the fetus in relation to the pelvis and birth canal of the dam, and of its extremities to its body. It is described as a normal or abnormal (faulty) disposition. During the first stage of parturition, the fetus assumes a normal disposition, which is the optimum to allow unimpeded passage through the birth canal. The disposition changes slightly during the expulsion of the fetus through the birth canal. Three terms are used to describe disposition in greater detail: presentation, position and posture.

Presentation
Fetal presentation describes the relationship between the longitudinal axis of the fetus and the maternal birth canal; it can be longitudinal or transverse. The description can be more specific if the part of the fetal body entering the pelvis first is determined, e.g. cranial longitudinal presentation, caudal longitudinal presentation, dorsal transverse presentation or ventral transverse presentation.

Position
Fetal position indicates the surface (quadrant) of the maternal birth canal to which the vertebral column of the fetus is apposed. The position can be described as dorsal, ventral, left lateral or right lateral.

Posture
Fetal posture refers to the disposition of the moveable appendages of the fetus, and involves flexion or extension of the neck and limb joints.

Normal
Normal fetal disposition is described as cranial longitudinal presentation, dorsal position, extended posture. Interestingly, caudal presentation (more accurately caudal longitudinal presentation, dorsal position, extended posture) is considered normal in dogs, and occurs in 30–40% of fetal deliveries. However, caudal presentations have been related both to higher neonatal mortality and to a predisposition for dystocia. This is because mechanical dilatation of the cervix may be inadequate, particularly when the caudal presentation involves the first fetus to be delivered. In addition, expulsion is rendered more difficult because the fetus is being delivered against the direction of its hair coat, and because the fetal chest, instead of being compressed, becomes distended by the pressure of the abdominal organs through the diaphragm.

Manual correction

A healthy fetus is active during the period when it is expelled, extending its head and limbs, twisting and rotating to help itself get through the birth canal. In most breeds the greatest bulk of the fetus is its abdominal cavity, whereas the bony parts (the head and the hips) are comparatively small. The limbs are short and flexible and rarely cause serious obstruction to delivery of a fetus of normal size. Further details of the diagnosis, treatment and management of difficult birth are given in Chapter 14. It is sufficient to state here that abnormalities of fetal disposition are quite common and that simple manual correction of a malpresented fetus may enable the bitch to deliver the remaining litter without further problems.

The postpartum period

In the postpartum period the bitch has often endured several hours of intensive labour and may be suffering from physical exhaustion, dehydration, acid–base disorders, hypotension, hypocalcaemia, hypoglycaemia, or a combination of these conditions. It is normal for the bitch to have a slightly elevated rectal temperature, up to 39.2°C (102.5°F), for a few days after parturition. However, the temperature should not exceed 39.5°C (103.1°F). Fever during this period usually results from a condition of the uterus or the mammary glands.

In the postparturient period it is normal for the bitch to have a serosanguineous vaginal discharge for 3–6 weeks. Normally uterine involution is complete within 12–15 weeks after whelping. Subinvolution of placental sites should be suspected if a sanguineous vaginal discharge persists for longer than 6 weeks. True haemorrhage should be distinguished from normal vaginal postparturient discharge. Some haemorrhage from the genital tract after parturition is normal, but maternal blood loss should never exceed a scant drip from the vulva. Excessive haemorrhage after parturition may indicate uterine or vaginal tears or vessel rupture, or may be evidence of a coagulation defect.

Retained placentas may cause severe problems in the bitch, especially when accompanied by retained fetuses or infection. Clinical signs of a retained placenta include a thick dark vaginal discharge. Retained fetuses can be identified by palpation or radiographic and ultrasonographic examination. The examination should also encompass the corpus uteri and the vagina in a search for partly expelled fetuses or fetal membranes. A retained placenta is often palpable in the uterus, depending on the size of the bitch and the degree of uterine involution. Extraction of retained tissue, by careful 'milking' of the uterine horn or by using forceps, is sometimes possible. Treatment with 1–5 IU/dog oxytocin s.c. or i.m. q6–12h for up to 3 days can help to expel retained placentas. The long-acting ergot alkaloids should not be used because they may cause closure of the cervix. Antibiotic treatment is advisable if the bitch is showing signs of illness.

Maternal behaviour

Maternal behaviour in the postpartum period is determined by hereditary factors, several hormones (e.g. oestrogens, progesterone, oxytocin and prolactin), previous experience as a mother, and the stimulus of the neonate. Good maternal behaviour includes nest building, and nursing and protecting the puppies. The period during which the bitch will form a bond with her puppies is probably <24 hours long. The bitch should spend most of the time with the litter for at least the first 2 weeks.

Most bitches have strong maternal instincts, but their behaviour depends upon their hormonal balance, general health and the environment. A higher incidence of bad mothering exists in some breeds, and in other species several genes for maternal behaviour have been identified.

Sublimation (close emotional attachment to a human) can cause problems at parturition, when the bitch may demonstrate a response akin to panic and reject her pups. In contrast, other bitches may resent human intervention, fail to accept assisted parturition and offspring delivered by a Caesarean operation, and may sometimes even kill the neonates. Major disturbances during and after parturition, mental instability or pain may also cause the mother to kill her neonates. Hypocalcaemia should always be ruled out in cases of aggression in lactating bitches. Good health, quiet and familiar surroundings and, most important of all, the presence of her young will promote normal maternal behaviour.

Perinatal loss

Based on a number of surveys, losses of puppies in the period up to weaning appear to range between 10% and 35%, and average around 12%. More than 65% of mortality in puppies occurs at parturition and during the first week of life; few neonatal deaths occur after 3 weeks of age. Breed differences in neonatal mortality patterns have been reported. The principal cause of mortality in puppies has been attributed to fetal asphyxia, which accounts for 42.5% of deaths. The majority of these puppies (82.2%) died during whelping or in the first 24 hours after birth. The death of just over half of these puppies could be attributed directly to dystocia. The remaining puppies were compromised during what appeared to be a normal whelping.

References and further reading

Bobic Gavrilovic B, Andersson K and Linde Forsberg C (2008) Reproductive patterns in the domestic dog – a retrospective study of the Drever breed. *Theriogenology* **70**, 783–794

Concannon PW (2003) Canine pregnancy: predicting parturition and timing events of gestation. In: *Recent Advances in Small Animal Reproduction*, ed. PW Concannon *et al.* International Veterinary Information Service (www.ivis.org), Document No. A1202.0500

Darvelid AW and Linde Forsberg C (1994) Dystocia in the bitch: a retrospective study of 182 cases. *Journal of Small Animal Practice* **35**, 402–407

Davidson AP (2003) Approaches to reducing neonatal mortality in dogs. In: *Recent Advances in Small Animal Reproduction*, ed. PW Concannon *et al.* International Veterinary Information Service (www.ivis.org), Document No. A1226.0303

Eneroth A, Linde Forsberg C, Uhlhorn M and Hall M (1999) Radiographic pelvimetry for assessment of dystocia in bitches: a clinical study in two terrier breeds. *Journal of Small Animal Practice* **40**, 257–264

Hollinshead FK, Hanlon DW, Gilbert RO *et al.* (2010) Calcium, parathyroid hormone, oxytocin and pH profiles in the whelping bitch. *Theriogenology* **73**, 1276–1283

Jones DE and Joshua JO (1982) *Reproductive Clinical Problems in the Dog*. Wright Group, Texas

Kutzler MA, Mohammed HO, Lamb SV and Meyers-Wallen VN (2003) Accuracy of canine parturition date prediction from the initial rise in preovulatory progesterone concentration. *Theriogenology* **60**, 1187–1196

Linde Forsberg C, Wikström C and Lundeheim N (2008) Differences between seasons of the year and breeds in mating frequency, gestation length and litter size in 13 breeds of dogs. *Proceedings of the 6th International Symposium on Canine and Feline Reproduction (ISCFR)*, Vienna, Austria, pp.132–133

Moore AH and Wotton PR (1993) Preparturient hypoglycemia in two bitches. *Veterinary Record* **133**, 396–397

11

Pregnancy diagnosis, normal pregnancy and parturition in the queen

Xavier Levy and Gary C.W. England

Introduction

The breeding of cats is a common pastime and there is a large industry of professional breeders producing pedigree cats for the show and pet markets. In addition, cat breeding within research colonies is not uncommon, providing a reproductive model for a number of exotic feline species that are endangered. It is important that the veterinary surgeon understands the normal physiology of pregnancy and parturition and is able to recognize when intervention is necessary.

Cyclicity

The reproductive cyclicity of the queen is interesting. Most queens do not cycle between October and January. As day length increases in spring, there is reduced secretion of melatonin from the pineal gland and therefore a removal of its inhibitory effect upon the hypothalamic–pituitary–gonadal axis; anoestrus is terminated and there is a return to cyclical activity.

During the breeding season, queens that are not mated come into oestrus approximately every 14–21 days, of which 3–7 days comprise pro-oestrus/oestrus, followed by 8–15 days of interoestrus, when hormone concentrations are generally low. When ovulation occurs it is followed by the formation of corpora lutea and increased concentrations of progesterone, and this can occur in the presence or absence of pregnancy. The non-pregnant luteal phase lasts approximately 30–40 days and is associated with an absence of oestrous behaviour during that time. The pregnant luteal phase lasts on average 65 days.

As a result of this reproductive cyclicity, in free-roaming queens there are often two peak times for pregnancy and parturition, with most kittens being born in mid-spring or late summer. Many queens have two pregnancies per year; the mean interval between litters is 5.2 months. On average the litter size for young queens is between three and five kittens.

Pregnancy diagnosis

The peculiarities of the oestrous cycle of the queen mean that endocrinological methods of pregnancy diagnosis cannot simply be adapted from other species. However, imaging technologies such as diagnostic B-mode ultrasonography and radiography are equally applicable to all domestic species.

Absence of oestrus

In many species the absence of a return to oestrus at 21 days can be used as an indicator of pregnancy. However, in the queen the absence of a return to oestrus indicates only that ovulation has occurred; it is not specific for pregnancy (Figure 11.1). A failure to return to oestrus after day 45 may

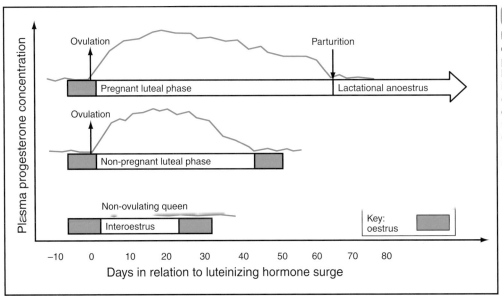

11.1 Plasma progesterone concentrations and periods of behavioural oestrus in the non-ovulating queen (bottom), ovulating but non-pregnant queen (middle) and pregnant queen (top).

be used as a positive sign of pregnancy, although by this time the pregnant queen is usually easily identified by changes in her physical appearance. Some domestic cats occasionally mate spontaneously during pregnancy.

Behavioural changes

Behavioural changes typical of pregnancy are not particularly useful for the diagnosis of pregnancy in queens, because such signs usually only develop during late pregnancy, by which time the non-pregnant queen will have returned to oestrus, and pregnant queens can be identified easily by their physical appearance. However, signs that typically occur during pregnancy include increased docility, excessive grooming of the perineal and mammary gland areas, and nesting behaviour.

Physical changes

In the queen bodyweight increases during pregnancy from an early stage (see Figure 11.6). As a result the abdomen may take on a rounded and swollen appearance, although the changes are less obvious than in the bitch and may not be noticed until after day 50. An increase in the size and degree of reddening of the nipples is commonly seen from day 21 onwards. Mammary gland enlargement occurs from day 58, and colostrum may be present in the teats in the last 7 days of pregnancy. Mammary gland enlargement and the secretion of milk are not normally features of pseudopregnancy in the cat, and therefore these signs may be used as indicators of pregnancy. During the last 2–3 weeks of pregnancy, fetal movements may be visible through the abdominal wall in the relaxed queen, especially when she is recumbent.

Manual palpation

The technique of detecting conceptuses by trans-abdominal palpation can be highly accurate and is not normally difficult in the queen. The optimum time for the diagnosis of pregnancy is between 21 and 25 days after mating, when the embryonic vesicles are approximately 2.5 cm in diameter. At this time an experienced clinician should achieve an accuracy of nearly 90%. Before this time the conceptuses are small tense swellings within the uterine horns, and although they can be detected as early as day 17 they may be overlooked because of their small size. From days 26–30 the conceptuses are tense spherical fluid-filled structures and can be readily palpated. However, from day 35 the conceptuses become elongated, enlarged and tend to lose their tenseness. They may be less easy to palpate at this time. After day 45 the uterine horns tend to fold upon themselves, resulting in the caudal portion of each horn being positioned against the ventral abdominal wall, and the cranial portion of the same horn being positioned dorsally. After day 55 the fetuses can often be identified easily, especially the fetal heads, which are particularly hard. It is difficult to count the number of conceptuses accurately by palpation except when performing an examination at approximately day 28. Adverse sequelae to the pregnancy caused by palpation have not been reported in the cat, but have been reported in other species.

Ultrasonography

B-mode ultrasonography can be used for early pregnancy diagnosis. The most accurate time to perform an examination is generally 3 weeks after the last mating.

In the non-pregnant queen the homogenous hypoechoic uterus can be identified dorsal to the bladder, and is relatively easily differentiated from the colon, which is hyperechoic. A small degree of uterine enlargement occurs within the luteal phase, whether the queen is pregnant or not, and therefore this is not diagnostic for pregnancy. Confirmation of pregnancy relies upon detection of fluid-filled (anechoic) conceptuses. Free fluid within the uterine lumen is not normal at any stage of pregnancy.

Accuracy

It is unlikely that queens undergoing an ultrasound examination in late pregnancy would be incorrectly diagnosed as non-pregnant. However, early examinations may also be inaccurate, especially if conducted too early, before fluid accumulation occurs within the conceptus. For this reason it is often recommended that the first examination is delayed until later than day 16 after mating. False-positive diagnoses may result from the confusion of empty loops of small intestine with early pregnancy. Fetal resorption may also produce a disparity between the number of conceptuses seen on the ultrasonogram and the number of offspring born.

The accuracy of detection of absolute fetal number is poor: the greatest accuracy is at the first examination and the lowest accuracy is in late pregnancy. Most commonly the number of fetuses is underestimated. The great advantage of using ultrasonography for pregnancy diagnosis is that at the time of the examination confirmation can also be made of fetal viability by assessing fetal movements and fetal heart rate. A small number of fetal abnormalities have also been detected by ultrasonography.

Radiography

Uterine enlargement can be detected from day 20 of pregnancy, when the enlarged uterus can be readily identified in the caudal abdomen, originating dorsal to the bladder and ventral to the rectum; it frequently produces cranial displacement of the small intestine. In some queens it is possible to detect uterine swellings but this is dependent upon the inherent contrast within the abdomen and the ability to identify the uterus. However, the early pregnant uterus has only soft tissue opacity and therefore any increase in size cannot be differentiated from other causes of uterine enlargement that occur at the same stage of the oestrous cycle, such as pyometra. A positive diagnosis of pregnancy is not possible until after day 40 when mineralization of the fetal skeleton is detectable radiographically.

There is sometimes concern about the use of radiography during pregnancy. However, it is unlikely that the fetuses will be damaged by the ionizing radiation after day 45 because organogenesis is largely complete by the fourth week of pregnancy. However, it is true that sedation or anaesthesia of the dam may be required and this is, of course, a potential risk. Importantly though, radiography is the method of choice to determine the number and size of fetuses in late pregnancy (when the technique is more accurate and useful than ultrasonography).

Detection of fetal heartbeats

Although no longer in common use, it is possible to detect fetal heartbeats using Doppler ultrasonography from day 30 onwards. Whilst this method has been superseded by diagnostic real-time ultrasonography for pregnancy diagnosis, there is some use of Doppler ultrasonography for fetal monitoring at the time of parturition, as described in the bitch (see Chapter 10).

In late pregnancy it is possible to auscultate fetal heartbeats in the queen using a stethoscope. Fetal heart rates may also be detected by recording an electrocardiogram (ECG). Both methods are diagnostic of pregnancy. Fetal heartbeats are not difficult to detect because the heart rate is usually more than twice that of the dam. In the normal fetus the heart rate exceeds 200 beats per minute.

Relaxin assay

In the queen the plasma progesterone concentration is elevated in both pregnancy and pseudopregnancy; therefore, measurement of this hormone before day 45 is not diagnostic of pregnancy. However, a pregnancy-specific increase in the concentration of plasma relaxin occurs from day 25 onwards. Relaxin is a peptide hormone that is produced by the fetoplacental unit, which as well as being luteotrophic is important for causing softening of the fibrous connective tissue of the dam's pelvis prior to birth. Relaxin concentrations can be measured easily in the practice laboratory using an enzyme-linked immunosorbent assay (ELISA) test kit. Given that relaxin is specific for pregnancy, it cannot be detected in the plasma after resorption of fetuses or following abortion with expulsion of the placentas.

Recently, a radioimmunoassay has been used to measure urinary relaxin and it was found that the concentrations mirrored those in plasma; urinary relaxin is first detectable between days 21 and 28 of pregnancy; the concentrations plateau between day 42 and day 50, and subsequently decline over the next 2 weeks towards parturition. It is possible that measurement of urinary relaxin may become a useful home-testing tool for pregnancy diagnosis by cat owners.

Normal pregnancy

During pregnancy a number of haematological changes occur in the queen; a normochromic normocytic anaemia develops in late pregnancy and results in a decrease in haematocrit. Interestingly, and in contrast to the bitch, leucocytosis is rarely observed. Other physiological changes include an increased blood volume and an increase in cardiac output by approximately 40% to compensate for the volume of the placenta and perfusion pressure. Furthermore, there is an increased respiratory rate to compensate for the metabolic needs of the fetus; oxygen consumption is increased by 20%. For information on the physiological and endocrinological changes that occur during pregnancy, see Chapter 1.

Gross observation of the queen will reveal noticeable signs of pregnancy in weeks 6 and 7. The most obvious signs include increased size of the mammary glands, increased firmness of the nipples and enlargement of the abdomen (see above).

Duration of pregnancy

Pregnancy in the queen is approximately 65 days (with a range of 52–74 days from mating to the onset of parturition). The variability in the observed length of pregnancy probably relates to the potentially long period of sexual receptivity and the uncertainty regarding the timing of ovulation in many queens, although larger litter sizes are associated with a slightly shorter duration of pregnancy.

Diagnostic imaging
Ultrasonography

In addition to early pregnancy diagnosis, B-mode ultrasonography can also be used to monitor the progress of the pregnancy and detect some fetal abnormalities. The conceptus may first be identified on day 8 after ovulation (approximately 9 days after mating), when it is 3 mm in diameter and contains only yolk sac fluid that is completely anechoic; the embryo proper is not yet visible. The embryo itself can be identified as a relatively homogenous mass at one pole of the conceptus by day 14 after ovulation, and usually on that day or 1 day later it is possible to identify cardiac activity within the embryo (noted as a rapid flickering of some of the pixels on the screen originating from an area within the embryo).

The conceptus increases in size rapidly and may lose its spherical outline, becoming oblate in appearance. At day 15 the embryo proper is approximately 3 mm in length and from day 20 after ovulation the embryo has enlarged and is approximately 8–9 mm in length. The yolk sac has also increased in length, becoming an elongate structure that extends along the longitudinal axis of the conceptus; it is often more than three times the length of the embryo at this early phase. At approximately day 20 it is possible to identify limb buds. The embryonic/fetal membranes are complicated and may appear confusing ultrasonographically; these are normally first identifiable from approximately day 21 onwards. It becomes possible to identify gross morphological features of the embryo (the head and body regions) from approximately day 25 onwards.

The conceptus is approximately 25 mm in diameter at day 30, and at day 35 the embryo is approximately 45 mm in length. The most rapid growth of the fetus occurs between days 32 and 55; during this time the limbs become even more apparent and it becomes possible to identify fetal movement, which often includes a spasmodic twitching of the fetus described as 'hiccoughing'. There is also substantial development of the embryonic membranes and, interestingly, fusion of the embryonic membranes of adjacent conceptuses has been noted in some queens. The allantois is now the prominent fluid-filled sac and entirely envelopes the amnion. The amnion surrounds the embryo but remains a poorly vascularized membrane.

The zonary placenta can usually be identified easily from this stage of pregnancy onwards and the fetal skeleton becomes evident from 40 days, when fetal bone appears hyperechoic and casts acoustic shadows. The heart can now be easily identified (Figure 11.2) because the hyperechoic valves can be seen to move. The large arteries and veins can be seen cranially and caudally. Lung tissue surrounding the heart is hyperechoic with respect to the liver, and the region of the forming diaphragm can be easily identified. During this time it is possible to determine the sex of individual fetuses by detection of the genital tubercle and scrotum in the male, and the genital tubercle only in the female. From 45 days onwards it is possible to identify the fluid-filled (anechoic) stomach, and a few days later the bladder can be seen. At this time the embryo is approximately 90 mm in (crown–rump) length and continues to increase in size and is almost completely developed by day 53. In late pregnancy the head, spinal column and ribs produce intense reflections and become more easily identifiable. In the last 20 days of gestation the kidneys can be seen and in late pregnancy the small intestine may be detected. The fetus is approximately 150 mm in length at the time of birth.

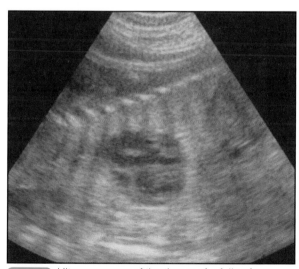

11.2 Ultrasonogram of the thorax of a feline fetus demonstrating the hyperechoic vertebrae casting acoustic shadows, and the centrally positioned hypoechoic heart with anechoic chambers.

In several domestic species ultrasonography is used to determine gestational age. This may also be of value in cats with uncertain mating times. Conventionally this is achieved using measurements of fetal size, including gestational sac diameter, crown–rump length, head diameter and body diameter; formulae are available using some of these measurements to allow estimation of fetal age in the cat (Figure 11.3).

Measurements to be used in the first half of pregnancy
Number of days of pregnancy = (Internal gestational cavity diameter in mm + 11.566) / 1.368
Number of days of pregnancy = (External gestational cavity diameter in mm + 12.130) / 1.602
Number of days of pregnancy = (Embryonic length in mm + 31.430) / 2.0087

Measurements to be used in the second half of pregnancy
Fetal age = [log (Fetal abdominal diameter in cm / 0.405565)] / 0.0372141
Fetal age = [log (Fetal biparietal diameter in cm / 0.483873)] / 0.02756
Fetal age = [log (Fetal stomach diameter in cm / 0.115113)] / 0.0388901

11.3 Estimation of the stage of pregnancy in the queen using ultrasonographic measurement of various embryonic and fetal structures. (Data from Zambelli *et al.*, 2002; 2004)

Whilst these measurements work well, for some clinicians they are cumbersome, and as yet are not programmed into the majority of ultrasound machines. It is often more simple to estimate gestational age by noting the time at which specific aspects of pregnancy can be identified by ultrasonography (Figure 11.4):

- Embryo proper – day 14
- Limb buds – day 20
- Urinary bladder – day 30
- Kidneys – day 40
- Renal cortex and medulla differentiation – day 50.

Days after mating	Ultrasonographic feature
16–17	Fetal heart beat
18 (range 17–19)	Limb buds
20 (range 19–21)	Yolk sac cavity has the same volume as the allantoic cavity
26 (range 24–27)	Fetus has definitive form
30 (range 29–32)	Identification of stomach, bladder, lung and liver
33 (range 30–34)	Detection of fetal limb movement
35 (range 35–39)	Detection of fetal eyes
37 (range 37–40)	Detection of fetal neck and head movement
39 (range 38–41)	Detection of kidneys
40	Fetal sexing possible
42 (range 35–45)	Detection of fetal skeleton
50	Differentiation between renal cortex and renal medulla is possible

11.4 Timing of ultrasonographic appearance of fetal and extra-fetal structures during pregnancy.

Radiography

Progressive mineralization results in an increasing number of bones that can be identified by radiography. It is generally possible to estimate the approximate stage of pregnancy by evaluating which bones are mineralized. Although there are variations depending upon the radiographic exposure a reasonable degree of accuracy can be obtained; in fact using these data it is possible to predict when parturition is likely to occur with an accuracy of ± 3 days in 75% of cases. The typical appearance of mineralization of specific bones at various stages of pregnancy is given in Figure 11.5.

Stage of pregnancy	Radiographic feature
From day 38–40 onwards	Mineralization of the skull, scapula, humerus, femur, vertebrae and ribs
From day 43 onwards	Mineralization of the tibia, fibula, ilium and ischium
From day 49 onwards	Mineralization of the metatarsals and metacarpals
From day 52–53 onwards	Mineralization of the digits and sternum
From day 56–63 onwards	Mineralization of the molar teeth

11.5 Stages of pregnancy in relation to the radiographic appearance of particular fetal components.

Care of the pregnant queen

Routine vaccination should be planned prior to breeding. The aims are to protect the queen and to ensure appropriate levels of antibodies in the colostrum. Similarly, it is best to ensure that treatments for endo- and ectoparasites are given prior to breeding.

Whilst some vaccinations and parasite treatments may be given during pregnancy, specific advice should be sought from the manufacturer before these are used, and in general it is best to avoid administration of any medications during pregnancy.

During early pregnancy there are no particular precautions that need to be taken with the pregnant queen, although it is always wise to reduce potential contact with unfamiliar cats of unknown disease status. Most queens will eat and exercise normally, although occasionally there is a short period of reduced appetite and vomiting, which appears to occur around the time of implantation of the conceptuses. Queens usually increase their food intake by approximately 10% per week throughout pregnancy. This increased appetite starts in early pregnancy and results in an initial and early gain in weight (Figure 11.6). The purpose of this basic increase in fat storage is to ensure the availability of sufficient energy later in pregnancy and throughout lactation. Usually, in later pregnancy (after day 45) the rate of weight gain parallels the gain in weight associated with the growth of the fetuses. As pregnancy progresses it may be advisable to feed several smaller meals per day as the enlarged uterus encroaches upon the stomach. At the end of pregnancy the queen will usually be consuming 70% more energy than in her non-pregnant state.

Preparation for parturition

In the last few weeks of pregnancy the queen is best isolated from other cats and kept in a quiet environment with a room temperature of approximately 22°C. The queen should be introduced to an easily cleanable kittening box with sides that are high enough to contain the kittens, and which is lined with absorbent bedding material. Some prediction of the estimated time of kittening may be given if the litter size is known (larger litters are associated with a shorter

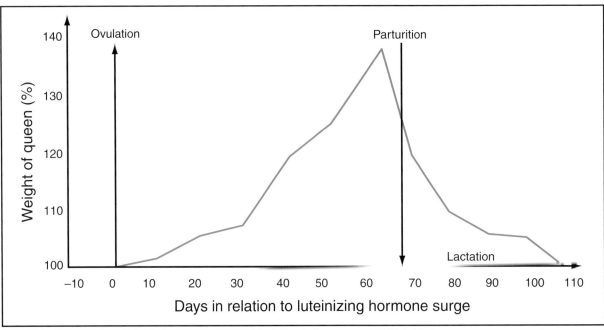

11.6 Changes in weight during pregnancy and lactation in the queen. NB Weight changes are expressed as a percentage change from the non-pregnant weight.

pregnancy length), if fetal measurements have been made with ultrasonography during pregnancy, or if the rectal temperature is recorded on a daily basis (although a dramatic reduction in rectal temperature is not noted as consistently in the queen as it is in the bitch). Some indication of impending parturition may also be given when there is a dramatic increase in the presence of milk in the mammary glands. Owners should be advised about the preparations they should make to assist in the kittening, including the collection of small towels to clean and dry kittens, thread to tie off umbilical cords, syringes to allow suction of fluid from the mouth, and a method for accessing veterinary assistance if necessary.

Parturition

Physiology of parturition

There is surprisingly little information concerning the physiology of the end of pregnancy and parturition in the queen; indeed, most information is inferred from studies in other species. However, it is clear that whilst there is a reasonable range in the observed length of pregnancy (62–70 days), with an average of 65 days, this can be influenced both by litter size and by breed. The effect of breed may relate to the variation in litter size. Interestingly, there are also variations in the successful outcome of pregnancy, with the best success appearing to be for smaller litters, when pregnancy length is longer, and for some particular breeds (Sparkes *et al.*, 2006). It is generally believed that in cases of abnormal parturition (dystocia), the rapid identification of a problem and prompt intervention improves the outcome, and for this reason it is important that the veterinary surgeon has a clear understanding of the events and time course of normal parturition (eutocia).

Stages of parturition

Generally parturition in the queen progresses unaided; indeed queens appear quite susceptible to excessive intervention, which can disrupt or even halt kittening and may promote cannibalism of the kittens. It is always best therefore to isolate the queen prior to parturition and to house her in a quiet and peaceful room (see above).

Preparation

It is proposed that the queen, like many other placental mammals, enters parturition by a cascade of events that commences with placental stress of the fetus (reduced ability to provide adequate nutrition), which stimulates the hypothalamic–pituitary–adrenal axis and results in the release of adrenocorticosteroid hormone from the fetus. An increase in cortisol from both the fetus and the dam results in:

- A shift in placental hormone production towards oestrogen rather than progesterone (which causes increased secretion of mucus within the caudal reproductive tract and also increases myometrial contractions)

- The release of prostaglandin F2α (which is luteolytic and reduces the concentration of progesterone, thereby also increasing myometrial contractions).

This early stage of parturition is called the phase of preparation, and although in the queen this is associated with a decline in progesterone concentration, progesterone often does not fall precipitously as it does in the bitch, but remains in the region of 3–5 ng/ml. The slight fall in progesterone noted during this stage tends not to result in a significant reduction in rectal temperature. Queens that have not settled into a kittening area will seek an isolated place, and there may be a transient anorexia. In addition, there may be increased milk production noted within the mammary glands.

First stage

The duration of the first stage of parturition is normally between 6 and 12 hours, although it may last longer in the nervous primiparous animal. This stage is characterized by the onset of coordinated myometrial contractions, during which the fetuses rotate from a ventral position (as normally seen on radiographs) to a dorsal position. It is not possible to observe the initial (intermittent) uterine contractions, but during this phase the queen may appear uncomfortable; usually she is very restless, often vocalizes and licks the genital area compulsively. The queen may appear to make a nest by turning around in the box, smelling the floor and pawing at the bedding. Late in the first stage of parturition, a slight reddish brown vulvar discharge is sometimes noted in the hours preceding the expulsion of the first kitten. The discharge originates from the marginal haematoma and is distinctly different from that noted in the bitch (which is greenish in colour).

Second stage

The onset of the second stage of parturition commences with abdominal contractions and ends with expulsion of a fetus. In the first instance these abdominal contractions are not part of 'Ferguson's reflex' (which is related to the release of oxytocin) but are a combination of uterine and abdominal contractions which together force the fetus against the cervix. This results in a neurohormonal reflex that stimulates (via the paraventricular nucleus) the release of oxytocin, which in turn stimulates further myometrial contractions (this latter component is 'Ferguson's reflex').

The second stage of parturition is often quite short in duration (normal length is 4–16 hours, with the majority complete within 6 hours) but can be up to 42 hours, and in rare cases if the queen is disturbed early in the second stage, the delivery may appear to be suspended for 2–3 days without apparent complication.

When the abdominal contractions begin the queen may assume a semi-squatting position. Contractions subside when the queen lies down on her side, and at this time she may keep purring. Kittens appear to be presented from one uterine horn and then the other; entrance of a kitten into the caudal reproductive tract results in a longer duration of abdominal contraction. Contractions may be associated with rupture of the allantoic–amniotic membrane and expulsion of the translucent allantoic fluid a few minutes before the appearance of a portion of the kitten at the vulval lips. Most kittens are born within the amnion, although this may also rupture during birth. If intact it is torn open by the queen, and this is followed by intensive licking of the kitten until spontaneous breathing occurs. Complete expulsion of each individual kitten often takes only 3–5 minutes.

Diagnosis: It is important that the clinician is able to determine whether the queen is in the second stage of parturition or still in the first stage. Inexperienced breeders tend to be nervous and do not understand the purpose of the preparatory stages, but once the second stage has commenced a critical feature of dystocia is failure of this stage to progress. The signs that are diagnostic for the onset of the second stage of parturition are:

* Visible abdominal straining
* Passage of fetal fluids.

Failure of the second stage of parturition to progress is an important clinical sign that necessitates veterinary examination (see Chapters 13 and 14). In the queen, as in other polytocous species, it is not possible to separate the second and third stages of parturition (which is marked by passage of the placenta).

Fetal disposition: This is the term used to describe the spatial arrangement of the fetus in relation to the pelvis and birth canal of the dam, and of its extremities to itself. Kittens are normally delivered in either: (a) the cranial longitudinal presentation, dorsal position, with extended posture; or (b) the caudal longitudinal presentation, dorsal position, with extended posture. The latter is not a breech disposition (which technically is caudal longitudinal presentation, dorsal position with hindlimbs flexed cranially at the hips – and is abnormal and results in dystocia). Approximately 70% of kittens have cranial and 30% have caudal presentation; the latter does not appear to predispose to dystocia.

Third stage
The third stage of parturition is characterized by the passage of the placenta. In the queen this generally occurs within the second stage of parturition; a placenta may be expelled after each kitten (usually 10–15 minutes later), although it is possible for several kittens to be delivered without placentas, and for these to be expelled together. At each placental site there is arterial vasoconstriction and therefore exsanguination and separation of the placenta, which is then expelled by uterine contractions. It is normal for the queen to consume a portion or all of the placentas, probably because they contain significant amounts of protein; this does not appear to cause any adverse effects. Parturition is followed by the puerperium, during which the reproductive tract reduces in size to one similar to that before pregnancy.

The postpartum period

Maternal behaviour
The majority of queens demonstrate strong maternal instincts, although this will depend upon environmental factors (including the level of human interference), the general health of the queen, hormonal changes (e.g. oestrogens, progesterone, oxytocin and prolactin), previous experience as a mother and the stimulus of the kittens. Sometimes young and inexperienced queens may neglect to clean, dry and stimulate kittens adequately in the interval between births, and this can be a cause of perinatal loss unless prompt attention is given. Where subsequent poor mothering behaviour is observed, careful attention should be paid to human interference but the possible presence of hypocalcaemia should also be investigated.

Perinatal losses
Approximately 8% of kittens are dead at birth, and there is some influence of breed upon this; a higher prevalence of stillbirths occurs in Exotic Shorthaired and Persian breeds. A higher incidence of stillbirths is also seen when litter size is greater. After birth, up to 10% of kittens that are born alive are lost before weaning; the majority of these die within the first week.

The puerperium
After normal parturition there is continued contraction of the uterus, followed by a decrease in muscle mass and a reduction in the volume of glandular and connective tissue. These events are collectively called 'involution'. During this phase phagocytic cells invade the uterine cavity and remove remnants of tissue and contaminant bacteria; this is associated with a characteristic lochial discharge from the vulva, which is normally present for a few weeks after parturition.

Onset of lactation
In the queen, prolactin is secreted in increasing amounts at the end of pregnancy and particularly at parturition; this is the most important hormone responsible for initiation of lactation. Shortly after parturition there is a rapid increase in the size of all eight thoracoabdominal mammary glands, and milk may be seen to drip from the nipples. Once milk production has been initiated, the continuation of lactation is maintained by the sucking activity of the kittens. The first milk produced is the colostrum and this is a critical source of immunoglobulins for the kittens (it is important to recognize that the endotheliochorial placenta of the queen allows relatively poor exchange of immunity, indeed only 5% of the antibodies needed by the kitten are present at the time of birth as a result of transplacental migration).

During the first 24 hours of life low concentrations of enzymes are produced from the enterocytes, thus allowing transfer of immunoglobulins directly into the blood of the kitten. Thereafter the enterocytes mature, 'gut closure' occurs and antibodies can no longer be absorbed directly.

Throughout lactation milk continues to be produced; it is secreted constantly during the day, but the act of sucking results in a neuroendocrine reflex resulting in the release of oxytocin, which elicits contraction of myoepithelial cells surrounding the mammary gland alveoli and the 'let-down' of milk into the excretory duct system. As mentioned above, reserves of fat are built up in early pregnancy (the average queen will weigh approximately 20% more after parturition than before she was pregnant; see Figure 11.6), and these are important to balance the energy requirements for lactation. Importantly though, during this time most queens should be offered a high energy (high fat) and high protein diet to enable effective lactation. The requirements for food intake are greatest during the third week after parturition, when milk production is at its peak. Thereafter, sucking activity reduces, and as weaning commences and prolactin concentrations decline, lactation ultimately ceases.

Resumption of cyclical activity

Queens are essentially within anoestrus in the early phase of lactation, presumably because there is suppression of gonadotrophin hormone secretion. However, many queens return to oestrous behaviour 2–4 weeks after parturition, whilst they are still lactating. Some queens do not return to oestrus until 2–8 weeks after weaning. However, if parturition has occurred late in the year, it is feasible that oestrus does not return but that the queen enters seasonal anoestrus.

References and further reading

Boyd JS (1971) Radiographic identification of the various stages of pregnancy in the domestic cat. *Journal of Small Animal Practice* 12, 501–506

Johnston SD, Root Kustritz MV and Olson PNS (2001) Feline pregnancy. In: *Canine and Feline Theriogenology*, pp.414–419. WB Saunders, Philadelphia

Root MV, Johnston SD and Olson PN (1995) Estrus length, pregnancy rate, gestation and parturition lengths, litter size, and mortality in the domestic cat. *Journal of the American Animal Hospital Association* 31, 429–433

Sparkes AH, Rogers K, Henley WE *et al.* (2006) A questionnaire-based study of gestation, parturition and neonatal mortality in pedigree breeding cats in the UK. *Journal of Feline Medicine and Surgery* 8, 145–157

Tsutsui T and Stabenfeldt GH (1993) Biology of ovarian cycles, pregnancy and pseudopregnancy in the domestic cat. *Journal of Reproduction and Fertility* 47(Suppl), 29–35

Zambelli D, Castagnetti C, Belluzzi S and Bassi S (2002) Correlation between the age of the conceptus and various ultrasonographic measurements during the first 30 days of pregnancy in domestic cats (*Felis catus*). *Theriogenology* 57, 1981–1987

Zambelli D, Castagnetti C, Belluzzi S and Paladini C (2004) Correlation between fetal age and ultrasonographic measurements during the second half of pregnancy in domestic cats (*Felis catus*). *Theriogenology* 62, 1430–1437

Zambelli D and Prati F (2006) Ultrasonography for pregnancy diagnosis and evaluation in queens. *Theriogenology* 66, 135–144

12

Clinical approach to unwanted mating and pregnancy termination

Alain Fontbonne

Introduction

A wide range of drugs and protocols have been used for pregnancy termination in the bitch and queen. From the point of view of the practitioner the ideal treatment must be safe, with good efficacy, an easy-to-follow protocol and, ideally, applicable as soon as the unwanted mating has occurred (Fieni *et al.*, 1998). In fact, none of the drugs available displays all these characteristics, and veterinary surgeons will have to decide on the most suitable product depending on the clinical situation. Anti-progestogens have recently become the treatment of choice in Europe. However, they may be very expensive to use in large bitches and this is why alternative techniques are also discussed in this Chapter.

Indications

Although most bitches are kept under close control by their owners, unwanted mating is still a common clinical problem. Bitches are frequently presented when they are still in oestrus and the owners may have observed mating or are suspicious that mating has taken place. In the queen, mating is often not observed and the owner notices changes during pregnancy.

Reasons for termination of pregnancy include:

- Non-breeding females – where the owners do not wish to have any offspring from their pet
- Breeding females – where the female has been mated by the wrong male (different breed, wrong sire in the kennel)
- Size – where a small bitch is mated by a much bigger dog
- Age – where females are either <12 months or >8 years of age (higher risk of dystocia and hypogalactia)
- High risk of dystocia or Caesarean operation – such as bitches with vaginal abnormalities, previous pelvic fractures or infantilism; single pup syndrome in a large or giant breed bitch
- Medical indications – such as endocrine conditions (diabetes), metabolic problems (pregnancy ketosis) or any disease that may deteriorate during pregnancy (e.g. renal failure, cardiac insufficiency, mammary gland fibroadenomatosis in the queen).

Whatever the reason for a client to approach the clinician for advice on abortion, it is important to remember that certain aspects of this procedure may have ethical and moral implications for the owners. Especially in cases of late pregnancy termination, when live fetuses can first be visualized by ultra-sonography and may later be expelled dead or alive after medical treatment, hospitalization during the abortion period should be considered and discussed.

Treatment

Surgical approach

Neutering in oestrus and early dioestrus
Surgical neutering in oestrus and early dioestrus may be used for non-breeding females and can be achieved with ovariectomy or ovariohysterectomy. Contrary to common belief, a study conducted at the author's clinic and based on more than 100 bitches showed that ovariectomy performed on oestrous bitches or at the beginning of dioestrus is well tolerated (Reynaud *et al.*, 2005).

Ovariectomy in mid-dioestrus
Performing an ovariectomy in mid-dioestrus may induce abortion, but the sudden removal of the ovaries and with it the progesterone supply will induce lactation. However, this may not be a real problem due to the high efficacy of dopaminergic drugs, especially cabergoline which can be used to treat iatrogenic pseudopregnancy.

Neutering in late dioestrus
In pregnancies of more than 45 days, surgical neutering should be performed using ovariohysterectomy rather than ovariectomy, owing to the risk of dead fetuses being retained in the uterus. Lactation is very likely to be induced and can be treated with cabergoline.

Pharmacological approach
Pharmacological termination of pregnancy can be achieved using drugs that act primarily on the uterus, ovary or pituitary gland. Of the products discussed below only those which act upon the uterus are licensed in Europe for pregnancy termination in the bitch.

Drugs that act on the uterus

Anti-progestogens: The first studies that used the human drug mifepristone (usually known as RU 486) for pregnancy termination in the bitch were conducted in the late 1980s. The most common drug used today for this indication in Europe is aglepristone (RU 534), a derivative, which is an oil–alcohol solution containing 30 mg/ml.

Anti-progestogens are synthetic steroids that compete with progesterone at the receptor level. They have strong binding affinity to progesterone receptors in the uterus of the bitch, cat and rabbit (Hoffmann and Schuler, 2000). They act as true receptor antagonists, devoid of any agonistic activity, although they do have slight anti-glucocorticoid activity. The affinity of aglepristone for uterine receptors in the bitch is three times greater than that of progesterone itself (nine times in the queen). These drugs prevent the effects of progesterone on the uterus initially without decreasing serum progesterone concentrations.

Bitches: Aglepristone is administered at a dose of 10 mg/kg s.c. twice, 24 hours apart. It can be used from day 1 to day 45 following mating. The intramuscular route should be avoided because the drug may cause local necrosis. Even careful subcutaneous injections can cause irritation, and a maximum volume of 5 ml per injection site is recommended. In larger dogs several injection sites will be necessary and these should be recorded and avoided for the second dose. Massaging the injection site after administration of aglepristone may also be beneficial. Interscapular subcutaneous injections seem to work well for most clinicians (Gogny and Fieni, 2004). The efficacy of aglepristone administered in the first 22 days after mating is almost 100%, and 95% later in pregnancy.

Queens: Aglepristone is administered at a rate of 15 mg/kg s.c. twice, 24 hours apart – this is off-label use in the UK. Although affinity to the progesterone receptor is higher in the queen than in the bitch, the bioavailability is lower and metabolism is faster. This means that the dose has to be increased in order to be effective. The efficacy of aglepristone in the queen is lower than in the bitch, at about 85%, and early intervention shows better results than administration at more advanced stages of pregnancy.

Contraindications: Small amounts of aglepristone are eliminated through the kidneys and may also modify the elimination of other drugs such as digoxin or vincristine (Gogny and Fieni, 2004).

Oestrogens: Oestradiol esters, especially estradiol (oestradiol benzoate), have been used to terminate unwanted pregnancies in bitches for many years. In the United States, oestradiol cypionate and valerate are available. Oestrogen alters transit time of the zygote in the oviduct, changes the biochemical environment and causes degeneration of the zygote, as well as having direct embryotoxic effects (Wanke et al., 2002). This is caused not just by the oestrogen

itself, but also by altering the ratio of progesterone to oestrogen in the plasma (Johnston et al., 2001). Oestrogen administration may also stimulate uterine contractions and induce cervical relaxation.

Bitches: Different protocols have been described in the literature. It is important to use as low a dose as early as possible after the unwanted mating in order to minimize side effects. One of the most common protocols comprises three injections of 0.01 mg/kg of estradiol 3, 5 and 7 days after mating. The efficacy is around 95% (Sutton et al., 1997). A more recent study recommends a single injection of 0.2 mg/kg 2 days after mating (Tsutsui et al., 2006). According to these authors, the efficacy is unchanged.

Oestrogen has not been used very commonly since the introduction of aglepristone. The requirement for administration close to the time of mating and the more severe side effects have made it less attractive. In some practices it may still be the drug of choice in very large breeds owing to cost considerations. Oestrogen given during this stage of the cycle may lead to a prolonged oestrus of 7–15 days. When bitches are treated during pro-oestrus it may even induce ovulation and therefore facilitate conception (Fieni et al., 1998).

Side effects: Oestrogens may have a negative effect on haemopoiesis, leading to bone marrow suppression with subsequent aplastic anaemia.

Uterine diseases such as cystic endometrial hyperplasia and pyometra have been reported post treatment. This may be caused by the artificially induced presence of both progesterone and oestrogen in the uterus at the same time. Oestrogen increases the number of progesterone receptors, which can then be used by the increasing amount of progesterone produced from the ovaries at this stage of the cycle. Pyometra may result from pregnancy termination using oestrogen treatment in 7–15% of cases (Sutton et al., 1997; Eilts 2002). Some authors (Fieni et al., 1998) have reported a negative effect on the secretion of gonadotrophins from the pituitary gland, which causes further long-term fertility problems.

Recommendation: Owing to the high incidence of cystic endometrial hyperplasia and pyometra post treatment, and given the availability of a better alternative, oestrogens should not be used routinely for this indication.

Drugs that act on the ovaries

Prostaglandins: These drugs lyse the corpora lutea and induce a decrease in progesterone. The corpora lutea in bitches are quite resistant to prostaglandins and repeated treatments are necessary to achieve lysis. However, the bitches themselves are very sensitive to the smooth muscle contractions also caused by prostaglandins, and dose rates must be monitored very carefully.

Prostaglandin F2α (PGF2α) is available in two forms. Naturally occurring PGF2α is produced in the form of a tham-salt (dinoprost tromethamine). It has a very short half-life, especially after intramuscular injection, and requires two to three daily applications. Subcutaneous administration can prolong the activity for slightly longer. Depending on the protocols that are used (see below) the recommended dose varies between 30 and 250 µg/kg.

Synthetic PGF2α analogues, such as cloprostenol, luprostiol and alfaprostol, often show a better luteolytic effect in cattle (Wanke *et al.*, 2002). They have a longer half-life and require administration only once daily. It is important to remember that the dose rate for synthetic PGF2α is much lower than for natural PGF2α. The dosage for synthetic PGF2α is 2–10 µg/kg, depending on the compound used. A mix up between the dosages of natural and synthetic PGF2α may lead to a fatal overdose in treated bitches.

Mode of action: PGF2α terminates pregnancy through its luteolytic action as well as inducing uterine contractions and cervical dilation. Administration of PGF2α arrests the synthesis of progesterone in the corpora lutea, although its specific mechanism of action in carnivores is poorly understood. Contraction of the myometrium in combination with dilation of the cervix causes the expulsion of intra-uterine fetal debris.

Uses: PGF2α is not authorized for use in domestic carnivores, although it has been used in many countries to terminate pregnancies in bitches. Owing to its high efficacy, PGF2α must be handled with care when used in carnivores. Pregnant women and people with asthma should also take extra precautions.

Side effects: Preparations of PGF2α have short-acting and dose-dependent secondary effects, which never last for more than 2 hours (Fieni *et al.*, 1997). These effects tend to be more severe after the first administration of PGF2α and decrease in intensity with every subsequent injection, because most bitches seem to respond less acutely to the drug over time (Wanke *et al.*, 2002). Gastrointestinal side effects are most common and result in hypersalivation, emesis, reflex defecation and/or urination. However, cardiac (bradycardia) or respiratory (bronchoconstriction) side effects may also occur. In the author's experience, administration of PGF2α to brachycephalic breeds, especially English Bulldogs, is not advisable because sudden death may occur. To minimize side effects, some authors recommend starting off with low doses and progressively increasing them (Wanke *et al.*, 2002) or using premedication (Figure 12.1). This will eliminate the side effects in 58% of cases (Fieni *et al.*, 1997).

Drug	Dosage (administer 15 minutes before PGF2α)
Atropine	0.025 mg/kg s.c.
Prifinium bromide	0.75 mg/kg s.c.
Metopimazine	0.5 mg/kg s.c.

12.1 Suggested premedication to reduce the incidence of side effects of PGF2α administration. (Data from Fieni *et al.*, 1997)

Drugs that act on the pituitary gland

Dopamine agonists: These drugs can be divided into true dopamine agonists, such as bromocriptine and cabergoline, which have a high affinity for D2 dopaminergic receptors, and anti-serotonergic drugs, such as metergoline, which possess a dopaminergic effect when used at high doses.

Mode of action: In the bitch prolactin is essential to maintain luteal function from mid-pregnancy (day 30) onwards. During the second half of pregnancy dopamine agonists can be used to block prolactin, thus causing luteolysis and termination, although they are not authorized for this use.

Uses: In practice dopamine agonists are mostly used in combination with other drugs such as PGF2α, or in cases where other protocols involving anti-progestogens have failed in earlier pregnancy (see below).

Side effects: Cabergoline is authorized for use in the dog in the UK for the treatment of pseudopregnancy and has few side effects because it is less able than other dopamine agonists to cross the blood–brain barrier (Wanke *et al.*, 2002). Bromocriptine causes emesis immediately after administration in many dogs, and metergoline may cause restlessness.

Gonadotrophin-releasing hormone (GnRH) antagonists: These are synthetic peptides that compete with GnRH at the receptor level and decrease or block GnRH action. GnRH antagonists are effective when used for termination in very early pregnancy, ideally 1 week before implantation (Vickery *et al.*, 1989). At this early stage the drugs induce a decrease in plasma progesterone concentration within 2 hours of administration. Their efficacy decreases to 50% when used in mid-pregnancy.

Uses: GnRH antagonists are used currently in human medicine, but owing to their high costs are not available in veterinary medicine at present.

Side effects: GnRH antagonists may induce an inflammatory reaction caused by degranulation of mast cells (Fieni *et al.*, 1998).

Corticosteroids

The corticosteroid that is used most commonly for pregnancy termination is dexamethasone. The mode of action is poorly understood; some authors suggest that it possesses a direct anti-progesterone effect, whilst others believe that it increases uterine and placental prostaglandin synthesis, thus leading to the same events that precede normal parturition.

Uses: Dexamethasone is a relatively cheap drug and has been used extensively in some regions (such as South America) to terminate pregnancies in bitches. Use of dexamethasone for this indication in cats has not been studied. Corticosteroids are used from 30 days of pregnancy onwards; dexamethasone is administered at a rate of 200 µg/kg orally q12h. Abortion will occur between 7 and 13 days after the start of treatment.

Side effects: Corticosteroids have many side effects, ranging from vaginal discharge, prolonged polydipsia/polyuria, increased risk of diabetes or hyperadrenocorticism, transient weakness, lactation and the risk of live fetuses being aborted. Due to the many side effects corticosteroids are not recommended for pregnancy termination.

Practical approach to pregnancy termination in the bitch

Presentation: during oestrus following suspected unwanted mating

A summary of the management of the bitch presented soon after a suspected unwanted mating is shown in Figure 12.2.

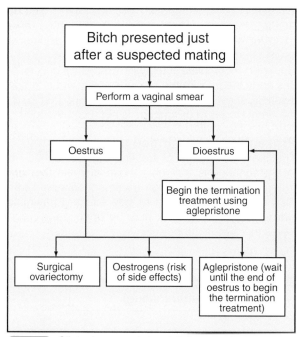

12.2 Clinical approach to a bitch presented soon after a suspected unwanted mating.

Treatment

- Non-breeding bitches may be safely neutered during oestrus or early dioestrus (Reynaud *et al.*, 2005).
- Aglepristone is very effective in terminating pregnancies at this early stage (see above) without compromising the future breeding potential of the bitch. It is short acting and should therefore be used when the bitch is in dioestrus to prevent pregnancy occurring as a result of mating after administration or prolonged survival of sperm in the female reproductive tract.

Vaginal cytology

The likelihood of the bitch being at the correct stage of the oestrous cycle to conceive should be investigated before embarking on surgical or pharmacological termination. A history should be taken to determine whether the bitch was actually seen mating/tied to a dog, and to identify the stage of the oestrous cycle.

Vaginal cytology can be helpful in both cases. Exfoliated vaginal cells may allow determination of the stage of the oestrous cycle, and samples from 65% of bitches that had been mated in the previous 24 hours showed spermatozoa under microscopic examination (Bowen *et al.*, 1985) (Figure 12.3). However, the absence of sperm does not indicate that a mating has not taken place.

12.3 Spermatozoa can often be visualized on a vaginal smear within 24 hours of mating. The arrows show an entire spermatozoon and a detached head. (Harris–Schorr stain; original magnification X400) (© A. Fontbonne)

Lennoz (2006) recommended that veterinary surgeons should monitor the onset of dioestrus to avoid conception after treatment. Aglepristone is active for an average of 6 days after subcutaneous administration. If a bitch is mated shortly before ovulation, when the plasma progesterone concentration is still low, and treated with aglepristone on the same day, she may still become pregnant owing to the prolonged survival of sperm in the genital tract. The bitch could also be mated again after treatment and conceive later. Aglepristone should be used at the end of oestrus or the beginning of dioestrus, as confirmed by vaginal cytology (Figures 12.4 and 12.5).

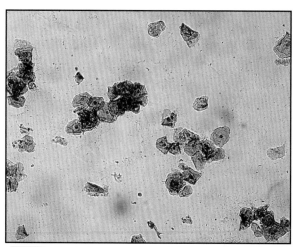

12.4 Vaginal smear obtained during oestrus. A high percentage of cornified superficial cells are present. Anti-progestogens should not be used until the early stages of dioestrus. (Harris–Schorr stain; original magnification X100) (© A. Fontbonne)

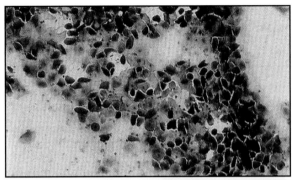

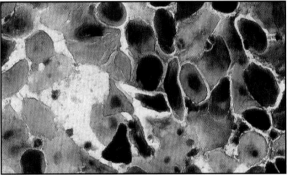

12.5 Vaginal smears obtained during early dioestrus. Numerous nucleated intermediate and parabasal cells, together with neutrophils, are visible. This is the best time to use anti-progestogens as abortive agents. (Harris–Schorr stain; original magnification X100 (top) and X400 (bottom)) (© A. Fontbonne)

Presentation: suspected early pregnancy, diagnosis not possible

The medical treatment of an animal of unknown pregnancy status is always difficult. Several studies have shown that many bitches presented at this stage are in fact not pregnant. Eilts (2002) reported that only 38% of bitches that had undergone an unwanted mating were truly pregnant. On the other hand, the efficacy of abortive drugs declines in later pregnancy and side effects may be more severe for the bitch. The birth of premature fetuses later in pregnancy, as opposed to resorption earlier on, must also be considered. All the options should be discussed with the owner and a decision taken with their full knowledge and consent.

Treatment

Aglepristone: This is the safest and most effective drug to terminate pregnancy at this early stage. However, especially in large or giant breeds, given the relatively high cost of the drug, clinicians will need to discuss with the owners whether to treat the bitch at once or wait until diagnosis of pregnancy becomes possible.

The sooner in early pregnancy aglepristone is used the better the efficacy. Studies have shown that aglepristone administered twice within 24 hours before day 21 of pregnancy causes abortion in 99.6% of treated bitches (Lennoz, 2006). Owing to this high success rate it does not seem necessary to perform a confirmatory ultrasound examination on day 25–30. At this early stage, it is thought that embryonic resorption occurs in <3 days.

The possible side effects include a serous vulval discharge, which may be seen in around 20% of cases between 24 and 96 hours following the first injection, transient anorexia, hypothermia and restlessness (Fieni *et al.*, 1996). In the long term it is important to inform the owner of the bitch that the interoestrous interval may be shortened. The interval was reduced to 158 ± 16 days from 200 ± 5 days in treated *versus* untreated Beagle bitches in one study (Galac *et al.*, 2004). There is no adverse effect on future fertility or on the risk of uterine infection (Fieni *et al.*, 1996).

PGF2α: This does not work well in early pregnancy as the corpora lutea are not very sensitive to prostaglandins at this stage. Therefore, either a high dosage is needed, which causes more side effects, or lower doses can be used with numerous repeated injections. A recommended protocol using natural PGF2α involves administration of 150–200 µg/kg q12h for 4–7 days, sometimes longer, beginning at least 15 days after mating (Wanke *et al.*, 2002). However, incomplete luteolysis frequently occurs at this early stage, with the risk of subsequent abortion failure (Lange *et al.*, 1997). Thus, use of PGF2α in the early stages of pregnancy is not recommended

Presentation: confirmed mid-pregnancy

The main advantage of treating bitches in mid- to late pregnancy is that only those animals that are truly pregnant will undergo treatment (Figure 12.6). This avoids the induction of side effects in animals where medical treatment may be unnecessary and saves the owners money.

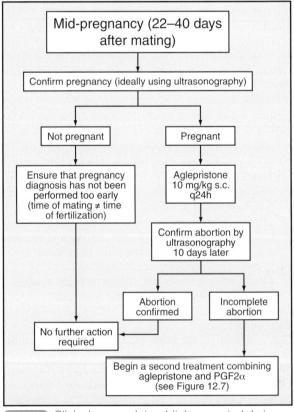

12.6 Clinical approach to a bitch presented during mid-pregnancy.

Treatment

Aglepristone: The efficacy of aglepristone at mid-pregnancy is good (it varies between 94.4 and 96.2%), although slightly lower than at an earlier stage. Prolific bitches may be predisposed to have only a partial abortion at this later stage, compared with treatment earlier in pregnancy (Lennoz, 2006).

At this stage aglepristone induces not resorption but abortion, which occurs 60–132 hours (maximum 7 days) after treatment. In 50% of cases expulsion is complete within 4 days (Fieni *et al.*, 1996). There is no change in the plasma progesterone concentration, and abortion occurs with high levels of circulating progesterone, because aglepristone blocks the uterine receptors. The subsequent interoestrous interval is significantly reduced in most bitches, and the next oestrous period may occur as early as 1–3 months after the abortion (Fieni *et al.*, 2001).

Side effects are more common in late abortions than with earlier intervention. In 32% of cases a brownish vulval discharge is observed, which may continue for 3–5 days after the first injection of aglepristone. This may be mistaken for the development of a uterine infection, but it is caused by the expulsion of embryonic debris. Antibiotics are not required unless other signs of infection develop. The increase in prolactin following the termination can induce lactation, maternal behaviour and sometimes slight depression and/or transient anorexia (Fieni *et al.*, 1996).

Most authors recommend performing an ultrasound examination 10–20 days after the second injection of aglepristone to ensure that complete expulsion has taken place. If the uterus does not appear to be greatly reduced in size and empty during this examination, a further dose of aglepristone should be administered (Lennoz, 2006).

In cases of complete or partial failure of termination it is recommended that the bitch is hospitalized and a second course of aglepristone in combination with injections of PGF2α begun (Figure 12.7). Some authors suggest that a combination of aglepristone and PGF2α should be used for any termination after 35 days of pregnancy (Lennoz, 2006). The PGF2α should be given repeatedly until complete expulsion has been achieved, as monitored by ultrasonography. For example, faced with failure of abortion, with two live fetuses remaining 1 week after the first administration of aglepristone, Gogny (2008) obtained

good results within 2 days using aglepristone and cloprostenol (with premedication using metoclopramide 0.5 mg/kg i.v. q12h). The vulval discharge stopped within 3 days and the following oestrus occurred 2 months later.

PGF2α: The efficacy of PGF2α is much greater at this stage (especially 25–30 days post ovulation) than earlier in pregnancy, owing to the increased sensitivity of the corpora lutea to the luteolytic effect of prostaglandins. Several protocols have been published (Figure 12.8). To minimize secondary effects when using natural PGF2α, it is recommended to use progressively increasing doses: 2 days at 25 μg/kg q12h; 2 days at 50 μg/kg q12h; and >4 days at 100 μg/kg q12h (Wanke *et al.*, 2002). In some bitches, a minimum of 9 days of treatment is necessary to produce complete embryonic or fetal death, whatever the dose used (Wanke *et al.*, 2002). Fieni *et al.* (1997) obtained 92.5% efficacy using only three injections of cloprostenol (2.5 μg/kg) at 48-hour intervals.

Drug	Dosage	Duration of treatment
Natural PGF2α	20–30 μg/kg s.c. q8–12h	4–7 days or longer
Natural PGF2α	100 μg/kg s.c. q12h	4–7 days or longer
Natural PGF2α	25 μg/kg s.c. q8–12h 50 μg/kg s.c. q8–12h 100 μg/kg s.c. q8–12h	Days 1 and 2 Days 3 and 4 >Day 4
Cloprostenol (PGF2α analogue)	2.5 μg/kg s.c.	Day 1, day 3, day 5

12.8 Protocols using PGF2α to induce termination at mid-pregnancy (>25 days) in the bitch. In order to use cloprostenol, it is recommended to dilute 1 ml of cloprostenol in 9 ml of isotonic saline solution. This solution is used at 0.1 ml/kg s.c. and may be kept at +4°C for the entire duration of treatment. (Data from Fieni *et al.*, 1997, Wanke *et al.*, 2002)

Most authors recommend performing an ultrasound examination to ensure that the pregnancy has been terminated and the uterus is completely empty. Partial abortions may occur if treatment is discontinued too early (Wanke *et al.*, 2002). There have been reports of a transitory depression of progesterone concentrations followed by a renewed increase, permitting the survival of some fetuses (Wanke *et al.*, 2002). The decrease in plasma relaxin after abortion is not sufficiently fast to be of any use as a control parameter.

In order to minimize the side effects of PGF2α and to apply protocols earlier in pregnancy, some authors recommend a combination of PGF2α and dopaminergic agonists (Figure 12.9). These protocols should never begin before 25 days of pregnancy and a further ultrasound examination at the end of treatment should always be performed to ensure that failure has not occurred.

Drug 1	Drug 2
Aglepristone 10 mg/kg s.c. two injections 24 hours apart	Natural PGF2α (dinoprost) 50 μg/kg s.c. q12h until complete abortion
Aglepristone 10 mg/kg s.c. two injections 24 hours apart	Cloprostenol 1.5 μg/kg s.c. q24h until complete abortion
Aglepristone 10 mg/kg s.c. two injections 24 hours apart	Cloprostenol 1 μg/kg s.c. 10 times/ day for 2 days (Gogny, 2008)

12.7 Protocols combining aglepristone and PGF2α for treatment of the bitch in cases of failure of termination at mid-pregnancy using aglepristone alone.

Cloprostenol	Dopaminergic drug
1 μg/kg s.c. every other day (up to 9 days)	Cabergoline: 5 μg/kg (0.1 ml/kg) orally q24h (for 9 days)
1 μg/kg s.c. days 1 and 5 of treatment	Cabergoline: 5 μg/kg (0.1 ml/kg) orally q24h (for 10 days)
2.5 μg/kg s.c. day 1 of treatment	Cabergoline: 5 μg/kg (0.1 ml/kg) orally q24h (for 10 days)
2.5 μg/kg s.c. day 1 of treatment	Bromocriptine: 30 μg/kg orally q8h (for 10 days)
1 μg/kg s.c. days 1 and 5 of treatment	Bromocriptine: 30 μg/kg orally q8h (for 10 days)
1 μg/kg s.c. days 1 and 3 of treatment	Cabergoline: 5 μg/kg (0.1 ml/kg) orally q24h (for 7 days). This protocol must begin a minimum of 35–45 days from the first mating

12.9   Protocols combining cloprostenol and dopaminergic drugs to induce termination at mid- to late pregnancy in the bitch. (Data from Onclin and Verstegen, 1999; Wanke *et al.*, 2002; Corrada *et al.*, 2006)

Dopamine agonists: Suggested protocols for termination in mid- to late pregnancy are detailed in Figure 12.10. However, these drugs show nearly no activity before 30 days of pregnancy, and only moderate activity between 30 and 40 days. Treatment should never be discontinued until complete termination has been achieved. In the author's experience dopamine agonists are useful in complementing other forms of treatment, or as a last resort when other treatments have failed.

Drug	Protocol	Duration of treatment
Cabergoline	5 μg/kg (0.1 ml/kg) orally q24h	>7–9 days
Metergoline	0.6 mg/kg orally q12h	>3–23 days

12.10   Protocols using dopaminergic compounds alone to induce termination in mid- to late pregnancy (preferably after 40 days) in the bitch. Note: bromocriptine alone would have to be used at a higher dose (100 μg/kg), which is likely to cause vomiting. (Data from Wanke *et al.*, 2002; Nöthling *et al.*, 2003)

Presentation: late pregnancy

Terminations at this stage (>40 days after mating) cause the bitch to give birth to live fetuses, which die shortly after parturition (Figure 12.11). Therefore, hospitalization is recommended to avoid stress to the owner. In most cases the bitch will produce milk post partum.

Treatment

Aglepristone: This is not recommended for use after 45 days of pregnancy, owing to the risk of retaining dead fetuses *in utero*. It has been used successfully in combination with PGF2α (see Figure 12.7) or dopamine agonists. The bitch should be hospitalized for treatment.

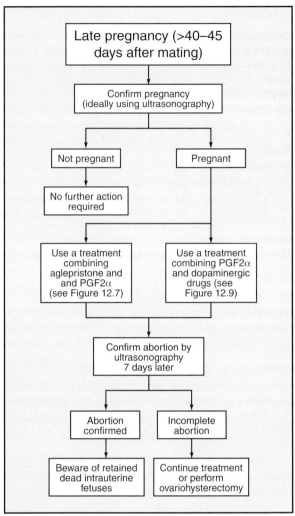

12.11   Practical approach to pregnancy termination when the bitch is presented in late pregnancy (>day 40–45 after mating).

Other drugs: PGF2α has good efficacy at this stage with repeat injections. Careful monitoring post treatment is required to ensure that all uterine contents have been expelled. Dopamine agonists may be used alone or in combination with PGF2α (see Figure 12.9), but they have a slower mode of action with a minimum of 5 days after the first administration before a response is seen. This carries the risk of inducing premature parturition rather than late abortion. In all cases, hospitalization of the bitch is highly recommended.

Practical approach to termination of pregnancy in the queen

Pharmacological termination of pregnancy is rarely performed in the queen and clinical data are not readily available. Surgical neutering is the most common procedure at any stage of unwanted pregnancy. After 45 days of pregnancy, ovariohysterectomy should be performed because there is placental secretion of progesterone that will maintain pregnancy if ovariectomy alone is performed (Verstegen *et al.*, 1993).

Treatment

Oestrogens

In cats fewer side effects are observed after oestrogen treatment than in dogs. Nevertheless, use of oestrogen is no longer advisable because treatments with fewer side effects are available (Verstegen *et al.*, 2003).

Aglepristone

Aglepristone is not authorized for use in cats, but data are widely available. Dose rates are higher in queens than in bitches because cats metabolize aglepristone more quickly. The recommended protocol is to give two injections at a rate of 15 mg/kg (0.5 ml/kg) s.c. within 24 hours for terminations between day 0 and day 45 of pregnancy. The reported efficacy before implantation (<14 days) is around 95% (Fieni *et al.*, 2006). Recently, Goericke-Pesch *et al.* (2010) reported a 100% efficiency when treating queens with aglepristone on days 5 and 6 after mating. The efficacy decreases to 87% when termination is performed later, and this increases the risk of dead fetuses being retained within the uterus (Georgiev and Wehrend, 2006). Termination in the queen should therefore be performed as early as possible. Many authors recommend combining aglepristone with PGF2α when used after day 25. The subsequent interoestrus interval is much reduced; sometimes oestrus may commence within a few days following termination.

PGF2α

PGF2α is better tolerated in cats than in dogs. However, as in bitches, the sensitivity of the corpora lutea to PGF2α is low during the first month of pregnancy. In order to minimize side effects combination with cabergoline is possible. Suggested protocols are listed in Figure 12.12.

Drug	Protocol	Time of treatment
Aglepristone	Two injections at a rate of 15 mg/kg (0.5 ml/kg) s.c. 24 hours apart	0–45 days of pregnancy
Estradiol [a]	Two to three injections at a rate of 0.01–0.1 mg/kg every other day	From mating. No later than day 10
Natural PGF2α	250–500 µg/kg s.c. q8–12h for 5–7 days	From pregnancy diagnosis
Cloprostenol	2.5–5 µg/kg s.c. q24h for 5–7 days	From pregnancy diagnosis
Cabergoline + Cloprostenol	Cabergoline 5 µg/kg (0.1 ml/kg) orally q24h (7–14 days) plus cloprostenol 5 µg/kg s.c. q24h for >7 days	From pregnancy diagnosis
Cabergoline + Cloprostenol	Cabergoline 5 µg/kg (0.1 ml/kg) orally q24h (7–10 days) plus cloprostenol 2.5 µg/kg s.c. on days 1, 3 and 5	From pregnancy diagnosis

12.12 Protocols to induce termination of pregnancy in the queen. [a] Although licensed in some countries estradiol is generally not recommended for use in cats. (Data from Wanke *et al.*, 2002; Verstegen, 2003)

Dopamine agonists

There are contradictory results concerning the use of dopamine agonists alone, which may be due to the placental secretion of progesterone at the end of pregnancy (Erünal-Malal *et al.*, 2004). Treatment fails after 45 days; therefore, use of prostaglandins seems preferable and permits the start of some termination protocols at around 25 days of pregnancy (Onclin and Verstegen, 1997; see Figure 12.9).

Acknowledgement

The author wishes to thank Felicity Leith-Ross for her great help with the English language.

References and further reading

Bowen RA, Olson PN, Behrendt MD *et al.* (1985) Efficacy and toxicity of oestrogens commonly used to terminate canine pregnancy. *Journal of the American Veterinary Medical Association* **186**(8), 783–788

Corrada Y, Rodriguez R, Tortora M *et al.* (2006) A combination of oral cabergoline and double cloprostenol injections to produce third-quarter gestation termination in the bitch. *Journal of the American Veterinary Medical Association* **42**, 366–370

Eilts BE (2002) Pregnancy termination in the bitch and queen. *Clinical Techniques in Small Animal Practice* **17**(3), 116–123

Erünal-Maral N, Aslan S, Findik M *et al.* (2004) Induction of abortion in queens by administration of cabergoline (Galastop™) solely or in combination with the PGF2α analogue alfaprostol (Gabbrostim™). *Theriogenology* **61**, 1471–1475

Fieni F, Dumon C, Tainturier D and Bruyas JF (1997) Clinical protocol for pregnancy termination in bitches using prostaglandin F2α. *Journal of Reproduction and Fertility Supplement* **51**, 245–250

Fieni F, Martal J, Marnet PG *et al.* (2001) Hormonal variation in bitches after early or mid-pregnancy termination with aglepristone (RU534). *Journal of Reproduction and Fertility Supplement* **57**, 243–248

Fieni F, Martal J, Marnet PG *et al.* (2006) Clinical, biological and hormonal study of mid-pregnancy termination in cats with aglepristone. *Theriogenology* **66**, 1721–1728

Fieni F, Mérot J, Tainturier D *et al.* (1998) L'interruption médicale de la gestation chez la chienne. *Recueil de Médecine Vétérinaire* **174**(7/8), 141–150

Fieni F, Tainturier D, Bruyas JF *et al.* (1996) Etude clinique d'une anti-hormone pour provoquer l'avortement chez la chienne: l'aglépristone. *Recueil de Médecine Vétérinaire* **172**(7/8), 359–367

Galac S, Kooistra SJ, Butinar J *et al.* (2000) Termination of mid-pregnancy in bitches with aglepristone, a progesterone receptor antagonist. *Theriogenology* **53**, 941–950

Galac S, Kooistra HS, Dieleman SJ *et al.* (2004) Effects of aglepristone, a progesterone receptor antagonist, administered during the early luteal phase in non-pregnant bitches. *Theriogenology* **62**, 494–500

Georgiev P and Wehrend A (2006) Mid-gestation pregnancy termination by the progesterone antagonist aglepristone in queens. *Theriogenology* **65**, 1401–1406

Goericke-Pesch S, Georgiev P and Wehrend A (2010) Prevention of pregnancy in cats using aglepristone on days 5 and 6 after mating. *Theriogenology* **74,** 304–310

Gogny A (2008) Gestion d'un échec de protocole abortif: l'intérêt de l'association aglépristone-cloprosténol. *Virbac Info* **106**, 6–7

Gogny M and Fieni F (2004) Principe actif: l'aglépristone. *Le Nouveau Praticien Vétérinaire* **Janvier**, 14–15

Hoffmann B and Schuler G (2000) Receptor blockers – general aspects with respect to their use in domestic animal reproduction. *Animal Reproduction Science* **60–61**, 295–312

Johnston SD, Root Kustritz MV and Olson PNS (2001) Prevention and termination of canine pregnancy. In: *Canine and Feline Theriogenology*, ed. SD Johnson *et al.*, pp.168–192. WB Saunders, Philadelphia

Lange K, Günzel-Apel A-R, Hoppen H-O *et al.* (1997) Effects of low doses of prostaglandin F2α during the early luteal phase before and after implantation in beagle bitches. *Journal of Reproduction and Fertility Supplement* **51**, 251–257

Lennoz M (2006) Practical uses of aglepristone: review of a recent expert meeting. Nice 17–18 June 2005. *Proceedings 5th Biannual EVSSAR Congress*, Budapest, Hungary, 7–9 April 2006, pp.152–159

Nöthling JO, Gerber D, Gerstenberg C *et al.* (2003) Abortifacient and endocrine effects of metergoline in beagle bitches during the second half of gestation. *Theriogenology* **59**, 1929–1940

Onclin K and Verstegen J (1997) Termination of pregnancy in cats using a combination of cabergoline, a new dopamine agonist, and a synthetic PGF2α cloprostenol. *Journal of Reproduction and Fertility Supplement* **51**, 259–263

Onclin K and Verstegen JP (1999) Comparisons of different combinations of analogues of PGF2α and dopamine agonists for the termination of pregnancy in dogs. *Veterinary Record* **144**, 416–419

Reynaud K, Fontbonne A, Marseloo N *et al.* (2005) *In vivo* meiotic resumption, fertilisation and early embryonic development in the bitch. *Reproduction* **130**, 193–201

Sutton DJ, Geary MR and Bergman GHE (1997) Prevention of pregnancy in bitches following unwanted mating: a clinical trial using low dose oestradiol benzoate. *Journal of Reproduction and Fertility Supplement* **51**, 239–243

Tsutsui T, Mizutani W, Hori T *et al.* (2006) Estradiol benzoate for preventing pregnancy in mismated dogs. *Theriogenology* **66**(6-7), 1568–1572

Verstegen J (2003) Estrus cycle regulation, estrus induction and pregnancy termination in the queen. *Proceedings of the SFT Meeting*, 16–20 September 2003, Colombus, Ohio, pp.334–339

Verstegen JP, Onclin K, Silva LD and Donnay I (1993) Abortion induction in the cat using prostaglandin F2 alpha and a new anti-prolactinic agent, cabergoline. *Journal of Reproduction and Fertility Supplement* **47**, 411–417

Vickery BH, McRae GI, Goodpasture JC *et al.* (1989) Use of potent LHRH analogues for chronic contraception and pregnancy termination in dogs. *Journal of Reproduction and Fertility Supplement* **39**, 175–187

Wanke MM, Romagnoli S, Verstegen J and Concannon PW (2002) Pharmacological approaches to pregnancy termination in dogs and cats including the use of prostaglandins, dopamine agonists and dexamethasone. In: *Recent Advances in Small Animal Reproduction*, ed. PW Concannon *et al.* IVIS (www.ivis.org)

13

Clinical approach to abnormal pregnancy

Autumn Davidson

Evaluation of pregnancy

Abnormal pregnancy is not an uncommon presentation in clinical veterinary practice and may be manifest as pregnancy loss or a variety of metabolic conditions that occur during pregnancy. The presence of normal luteal function in non-pregnant dioestrus, along with the inability to detect pregnancy before 21 days (16 in the queen), results in very early fetal resorption being indistinguishable from a failure to conceive. Later fetal loss can be documented by serial ultrasonography (Cain, 2001).

Pregnancy detection by abdominal palpation (at approximately 30 days gestation) or radiography (≥43–46 days after the peak in luteinizing hormone (LH)) can confirm the presence of fetuses. Prior to fetal skeletal mineralization, other causes of uterine enlargement (e.g. hydrometra, pyometra) cannot be ruled out radiographically. Radiography cannot be used to assess fetal viability in a timely fashion. Once profound post-mortem changes have occurred, radiography can detect intrafetal gas accumulation or abnormal skeletal arrangements suggesting fetal death. Early fetal loss resulting in eventual resorption or abortion can only be detected by ultrasonography. Ultrasound examination is the best method to evaluate early gestation pregnancies and to determine whether the lack of a normal pregnancy is the result of failure to conceive (see Chapters 6, 7 and 8) or the loss of conceptuses (Baker, 2007).

Fetal loss

Intrauterine death

Early fetal death is characterized ultrasonographically by the preservation of fetal anatomy without a heartbeat (Figure 13.1a). Over time, fetal demise results in the loss of recognizable fetal anatomy and the loss of normal fluid content and shape of the vesicle (Figure 13.1b) (Baker, 2007).

Premature parturition

Mid- to late term gestational loss attributed to preterm or premature parturition is a controversial topic in small animal reproduction. Both hypoluteoidism and inappropriate uterine activity accompanied by cervical changes have been implicated in the pathophysiology of preterm birth in veterinary medicine. Whilst the human literature on this topic is abundant, there are

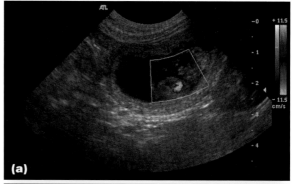

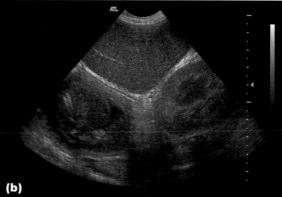

13.1 Fetal death. **(a)** Lack of detectable cardiac motion on Doppler ultrasonography (normal shown here) can be the earliest method of confirming fetal demise. **(b)** There is a loss of recognizable fetal anatomy and reduction in the amount of fluid within the vesicle. Note the adjacent (near-field) enlarged, fluid-filled uterine horn. (Courtesy of T. Baker)

few veterinary medicine publications and this syndrome is not well understood or even researched.

Causes

Premature parturition is defined here as uterine activity and cervical changes leading to the loss of the pregnancy via resorption or abortion before term, for which no metabolic, infectious, congenital, traumatic or toxic cause can be identified. Premature parturition is associated with progesterone concentrations of <2 ng/ml. Premature parturition is often a retrospective diagnosis after a thorough evaluation of the dam and fetuses following loss of pregnancy. This evaluation should include metabolic screening of the dam for systemic disease, infectious disease evaluation, histopathology of the expelled fetuses and placentas, and a review of kennel/cattery husbandry including nutrition, medications and environmental factors. All results are normal or negative in this condition. Dams that experience premature myometrial activity in one pregnancy may or may not exhibit it during subsequent pregnancies, but the syndrome can be a chronic cause of failure to reproduce.

In human medicine, preterm birth complicates 10–12% of pregnancies, but accounts for 80% of fetal morbidity and mortality. The diagnosis of preterm labour, placing the fetus at risk of premature delivery, is dependent upon evaluation of uterine contractility by tocodynomometry, measurement of fetal fibronectin, and transvaginal cervical length measurement determined via ultrasonography, which together have a high negative predictive value. Amniocentesis is also advocated in women as a method of evaluating fetal lung maturation and microbial invasion of the amniotic cavity. The presence of contractions alone does not warrant intervention. Multifetal gestations (i.e. litters) are associated with exaggerated physiological changes that promote premature parturition and complicate tocolytic therapy. Women with a history of preterm deliveries do appear to be at risk for the same occurrence in subsequent pregnancies (Newman *et al.*, 1998).

Treatment
If intervention is indicated, tocolytic agents have been commonly advocated. Antibiotics, rest and hydration do not appear to give any benefit. Contraindications to tocolytic therapy include severe pre-eclampsia, placental abruption, intrauterine infection, lethal congenital or chromosomal abnormalities, advanced cervical dilatation, and evidence of fetal compromise or placental insufficiency. Tocolytic agents inhibit myometrial contractions and include:

- Beta mimetics (terbutaline, ritodrine)
- Magnesium sulphate
- Calcium channel-blockers
- Prostaglandin synthetase inhibitors (indomethacin, ketorolac, sulindac).

In the future it is hoped that veterinary surgeons will be able to use anti-cytokine (interleukin-10) and anti-prostaglandin therapy to suppress the pathogenic process more completely at multiple sites along the pathway, rather than just treating the processes at the end of preterm parturition.

Maintenance of pregnancy: The maintenance of canine and feline pregnancy requires serum progesterone concentrations >1–2 ng/ml. Serum progesterone concentrations during pregnancy normally range from 15–90 ng/ml, declining gradually during the latter half of gestation, and falling abruptly at term (usually the day before or the day of parturition). Progesterone promotes the development of endometrial glandular tissue, inhibits myometrial contractility (causes relaxation of myometrial smooth muscle), blocks the action of oxytocin, inhibits the formation of gap junctions and inhibits leucocyte function in the uterus.

In several species, local changes in the progesterone concentration or in the ratio of progesterone to oestrogen in the placenta, decidua or fetal membranes is important in the initiation of parturition. Progesterone antagonists administered at term can result in an increased rate of spontaneous abortion. In the bitch, the corpora lutea are the sole source of progesterone, whilst in the queen placental production of progesterone occurs in the latter half of gestation. Canine luteal function is autonomous early in pregnancy but supported by luteotrophic hormones (LH and prolactin) after the second week of gestation.

Administration of progesterone to maintain pregnancy in cases of primary fetal abnormalities, placentitis or intrauterine infection can cause continued fetal growth with the possibility of dystocia and sepsis, requiring close monitoring. Administration of excessive progesterone to maintain pregnancy in a dam that does not actually require therapy can delay parturition and affect lactation, thus endangering the life of the bitch and her fetuses; it can also masculinize female fetuses.

Hypoluteoidism: This refers to primary luteal failure that occurs before term gestation. Hypoluteoidism is a potential but as yet not documented cause of late abortion in otherwise normal bitches. It has been reported that the induction of abortion in a normal but undesired pregnancy requires a reduction in plasma progesterone concentrations to <2 ng/ml (see Chapter 12). The diagnosis of gestational loss due to premature luteolysis is difficult, requiring documentation of inadequate plasma progesterone concentrations prior to abortion for which no other cause can be found. Precise measurement of progesterone concentrations, especially in the critical 1–3 ng/ml range, is not possible using the currently available rapid in-house enzyme-linked immunosorbent assay (ELISA) kits, necessitating the use of commercial laboratories in most practice situations. A few academic and human private laboratories provide a more rapid (<8 hours) turnaround, facilitating the diagnosis. Progesterone concentrations diminish in response to fetal death, thus documentation of a low progesterone concentration following abortion does not establish hypoluteoidism as the primary cause of reproductive failure.

Therapeutic intervention in primary hypoluteoidism can be accomplished with the administration of either injectable natural progesterone or oral synthetic progestogens. Total serum concentrations of progesterone can be monitored only when the bitch is supplemented with the natural product. Progesterone in oil is given at a rate of 2 mg/kg i.m. q72h. Altrenogest (a synthetic progestogen manufactured for use in horses) is given at a rate of 0.088 mg/kg orally q12–24h. Both forms of supplementation must be discontinued in a timely fashion so as not to interfere with normal parturition: within 24 hours of the date of parturition for the oral synthetic product; and within 72 hours for the natural injectable depot form. This requires accurate identification of gestational length via prior timing of ovulation (parturition is expected to occur 64–66 days after the LH surge or initial rise in progesterone, or 56–58 days from the first day of cytological dioestrus). Less accurate identification of gestational length can be made from the dates of breeding (58–72 days from the first mating), radiography and ultrasonography (see also Chapter 14).

Myometrial activity: Dams with documented low concentrations of progesterone and a history of late term loss of pregnancy with no apparent pathology can also be evaluated for premature myometrial activity mid-gestation, using uterine monitoring (Figure 13.2). Elaboration of prostaglandins from the endometrium and placenta, which are associated with premature myometrial activity, can result in secondary luteolysis. Premature uterine activity that endangers fetal survival can be identified before significant luteolysis occurs, and intervention is indicated if the pregnancy is otherwise normal. Pharmacological intervention to decrease myometrial activity, using progestational compounds and tocolytic agents alone or in combination, is recommended.

13.2 Tocodynomometry. Uterine sensor and recorder in place on a Labrador Retriever bitch in mid-pregnancy.

Terbutaline at a rate of 0.03 mg/kg orally q8h has been used to suppress uterine contractility in bitches and queens with a history of loss of otherwise normal pregnancies before term. The dose is ideally titrated to effect using tocodynomometry. Therapy is discontinued 24 hours before parturition (Plumb, 1999).

Infectious agents

Canine herpesvirus
Canine herpesvirus is a widely recognized and commonly cited cause of neonatal death. Exposure of a naïve bitch to canine herpesvirus during the last 3 weeks of gestation can result in either late term abortion, or neonatal death within the first few weeks of life, because inadequate maternal antibodies exist to allow passive immunity to be acquired by the neonates. Transmission of canine herpesvirus occurs subsequent to contact with infectious vaginal fluids or oronasal secretions.

Clinical signs: In the neonate clinical signs include anorexia (poor weight gain), dyspnoea, abdominal pain, incoordination, diarrhoea, serous/haemorrhagic nasal discharge and petechiation of the mucous membranes. The mortality rate in litters infected *in utero* or during birth is commonly 100%, with deaths occurring within the first 7 days of life. Puppies born to a naïve bitch may also come in contact with canine herpesvirus from another dog that is shedding the organism. Older puppies (>3–4 weeks of age) exposed to herpesvirus may have an unapparent infection, but central nervous signs including blindness and deafness have been observed later. The recently infected bitch generally has no clinical signs. Subsequent litters from the bitch are usually normal, having acquired maternal antibodies (Percy *et al.*, 1971; Carmichael, 1999).

Diagnosis: Both pre- and post-mortem diagnosis of canine herpesvirus infection in neonates can be challenging. Typical post-mortem changes that occur in the kidneys include multifocal petechial haemorrhage; however, this can also be present with bacterial septicaemia. Intranuclear inclusion bodies can be difficult to find. Diagnosis by virus isolation or canine herpesvirus-specific polymerase chain reaction (PCR) is confirmatory.

Treatment: This has been reported to be unrewarding and rarely successful, with recovery usually associated with residual cardiac and neurological damage. Treatment with immune serum from affected dams is reported to be ineffective in infected puppies. There is one case report of successful treatment with the antiviral drug aciclovir. A vaccine against herpesvirus is now available in the UK. The first vaccination is given at the time of mating or 7–10 days later, with the second vaccination administered 1–2 weeks before the expected date of whelping. The vaccination procedure has to be repeated during each pregnancy.

Brucellosis
Canine brucellosis, caused by *Brucella canis*, is the primary contagious infectious venereal disease of concern in canine reproduction; it is rare in the UK at present. *Brucella* causes reproductive failure in both dogs and bitches. Screening for *B. canis* is an important part of the pre-breeding evaluation of any dog in countries where it is endemic, and should be included in the initial diagnostic investigation for any case of canine abortion or apparent infertility.

Transmission: B. canis is a small, Gram-negative, non-spore forming aerobic coccobacillus. *B. abortus*, *B. melitensis* and *B. suis* have occasionally caused canine infections but are comparatively very rare. Transmission occurs through direct exposure to body fluids containing an infectious dose of the organism (semen, lochia, aborted fetuses/placentas, milk and, less commonly, urine that contains semen). There are 2×10^6 colony forming units in an infective dose. Transmission is therefore primarily venereal and oral (i.e. through the mucous membranes); the latter is associated with the ingestion of infectious materials.

Urine and indirect mucous membrane contact are not important routes of transmission. The aerosol route is only important if kennel conditions are very crowded. *B. canis* is short-lived outside the dog and is readily inactivated by common disinfectants, such as 1% sodium hypochlorite, 70% ethanol, iodine/alcohol solutions, glutaraldehyde and formaldehyde (Barr *et al.*, 1986).

Brucella organisms can be identified in the rough endoplasmic reticulum of placental trophoblastic giant cells in the gravid uterus of an infected bitch. Severe necrotizing placentitis with infarction of the labyrinth region, coagulation necrosis of the chorionic villi and necrotizing arteritis result in fetal death. The organism can be found in the gastric contents of the aborted fetuses. Necrotizing vasculitis in the male causes granulomatous lesions and results in epididymal and subsequent testicular and prostatic pathology.

Geographical distribution: Members of the Canidae family are the natural hosts of *B. canis*, and any mature, reproductively active dog of any breed is susceptible. Canine brucellosis occurs most commonly as an outbreak in a large commercial kennel, and less commonly in privately owned dogs. Outbreaks of canine brucellosis traditionally have a geographical orientation, with an increased incidence seen in the southernmost states of the United States, and more commonly in Mexico, Central and South America, China and Japan. The incidence in Europe is low. The increased practice and success of processing canine semen for exportation for use in artificial insemination (see Chapter 9) now makes canine brucellosis a concern worldwide, because direct mucosal contact amongst dogs is no longer necessary for transmission (Carmichael and Joubert, 1998).

Zoonotic potential: Humans can become infected with *B. canis*, but this is rare. Approximately 40 cases of human infection have been reported in several countries; however, the actual number is unknown as human cases are rarely diagnosed or reported. The transmission to humans occurs most commonly through contact with semen from an infected dog, vulvar discharge from an infected bitch, aborted fetuses or placentas, or through direct, accidental laboratory exposure.

Clinical signs: Canine brucellosis has high morbidity but low mortality. The clinical systemic signs are often subtle and include suboptimal athletic performance, lumbar pain, lameness, weight loss and lethargy. The primary clinical sign of canine brucellosis in the bitch is pregnancy loss, which can occur early in gestation (day 20) resulting in fetal resorption, or more commonly (75%) later in gestation (generally at 45–59 days) resulting in abortion. Bitches with pregnancy loss early in gestation may appear to be infertile, unless early ultrasonographic pregnancy evaluation has been performed. Non-gravid bitches can be asymptomatic, or can show regional lymphadenopathy (pharyngeal if orally acquired, inguinal and pelvic if venereally acquired; see Chapters 6 and 20).

Chronic infections can result in uveitis, granulomatous splenopathy, discospondylitis, granulomatous dermatitis, meningoencephalitis and nephritis. Bacteraemia can persist for years and asymptomatic dogs can remain infectious for long intervals. A large number of organisms are shed in the vulvar discharge of bitches 4–6 weeks after abortion. Urine can serve as a contaminated vehicle, owing to the proximity of the urinary and genital tracts in the dog. Milk can also serve as a vehicle for shedding.

Spontaneous recovery can occur 1–5 years post infection, but is difficult to document. Bitches can produce normal litters subsequent to multiple abortions, but can remain infectious to their offspring.

Diagnosis: Despite improvements in serological diagnostic methods, confirmatory blood cultures have classically been indicated when the disease is suspected. *B. canis* is readily isolated from the blood of bacteraemic individuals for several months after infection. A positive *B. canis* culture is a definitive diagnosis, and this has been advocated as the best diagnostic test in the first 2 months of the disease; however, dogs can become abacteraemic after 27–64 months.

Treatment: Infected dogs and bitches should be removed from breeding programmes and neutered. Antibiotic therapy has not been rewarding historically, probably because the organism is intracellular and bacteraemia is periodic. Antibiotic therapy may reduce antibody titres without clearing the infection, and relapses are common. Combination therapy with a tetracycline (doxycycline or minocycline 25 mg/kg orally q12h for 4 weeks) and dihydrostreptomycin (10–20 mg/kg i.m., s.c. q12h for 2 weeks, weeks 1 and 4) or an aminoglycoside (gentamicin 2.5 mg/kg i.m., s.c. q12h for 2 weeks, weeks 1 and 4) has been advocated as being the most successful. However, nephrotoxicity, the requirement for parenteral therapy and expense remain problematic (Johnson and Walker, 1992; Hollet, 2006).

One study recently reported an encouraging outcome of therapy with enrofloxacin (5.0 mg/kg orally q12h for 4 weeks) in a small group of infected dogs and bitches. Enrofloxacin was not completely efficacious in eliminating *B. canis*, but it maintained fertility and prevented the recurrence of abortions, transmission of the disease to subsequently whelped puppies, and dissemination of microorganisms during parturition (Wanke *et al.*, 2006).

Attempts to develop an appropriate vaccine that is capable of inducing immunity, but does not provoke serological responses that interfere with the diagnosis, have not been successful. Currently, the development of a vaccine is considered undesirable because the *Brucella* vaccines evaluated offered only moderate protection and immunized dogs developed antibodies that confounded serodiagnosis.

Adverse metabolic conditions during pregnancy

Gestational diabetes

Gestational diabetes occurs infrequently in the bitch and queen, and is attributed to the anti-insulin effect of progesterone (mediated by increased concentrations of growth hormone) during the luteal phase. Polydipsia, polyuria and polyphagia with weight loss are the clinical signs. Diets containing high levels of protein and low levels of carbohydrate may be helpful in the queen, while high-fibre diets promote euglycaemia in the bitch. Insulin may be indicated. Oversized fetuses can result from the increased production of insulin in response to maternal hyperglycaemia, and may cause dystocia due to fetal–maternal mismatch. The condition usually (in approximately 70% of cases) resolves with parturition. Termination of pregnancy may be indicated if the condition of the dam is grave and not otherwise manageable. Breed tendencies have been reported.

Pregnancy toxaemia

Pregnancy toxaemia is uncommon in the bitch and occurs as a result of altered carbohydrate metabolism in late gestation, resulting in ketonuria without glycosuria or hyperglycaemia. The most common cause is poor nutrition or anorexia during the latter half of gestation, typically in small breeds with a large litter size (Figure 13.3). Hepatic lipidosis can occur. An improved plane of nutrition can resolve the condition in most cases, but termination of the pregnancy may be indicated in severe cases.

Vasculitides

Human pregnancy is known to be a hypercoagulable condition. Procoagulability has also been recognized in pregnant dogs. Affected individuals have an increased tendency for thrombosis (as evidenced by an elevated concentration of D-dimers) which has a variable clinical appearance (Figure 13.4). The tendency for thrombosis in women is documented by specific testing in conjuction with ultrasonography. Anti-thrombotic therapy with low molecular weight heparin is well reported in women, but is undocumented in the bitch, and can easily result in congenital defects (aspirin-induced cleft palates) or loss of pregnancy due to placental or fetal haemorrhage. Warfarin is contraindicated in pregnancy because it crosses the placenta. The condition is believed to have a genetic predisposition in women. Affected bitches should be removed from the breeding pool.

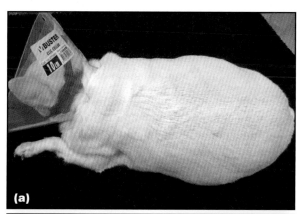

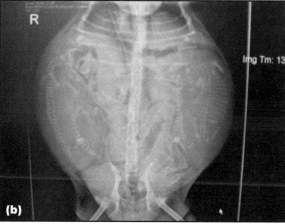

13.3 **(a)** Pregnancy ketosis occurs most commonly in small-breed bitches with large litters.
(b) Radiography can be used to document litter size in the later stages of pregnancy.

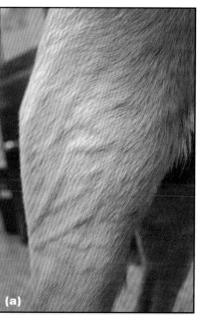

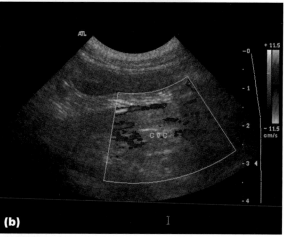

13.4 **(a)** Venous distension secondary to deep vein thrombosis in a bitch during pregnancy.
(b) Ultrasonogram showing almost complete thrombosis of the femoral vein. (Courtesy of T. Baker)

Oedema

Marked oedema of the perineum and pelvic limbs (Figure 13.5) has been observed, usually in large-breed bitches with large litters. Venous thrombosis should be ruled out using Doppler ultrasonography (see Figure 13.4b). Gentle exercise (walking or swimming) can be helpful.

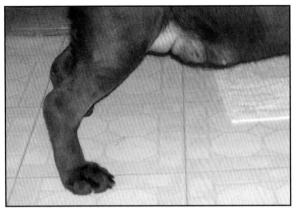

13.5 Pitting oedema of the distal pelvic limbs in late pregnancy.

Mild elevation of oestradiol towards the end of gestation in the bitch can induce recurrence of vaginal hyperplasia (Figure 13.6), which is seen more commonly in oestrus and causes compromise of the birth canal. An elective Caesarean operation is indicated.

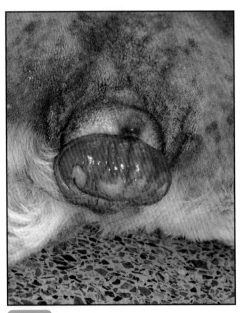

13.6 Vaginal hyperplasia occluding the birth canal.

Eclampsia

Eclampsia can occur during the last few weeks of pregnancy. For information see Chapter 14.

Vaccination

Rescue centres and veterinary surgeons may be faced with the dilemma of vaccinating a pregnant dog. Historically, vaccination during pregnancy has been advised against in small animal medicine, because of the paucity of data concerning vaccine safety and efficacy during gestation, and because it is accepted that no substance should be administered unnecessarily during pregnancy. However, when the immunity of the dog is unknown, the risk of maternal, fetal and neonatal infection must be weighed against that of vaccination. The Centers for Disease Control in the United States state that the risk to a developing fetus from vaccination of the mother during pregnancy is primarily theoretical, that no evidence exists of risk resulting from vaccination of pregnant women with inactivated viral or bacterial vaccines or toxoids, and that the benefits of vaccinating pregnant women usually outweigh potential risks when the likelihood of disease exposure is high.

Further research into vaccination of pregnant dogs is needed; extrapolating from the human field is advised at this time. Depending on the environment the pregnant bitch will be exposed to, likely diseases such as canine distemper virus, parvovirus and *Bordetella bronchiseptica* (intranasal vaccine) warrant vaccination. Inactivated subunit recombinant polysaccharide conjugate vaccines should be selected for use in the pregnant dog if available.

References and further reading

Baker TW (2007) Diagnostic imaging of the reproductive tract and adnexa in cats and dogs. *Proceedings, Norwegian Annual Congress in Small Animal Reproduction*, pp. 14–18. Oslo, Norway

Barr SC, Eilts BE, Roy AF and Miller R (1986) *Brucella suis* biotype 1 infection in a dog. *Journal of the American Veterinary Medical Association* **189**(6), 686–687

Cain JL (2001) A logical approach to infertility in the bitch. *Veterinary Clinics of North America: Small Animal Practice* **31**, 237–245

Carmichael L (1999) Neonatal viral infections of pups: canine herpesvirus and minute virus of canines (canine parvovirus-1). In: *Recent Advances in Canine Infectious Diseases*, ed. L Carmichael. IVIS (www.ivis.org)

Carmichael LE and Joubert JC (1998) Transmission of *Brucella canis* by contact exposure. *Cornell Veterinarian* **78**, 63–73

Concannon PW, McCann JP and Temple M (1989) Biology and endocrinology of ovulation, pregnancy and parturition in the dog. *Journal of Reproduction and Fertility Supplement* **39**, 3–25

Davidson AP, Grundy SA and Foley JE (2003) Successful medical management of neonatal canine herpesvirus: a case report. *Communications in Theriogenology* **3**(1), 1

Evermann JF (1989) Diagnosis of canine herpetic infections. In: *Current Veterinary Therapy X: Small Animal Practice*, ed. RW Kirk, pp.1313–1316. WB Saunders, Philadelphia

Freshman JL (2009) Pregnancy loss in the bitch. In: *Kirk's Current Veterinary Therapy XIV*, ed. JD Bonogura and DC Twedt, pp. 986–989. WB Saunders, Philadelphia

Hollet RB (2006) Canine brucellosis outbreaks and compliance. *Theriogenology* **66**(3), 575–587

Johnson CA and Walker RD (1992) Clinical signs and diagnosis of *Brucella canis* infection. *Compendium on Continuing Education for the Practicing Veterinarian* **14**, 763–772

Keid LB, Soares RM, Vasconcellos SA *et al.* (2007) A polymerase chain reaction for the detection of *Brucella canis* in semen of naturally infected dogs. *Theriogenology* **67**(7), 1203–1210

Newman RB, Campbell BA and Stramm SL (1998) Objective tokodynomometry identifies labour onset earlier than subjective maternal perception. *Obstetrics and Gynecology* **76**, 1089–1092

Percy DH, Carmichael LE, Albert DM, King JM and Jonas AM (1971) Lesions in puppies surviving infections with canine herpesvirus *Veterinary Pathology* **8**, 37–53

Plumb DC (1999) *Veterinary Drug Handbook, 3rd edn*. Iowa State University Press, Ames

Saballus MK, Lake KD and Wager GP (1987) Immunizing the pregnant woman. Risks *versus* benefits. *Postgraduate Medicine* **81**(8), 101–108

Wanke MM, Delpino MV and Balth PC (2006) Use of enrofloxacin in the treatment of canine brucellosis in a dog kennel (clinical trial). *Theriogenology* **66**(7), 1573–1578

14

Problems during and after parturition

Autumn Davidson

Introduction

Although many bitches and queens deliver in the home or kennel/cattery setting without difficulty, requests for veterinary obstetrical assistance are becoming more common. The financial and emotional value of stud dogs, breeding bitches, tom cats, queens and their offspring makes the preventable loss of even one neonate undesirable.

Veterinary involvement in canine and feline obstetrics has several goals, including:

- To increase live births (minimizing stillbirths resulting from difficulties in the birth process)
- To minimize morbidity and mortality in the dam
- To promote increased survival of neonates during the first week of life.

Neonatal survival is directly related to the duration and quality of parturition. There is no doubt that careful management of whelping/kittening will result in improved survival of both dam and fetuses. This requires an understanding of normal parturition and delivery in the bitch and queen, as well as the clinical ability to detect abnormalities in the process of parturition and to apply suitable treatment regimes quickly and effectively.

Predicting the time of parturition

Clinicians are commonly asked to ascertain whether a bitch or queen is at term and ready to deliver a litter physiologically, and then to intervene if parturition has not begun. An accurate determination of pregnancy length can be difficult, especially if numerous copulations occurred and the timing of ovulation was not established (i.e. the length of pregnancy will be more difficult to calculate). Prolonged gestation is a form of dystocia, which in itself can be difficult to manage without knowing when ovulation occurred. Determining the expected time of parturition in the bitch is more challenging than in the queen, because bitches are spontaneous ovulators, whereas in the queen ovulation occurs in response to mating, and usually mating has been observed.

Bitches
The normal duration of pregnancy in the bitch can be approximated from any of the following criteria:

- 56–58 days from the first day of cytological metoestrus/dioestrus (detected by serial vaginal cytology and defined as the first day that the percentage of cornified/superficial cells returns to <50%)
- 61–63 days from ovulation determined by progesterone concentrations of 5–7 ng/ml
- 64–66 days from the initial rise in progesterone from baseline concentrations (generally to values >2 ng/ml)
- 58–72 days from the first instance that the bitch permitted breeding.

Prediction of the length of pregnancy, without having established the time of ovulation during the preceding oestrus, is difficult in the bitch because of the disparity between oestrous behaviour and the actual time of conception, and the length of time for which semen can remain viable in the reproductive tract (often ≥7 days). Mating and conception dates do not correlate closely enough to permit a very accurate prediction of whelping date simply by counting the number of days from when mating was observed. In addition, the clinical signs of the end of pregnancy are not specific, and while some are suggested to be useful there is considerable variation both within and between bitches; for example, the radiographic appearance of fetal skeletal mineralization varies at term, fetal size varies with breed and litter size, and the characteristic decline in rectal temperature (typically to <37°C) may not be detected in all bitches. Furthermore, breed, parity and litter size can also influence the length of pregnancy (Eilts *et al.*, 2005).

Queens
Given that the queen is an induced ovulator (ovulation follows coitus by 24–36 hours), pregnancy length can be predicted more accurately from the date of mating. This assumes that copulation provided adequate coital stimulation for the surge in luteinizing hormone (LH) and subsequent ovulation (see Chapter 1), and that a limited number of copulations were permitted. The duration of pregnancy in queens ranges from 52 to 74 days when recorded from the last mating or first mating to parturition, although the mean pregnancy length is 65–66 days.

Ultrasonography
When data on the timing of ovulation are not available, estimation of the stage of pregnancy can be

made with real-time diagnostic ultrasonography. The canine blastocyst can first be detected within the uterus as a spherical hypoechoic structure 2–3 mm in diameter surrounded by a hyperechoic rim at approximately 19–20 days after the preovulatory LH peak. Ultrasonography permits the evaluation of early fetal cardiac motion (21–22 days after the LH peak), fetal movement (31–32 days after the LH peak) and fetal heart rate, which enables assessment of viability. By 30 days gestation (i.e. 30 days after the LH surge) pregnancy diagnosis with ultrasonography is straightforward. Definite ultrasonographic diagnosis of pregnancy in the queen, based on the appearance of a 'fetal pole', can be made at 15–17 days post coitus, although gravid uterine enlargement (4–14 days) and the presence of a gestational sac (11–14 days) can be detected even earlier (Figure 14.1).

The determination of fetal age by ultrasonography can be accomplished in two ways:

- By noting the first appearance of visible structures
- By the measurement of certain parameters.

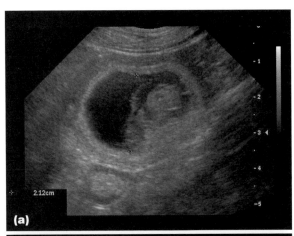

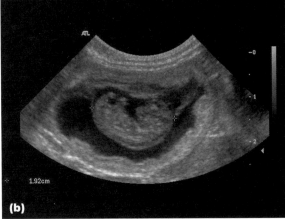

14.1   (a) Early feline pregnancy: 28 days gestation. The fetus has recognizable morphology within the gestational sac (cursors). Fetal membranes are evident. (b) Feline pregnancy: 35 days gestation. (Courtesy of I. Baker)

Prediction of fetal age by noting the first appearance of structures that correspond to a certain stage of pregnancy is often more accurate than making specific measurements of the fetus or uterus.

Measurements such as the diameter of the gestational sac, fetal occipitosacral (crown–rump) length and fetal skull (biparietal) diameter can be obtained ultrasonographically. These relate closely to fetal age, and permit estimation of pregnancy length and parturition date (Figures 14.2 and 14.3). However, there is considerable variation in breed size (especially in the dog) and individual variations in measuring technique, which are sources of inaccuracy when predicting fetal age with ultrasonography. Ultrasonography is less accurate than radiography in estimating litter

Bitches (gestational age ± 3 days)
Less than 40 days gestation
GA = (6 × GSD) + 20
GA = (3 × CRL) + 27
Greater than 40 days gestation
GA = (15 × HD) + 20
GA = (7 × BD) + 29
GA = (6 × HD) + (3 × BD) + 30
Days before parturition
DBP = 65 – GA
Queens (gestational age ± 2 days)
Greater than 40 days gestation
GA = 25 × HD + 3
GA = 11 × BD + 21
Days before parturition
DBP = 61 – GA

14.2   Formulas to predict gestational age and days before parturition in the bitch and queen. Gestational age (GA) is based on days after the LH surge in the dog and days post breeding in the cat. Gestational sac diameter (GSD), crown–rump length (CRL), head diameter (HD) and body diameter (BD) measurements are in centimetres. Days before parturition (DBP) is based on 65 ± 1 days post LH surge in the dog and 61 days post breeding in the cat. (Data modified from Nyland and Mattoon, 2002)

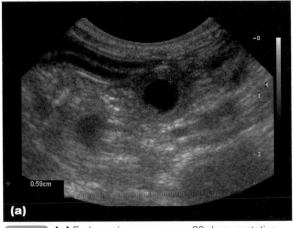

14.3   (a) Early canine pregnancy: 20 days gestation. The fetal pole is evident dorsally within the gestational sac. The diameter of the gestational sac was measured at 0.59 cm (cursors). (Courtesy of T. Baker) (continues)   ▶

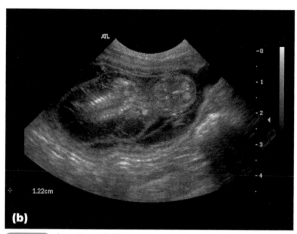

14.3 (continued) **(b)** Late gestation canine fetus. Cursors indicate proper positioning for measurement of the biparietal diameter, which was determined to be 1.22 cm. Ossification is evident. (Courtesy of T. Baker)

size, particularly later in gestation, owing to its dynamic nature. It should always be kept in mind that fetal resorption or abortion can alter litter size after early ultrasonographic estimates have been made.

Overdue parturition

Imminent or overdue parturition may be predicted in the bitch by:

- Counting the number of days from ovulation (if the time of ovulation was established)
- The metoestrus/dioestrus vaginal smear (as defined above)
- Ultrasonographic measurements of the fetuses.

It is also useful to instruct the owner to record the rectal temperature twice daily during the last third of pregnancy, because a decline in rectal temperature precedes parturition by approximately 12–36 hours in many but not all bitches.

In the queen, prediction of the time of expected parturition can be achieved by counting the number of days from mating. However, it is not uncommon for this information to be unknown or unclear and it is important, therefore, to perform a full clinical examination to ensure that the dam is clinically well and that she is pregnant, and to look for signs that indicate that parturition is imminent, or has commenced but has been overlooked by the owner.

Progesterone concentration

In the bitch, measurement of the plasma progesterone concentration can be used to assess whether parturition is imminent. Progesterone concentrations decrease approximately 24–36 hours before parturition. Demonstration of high plasma progesterone concentration therefore indicates that parturition is not imminent, while a low progesterone concentration indicates that parturition is imminent or should already have occurred. Intermediate values are difficult to interpret. Plasma progesterone can be measured easily in the practice laboratory by the use of enzyme-linked immunosorbent assay (ELISA) kits. It

can be seen that bitches that are still within their normal physiological pregnancy length will have high plasma progesterone concentrations, whilst bitches and queens that have had primary uterine inertia will have low plasma progesterone concentrations. Non-pregnant bitches may have high or low plasma progesterone concentrations, because the luteal phase of pregnancy and non-pregnancy is remarkably similar. Non-pregnant queens will have low plasma progesterone concentrations unless the queen has returned to oestrus and subsequently ovulated.

Dystocia

Dystocia is defined as difficulty with the normal vaginal delivery of a neonate from the uterus. Dystocia must be diagnosed in a timely fashion for medical or surgical intervention to improve outcome. In addition, the aetiology of dystocia must be identified for the best therapeutic decisions to be made.

Aetiology

Dystocia can result from maternal factors (uterine inertia, pelvic canal anomalies, intrapartum compromise), fetal factors (oversize, malposition, malposture, anatomical anomalies) or a combination of both (Figure 14.4). For effective management, the recognition of dystocia must be made in a timely manner, and the contributing factors must be identified correctly.

Cause	Bitch (%)	Queen (%)
Maternal	75.3	67.1
Primary complete inertia	48.9	36.8
Primary partial inertia	23.1	22.6
Birth canal too narrow	1.1	5.2
Uterine torsion	1.1	–
Uterine prolapse	–	0.6
Uterine strangulation	–	0.6
Hydrallantois	0.5	–
Vaginal septum formation	0.5	–
Fetal	24.7	29.7
Malpresentations	15.4	15.5
Malformations	1.6	7.7
Fetal oversize	6.6	1.9
Fetal death	1.1	1.1

14.4 The percentage of different presenting types of dystocia in bitches and queens. (Data from Ekstrand and Linde Forsberg, 2008 and Walett Darvelid and Linde Forsberg, 2008)

Maternal factors

Uterine inertia is the most common cause of dystocia. *Primary uterine inertia* results in the failure of delivery of any neonates at term, and is thought to be multifactorial, with the involvement of metabolic defects at the cellular level. In essence there is an intrinsic failure to establish a functional, progressive level of myometrial contractility, which results in failure of expulsion. A genetic component may be present, and it is not uncommon to identify dams from specific breeding lines in which primary uterine inertia is common. *Secondary uterine inertia* is the cessation of parturition after it has been initiated, which results in a failure to deliver the remaining fetuses. Secondary inertia can result from metabolic or anatomical (obstructive) causes, and it is also thought to have a genetic component.

Birth canal abnormalities such as vaginal strictures (Figure 14.5), stenosis from previous pelvic trauma or particular breed conformation, and intra-vaginal or intrauterine masses can cause obstructive dystocia. In most cases, birth canal abnormalities can be detected in the pre-breeding examination, and resolved or avoided by elective Caesarean operation. Causes of intrapartum compromise that render the dam unable to complete delivery include metabolic abnormalities such as hypocalcaemia and hypoglycaemia, systemic inflammatory reaction, sepsis and hypotension (caused by haemorrhage or shock).

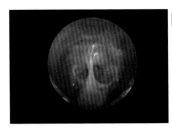

14.5

Septate band at the vestibulovaginal junction.

Fetal factors

Fetal factors that contribute to dystocia most commonly involve mismatch of fetal and maternal size, fetal anomalies and fetal malposition and/or malposture. Prolonged gestation with a small litter can cause dystocia owing to one or more oversized fetuses. Fetal anomalies such as hydrocephalus and anasarca (Figure 14.6) similarly can cause dystocia. Fetal malposition (ventrum of the fetus proximal to the dam's dorsum) and fetal malposture (most commonly flexed neck and scapulohumeral joints) promote dystocia because the fetus cannot traverse the birth canal smoothly.

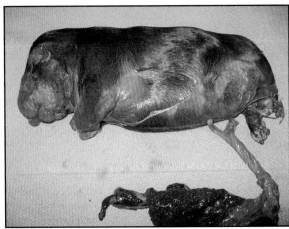

14.6 Oversized fetus due to anasarca, which was found obstructing the birth canal.

Diagnosis

When bitches and queens with presumed dystocia are examined by veterinary surgeons, there may be a variety of different presentations. In some cases, veterinary examination is late in the process and the dam may be distressed and debilitated and there may already be fetal death, whilst in others examination occurs early in the process of parturition. It is always best to engage with the owner, if possible, during pregnancy to provide support and education, and to ensure early examination and intervention if necessary.

History

An efficient diagnosis of dystocia is dependent upon taking an accurate history and performing a thorough physical examination in a timely manner. The clinician must quickly obtain a careful reproductive history, detailing breeding dates, any ovulation timing performed, historical and recent parturition as well as a general medical history.

Physical examination

The physical examination should address the general status of the patient and include: a digital and/or vaginoscopic pelvic examination for patency of the birth canal; evaluation of litter and fetal size (radiography is most useful); and assessment of fetal viability (ideally using Doppler or real-time ultrasonography) and uterine activity (tocodynomometry is most useful).

Clinical signs

Within the clinical history the following criteria are useful in indicating the likely presence of dystocia:

- There has been a reduction in rectal temperature (which may have returned to normal, be above normal or remained low) commencing more than 12 hours previously and there are no signs of parturition
- There has been passage of a green (bitch) or red–brown (queen) discharge from the vulva, which originates from the marginal haematoma and indicates that at least one placenta is beginning to become separated
- There has been expulsion of fetal fluids, commencing 2–4 hours previously, and no progression of parturition
- There has been cessation of signs of parturition for more than 2 hours, or infrequent signs of parturition for more than 2–4 hours
- There have been strong signs of parturition for more than 30 minutes but no signs of fetal expulsion
- There is other evidence of probable dystocia (e.g. a fetus is visibly stuck in the birth canal)
- The dam is depressed, lethargic or showing signs of shock, fluid loss or dehydration.

Non-obstructive dystocia: In females with non-obstructive dystocia the common presentation is with relaxation of the perineal musculature, dilatation of the cervix (this can only be assessed upon endoscopic examination and not digital palpation) and having already had a decline in plasma progesterone (and in bitches a subsequent decline in rectal temperature). When placental separation occurs, a green (bitch) or red–brown (queen) vulval discharge will be evident. Early in the course of the condition the fetuses will be alive, fetal movement may be palpated, and fetal movement and heartbeats may be detected ultrasonographically. Fetal death can be recognized immediately using ultrasonography by an absence of fetal movement and by a non-moving echogenic appearance to the heart. When these animals are presented later, fetal death can be detected radiographically by a change in the fetal posture, by the accumulation of gas within the fetus and/or uterus, and by the overlapping of the bones of the fetal skull, although these changes may take several days before they are evident.

Obstructive dystocia: In females with obstructive dystocia, whilst there may be similar confirmatory signs, there is usually a history of active parturition with significant straining that has been non-productive. Usually fetuses can be identified lodged within the birth canal as a result of fetal or maternal abnormalities.

Tocodynomometry

A novel approach to veterinary obstetrical monitoring in the United States involves external monitoring devices that use tocodynomometry (Healthdyne Inc.) and a hand-held Doppler instrument (Oxford Instruments) to detect and record uterine activity and fetal heart rates (Figure 14.7).

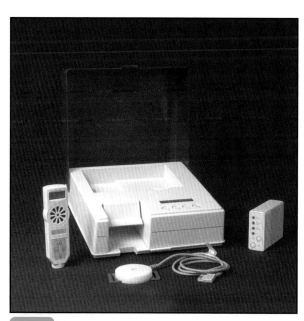

14.7 Tocodynomometer: sensor, monitor and recorder.

Equipment: These devices can be used either in the home or at the veterinary clinic. Their use requires the hair coat to be lightly clipped caudal to the ribcage, over the gravid area of the lateral flanks, to allow proper contact of the uterine sensor and fetal Doppler probe. The uterine sensor detects

changes in intrauterine and intra-amniotic pressures. The sensor is strapped over the caudolateral abdomen. The sensor's recorder is worn in a small backpack placed over the caudal shoulder area, and the animal is left to rest in the whelping/kittening box or in a crate/cage during the monitoring sessions. The monitoring equipment is well tolerated. Subsequent to each recording session, data are transferred from the recorder via a modem using a standard telephone. Fetal Doppler monitoring is performed bilaterally with a hand-held unit and the animal in lateral recumbency, using acoustic coupling gel. Directing the Doppler probe perpendicularly over a fetus results in a characteristic amplification of the fetal heart sounds, distinct from maternal arterial or cardiac sounds, which enables determination of the fetal heart rates (Davidson, 2003; 2009) (Figure 14.8).

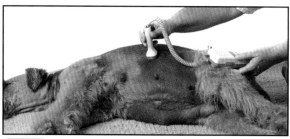

14.8 Fetal heart rate monitoring with a hand-held Doppler probe.

Interpretation of results: Interpretation of the contractile pattern on strips produced by the uterine monitor requires training and experience. In the USA the data that are transferred by the modem are interpreted at a central facility by experienced personnel who consult subsequently with the attending veterinary clinician and client. For home monitoring, recordings are made on a twice daily, hour-long basis, then intermittently as indicated during active parturition. In the veterinary clinic patients are monitored for shorter periods of time (a minimum of 20 minutes) whilst being evaluated for suspected dystocia.

The canine and feline uterus each has characteristic patterns of contractility, which vary in frequency and strength before and during the different stages of parturition. Serial tocodynomometry permits evaluation of the progression of parturition. During late term, the uterus may contract once or twice an hour before the actual first stage of parturition is initiated. During the first and second stages of parturition, uterine contractions vary in frequency from 0–12 per hour, and in strength from 15–40 mmHg, with spikes up to 60 mmHg. Contractions during active parturition can last 2–5 minutes. Recognizable patterns exist during preparturition and active (first, second and third stages) parturition. Abnormal, dysfunctional patterns of parturition can be weak or prolonged, and are often associated with fetal distress. In addition, the completion of parturition (or lack thereof) can be evaluated via tocodynomometry (Figure 14.9).

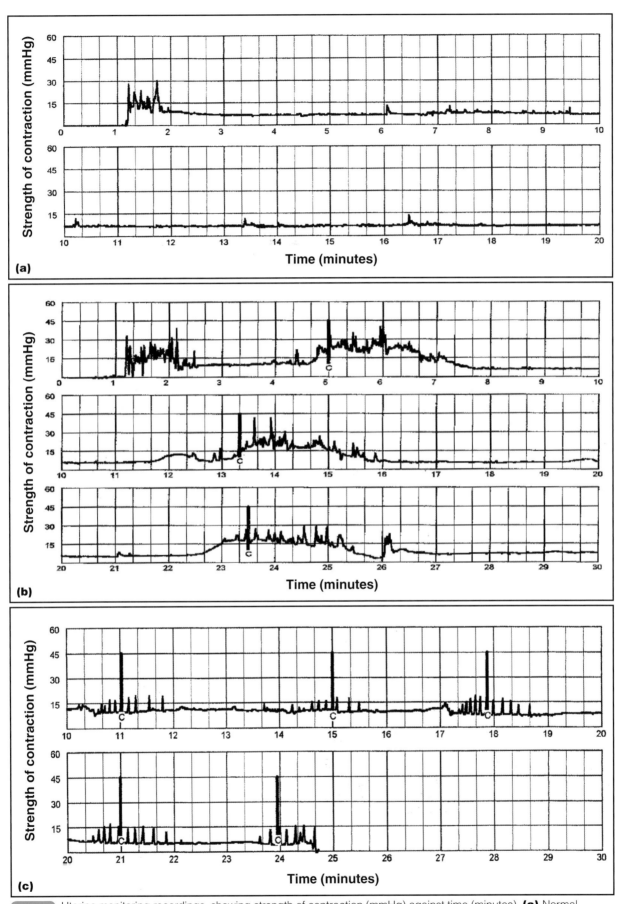

14.9 Uterine monitoring recordings, showing strength of contraction (mmHg) against time (minutes). **(a)** Normal baseline myometrial tracing (before parturition): no contractions. Variation from baseline seen at attachment of sensor (1–2 minutes). **(b)** Early active parturition (second stage): uterine contractions and abdominal pushing. **(c)** Active parturition: abdominal pushing with uterine inertia. Vertical spikes indicate abdominal efforts. C = contraction. (continues) ▶

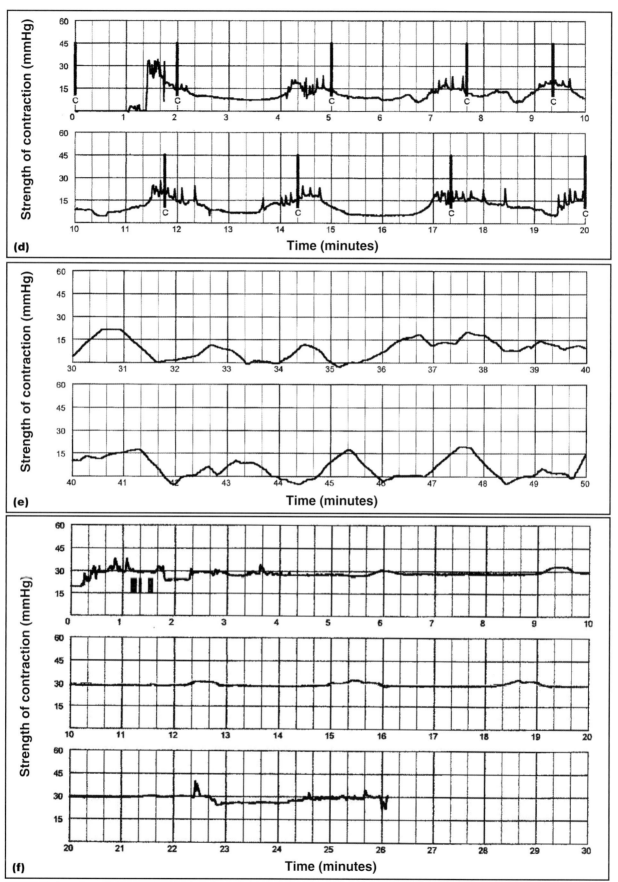

14.9 (continued) Uterine monitoring recordings, showing strength of contraction (mmHg) against time (minutes). **(d)** Same bitch as in Figure 14.3a, treated with 6 ml of 10% calcium gluconate s.c. and 0.50 IU oxytocin i.m. Abdominal pushing evident in conjunction with uterine contractions. The puppy was delivered in 26 minutes. **(e)** Uterine hyperstimulation: obstructed puppy and ecbolics contraindicated. Fetal distress was evident (persistent fetal bradycardia). **(f)** Empty postpartum uterus. C = contraction.

The presence of fetal distress is reflected by sustained deceleration of the heart rate. Normal canine and feline fetal heart rates at term are between 180 and 240 beats per minute (bpm), or at least four times the maternal heart rate. In the periparturient period the cardiac output of the fetus/neonate is mainly dependent on heart rate because the right ventricle is relatively stiff (low compliance) and the autonomic nervous system is immature (minimal inotropic response to catecholamines). Deceleration associated with uterine contractions suggests a mismatch in size between the fetus and dam, or fetal malposition/malposture. Transient accelerations occur with normal fetal movement. Fetal heart rates of ≤150–160 bpm indicate stress. Fetuses with heart rates ≤130 bpm have poor survival if not delivered within 2–3 hours, and fetuses with heart rates ≤100 bpm require immediate intervention to hasten delivery (medical or surgical) before their demise (Newman *et al.*, 1998; Davidson, 2001).

Treatment

Careful diagnostic work-up and prompt intervention are important to ensure a positive outcome in cases of dystocia. In cases that are presented early in the course of parturition, rather than as an emergency, the use of uterine and fetal monitors allows the clinician to detect and monitor parturition, as well as manage parturition medically or surgically with insight instead of guesswork. At the Guide Dogs for the Blind Inc., USA, the overall stillbirth rate declined from 9.2% to 2.5% with incorporation of uterine and fetal monitoring into the whelping process.

Manual intervention

In females with obstructive dystocia, the primary aim is to remove the obstruction. Manual obstetrical manipulations are very difficult in small animal patients due to the size of the pelvic canal. In large dogs manual relief of an obstruction can sometimes be achieved using retropulsion, realignment and traction techniques if the obstruction is the result of an abnormality of fetal presentation, position or posture. Given that the narrowest part of the birth canal is within the pelvis, it may be necessary for realignment to be performed by transabdominal palpation after the fetus has been pushed back into the uterus. These manipulations will need to be performed between periods of straining and not work against the uterine or abdominal contractions of the dam. When traction is to be applied there should be an initial generous application of lubrication and a grip placed around the fetal head or neck (a teaspoon placed over the head can be useful to help with traction when it is difficult to pass a finger beyond the dome of the skull), or the pelvis for a fetus in caudal presentation. It is best to avoid applying traction to the limbs, which are easily damaged. Squeezing transabdominally can also help in producing caudal movement of the fetus in small bitches and cats. Often it is useful to apply a gentle rocking movement from one side to the other at the same time as applying traction, and as the largest diameter of the fetus approaches the narrowest diameter of the pelvis, it

may be helpful to rotate the fetus slightly to one side. Great care must be taken not to rupture the uterus or physically damage the fetuses.

Obstetrical forceps should be used with extreme care because they can traumatize the vaginal wall (which may be pinched at the time the fetus is grasped) or cause serious damage to the fetus. They are best contemplated only when it is clear that removal of this particular fetus will enable completion of parturition relatively unaided; otherwise, and in cases of large litters, it is best to resort to a Caesarean operation. In all cases of manual correction of dystocia, care should be taken to ensure that parturition continues normally, because there is a risk of the development of secondary uterine inertia. However, in many cases, the dystocia is the result of a fetal or maternal abnormality that cannot be corrected to allow normal delivery, and in such circumstances it is necessary to resort to a Caesarean operation.

Medical intervention

Medical therapy for dystocia, based on the administration of oxytocin and calcium gluconate, can be empirical or preferably directed and tailored based on the results of monitoring. Generally, the administration of oxytocin increases the frequency of uterine contractions, whilst the administration of calcium increases their strength. Oxytocin is effective at mini-doses, starting at a rate of 0.25 IU/animal s.c. or i.m. and increasing to a maximum dose of 4 IU per bitch or queen. Higher doses of oxytocin or intravenous boluses can cause tetanic, ineffective uterine contractions that can further compromise fetal oxygen supply by placental compression. The frequency of oxytocin administration is dictated by the parturition pattern, and is generally not given more frequently than hourly.

Calcium gluconate (10% solution with 0.465 mEq Ca^{2+}/ml) is given at a rate of 1.0 ml/5.5 kg s.c. as indicated by the strength of uterine contractions, generally no more frequently than every 4–6 hours. Calcium is given before oxytocin in most cases, thus improving contraction strength before increasing frequency. In addition, the action of oxytocin appears to be improved when given 15 minutes after calcium. Most bitches and queens have normal serum calcium concentrations, suggesting that the benefit of calcium administration is at a cellular or subcellular level. It is interesting to note that, in relatively large studies, the manual and medical correction of dystocia is effective in only approximately 30% of cases, with the remainder ending in a Caesarean operation.

Surgical intervention

Surgical intervention (Caesarean operation) is indicated if a bitch or queen fails to respond to medical management, or if fetal distress is evident despite an adequate increase in uterine contractility (which suggests a mismatch of maternal birth canal to fetal size, or fetal malposition or malposture incompatible with vaginal delivery), or if aberrant contractile patterns are noted on uterine monitoring. A carefully planned and orchestrated Caesarean operation with suitable anaesthetic and neonatal resuscitation protocols should result in a positive outcome in most cases, unless there has been some significant delay in presenting the dam for veterinary examination. However, it should always be remembered that:

- The dam may be debilitated and require careful anaesthetic management
- There may be little time for pre-anaesthetic preparation
- The dam may have been fed recently.

Anaesthesia: Veterinary practices should plan anaesthetic regimes around the possibility of emergency cases as well as more carefully planned interventions. The general aims of anaesthesia are:

- To ensure adequate oxygenation (by pre-oxygenation by facemask, intubation and provision of inspired oxygen)
- To maintain blood volume and prevent hypotension (by the administration of intravenous fluid therapy)
- To minimize maternal and fetal depression during surgery and after delivery (by reducing the dose of anaesthetic agents used).

It is not possible to discuss all of the anaesthetic options for Caesarean operations in this text; however, there are a few points worth considering.

Premedication: For premedication, atropine is best not given routinely, because it crosses the placenta and blocks the normal bradycardic response of the fetus to hypoxia and causes relaxation of the lower oesophageal sphincter, making aspiration more likely. The use of an anticholinergic is indicated for the dam because of the anticipated vagal stimulation of the gravid uterus. Glycopyrrolate does not cross the placenta and is preferred. Most dams are tractable and do not need pre-anaesthetic tranquilization, which has a depressant effect on the fetuses. Phenothiazine tranquillizers are transported rapidly across the placenta. Alpha2-adrenoceptor agonists such as medetomidine and xylazine are contraindicated because of their severe cardiorespiratory depressant effects. Similarly, the respiratory depressant effect of opioids makes them unpopular prior to removal of the fetuses. Metoclopramide may be administered subcutaneously or intramuscularly prior to the induction of anaesthesia to reduce the risk of vomiting during the procedure.

Induction: For induction of anaesthesia, dissociative agents such as ketamine and the barbituates are best avoided because they produce profound depression of the fetuses. Propofol appears to be most useful because of its rapid redistribution and therefore limited effect upon the fetuses after delivery. Mask induction produces more maternal and fetal hypoxaemia than propofol induction.

Maintenance: For maintenance of anaesthesia, volatile agents are preferable, especially those with low partition coefficients such as isoflurane and sevoflurane. These agents show rapid uptake and elimination

by the animal, and it may have a better cardiovascular margin of safety than the more soluble agents such as halothane. Whilst nitrous oxide may be used to reduce the dose of other anaesthetic agents, it is transferred rapidly across the placenta and, although it has minimal effects upon the fetus *in utero*, it may result in a significant diffusion hypoxia after delivery. Using a local anaesthetic (bupivacaine 2 mg/kg) line block in the skin and subcutaneous tissues prior to incising permits a more rapid entry into the abdomen, whilst the dam is making the transition from propofol induction to inhalant maintenance, and helps with postoperative discomfort.

Fluid therapy: A number of factors are important when considering the most appropriate fluid for intravenous administration, including:

- There may be increased alveolar ventilation (an effect of progesterone), causing respiratory alkalosis
- The enlarged abdomen may produce a decreased tidal volume, causing respiratory acidosis
- There may be loss of acid because of vomiting
- There may be loss of blood as a result of the surgery.

The best choice agent is probably lactated Ringer's solution administered at a rate of 10–20 ml/kg/h.

Procedure: The surgical procedure is widely described in other texts (see *BSAVA Manual of Canine and Feline Abdominal Surgery*), but a few comments are made here. Operative speed is important because surgical delay and prolonged anaesthetic time are associated with fetal asphyxia and depression. However, care should be taken during incision of the linea alba to ensure that the gravid uterus is not also incised. Ideally the uterus should be exteriorized and packed off with moistened laparotomy sponges to prevent abdominal contamination with uterine fluid. This process should be undertaken carefully to ensure that the uterus and its broad ligament do not tear; it may be easier in some cases to exteriorize one horn at a time. The uterus should be penetrated in a relatively avascular area, and it is best to elevate the uterine wall from the fetus and to extend the incision with scissors to ensure that the fetus is not lacerated.

The fetuses may be brought to the incision by gently 'milking' them along the uterus, although in some cases or in large dams it may be necessary to make more than one incision. As the fetal fluid is released it is best to remove this by suction, and then to clamp the umbilicus before passing the fetus to an assistant for resuscitation. After each fetus is removed the associated placenta should be detached by gentle traction, but the placentas may be left *in situ* if they are firmly attached. It is essential that the uterine horns, the uterine body and the vagina are inspected thoroughly to ensure that all fetuses have been removed. Finally, after closure the uterus, its broad ligament and the vascular supply should be inspected carefully to ensure that any previously unnoticed tears have been identified before closure of the abdomen. If uterine viability is questionable an ovariohysterectomy should be performed. In the normal dam the uterus will begin to involute shortly after removal of the fetuses, but if this is not the case oxytocin may be administered (0.25–1.0 IU/dam) to facilitate involution and arrest any haemorrhage.

Analgesia: Postsurgical discomfort in the dam should be acknowledged. Non-steroidal anti-inflammatory drugs (NSAIDs) are not advisable due to their uncertain metabolism by the nursing neonates with immature renal and hepatic systems. Narcotic analgesia is preferable. Oral narcotics such as tramadol (10 mg/kg/day in divided doses) provide excellent analgesia for nursing bitches with minimal sedation of the neonates.

Postpartum conditions

The postpartum period can be associated with high morbidity and even mortality for the dam and neonates. The postpartum period is defined here as the period of 30–45 days after parturition. The effective diagnosis of postpartum problems first requires their recognition and differentiation from normal situations; effective treatment depends on both timely diagnoses and intervention.

Normal events
Normally, dams stay very close to their offspring during the first 2 weeks following birth, leaving the whelping/kittening box briefly (if at all) to eat and eliminate. They are alert and content to remain with their offspring. Some protective dams may show aggression towards other household animals or even people of whom they are normally tolerant; such behaviour tends to reduce after 1–2 weeks of lactation. Lactation typically presents the greatest nutritional and calorific demand of the female's life. Weight loss and dehydration may occur and will affect milk production if food and water are not made readily available; sometimes this entails leaving both in the nest box with a nervous dam. Partial anorexia can be exhibited during the last few weeks of pregnancy and again in the immediate postpartum period. The causes of this are not known, but could be associated with displacement of the gastrointestinal tract by the gravid uterus. Appetite should return to normal and then increase as lactation progresses. In some cases, poor appetite early in the postpartum period can occur secondary to digestive upset following the consumption of numerous placentas.

Diarrhoea can occur secondary to increased rations and rich food (bacterial overgrowth secondary to carbohydrate malassimilation). Marked hair loss (postpartum effluvium) is quite common and normal in the bitch, and usually occurs 4–6 weeks after whelping. In some severe cases the hair loss may spare only the head. This is usually more marked than the hair loss that occurs in conjunction with the typical oestrous cycle, and can be interpreted as abnormal by an owner, especially in conjunction with the weight loss typically associated with lactation.

The body temperature of the dam may be mildly elevated (≥39°C) in the immediate postpartum period, reflecting anticipation of the normal inflammation associated with parturition, but it should return to normal within 24–48 hours. If a Caesarean operation has taken place, it may be difficult to differentiate normal postsurgical inflammation from fever associated with pathology. This differentiation is best achieved by careful physical examination and evaluation of haematology. Normal postpartum lochia is a brick red colour, non-odorous, and diminishes over several days to weeks (uterine involution and repair occur for up to 16 weeks in the bitch). The mammary glands should not be painful; they should be symmetrical and moderately firm without heat, erythema or palpable firm masses. If expressed, normal milk is grey to white and of watery consistency.

Haemorrhage
Haemorrhage is an uncommon postpartum condition in the bitch and queen that may occur as a result of physical injury to the uterus or vaginal wall, or as a result of placental necrosis, surgical error, coagulopathy or subinvolution of placental sites. Blood loss is common at the time of parturition, although this is normally limited to the time of delivery of each fetus. A small volume of haemorrhagic fluid may be passed after the termination of parturition in the normal female, but this rapidly decreases in volume and changes in colour due to the release of uteroverdin from the marginal haematoma. Persistence of a haemorrhagic discharge is abnormal. The dam may be unsettled initially, but later may be depressed and have pale mucous membranes.

Inspection of the vaginal wall either digitally or using a speculum or endoscope may demonstrate the site of a physical injury. Uterine bleeding may be detected by the presence of haemorrhagic fluid exiting the cervix, or by transabdominal ultrasonography of the uterus. It is not normally necessary to measure the packed cell volume (PCV) of the fluid. The principal considerations are to prevent further blood loss and to maintain blood volume (as discussed for uterine prolapse, see below).

Cases of vaginal trauma can be treated either by direct pressure or by the application of a vaginal tampon, or occasionally by clamping of the bleeding tissue using artery forceps followed by subsequent ligation. With uterine bleeding, in the first instance, it is appropriate to attempt to speed uterine involution by the administration of oxytocin (0.25–1.0 IU/dam) or ergometrine (0.2–0.5 mg/kg). If there is no response, a laparotomy and ovariohysterectomy may be indicated, although a coagulopathy should always be suspected and assessments of clotting function should be performed prior to surgery.

Retained fetuses and placentas
Retained placentas are very uncommon in bitches and queens, despite the concern expressed by many breeders. Most commonly the owner's report is

mistaken, because placentas are not always expelled after each fetus and several may be delivered together some time later during parturition. Retained fetuses should be investigated as described for primary and secondary uterine inertia (see above).

The aetiology of placental retention is unknown, but appears to be more common in toy breeds of dog. In cases of true placental retention in the bitch, there is usually persistence of a green-coloured vulval discharge, and the bitch may be restless and not allow the puppies to suck. If the condition is not treated, the discharge will become malodorous and the bitch may become depressed, septicaemic and toxaemic. In general, queens have similar clinical signs, although in some cases there are no signs of vulval discharge until several weeks after parturition.

Diagnosis of the condition soon after parturition can be very difficult without ultrasonographic or endoscopic examination. Transabdominal palpation is often misleading because the involuting uterus may have sections that are dilated and adjacent regions that are smaller in diameter.

Removal of the infected material is imperative; however, stabilization of the patient with appropriate fluid and antimicrobial therapy is essential. In the early stages (within 24 hours of parturition), repeated administration of oxytocin may be sufficient to cause expulsion of the retained placentas. This results in resolution of the clinical signs, although the dam should be treated with broad-spectrum antimicrobial agents safe for nursing neonates such as amoxicillin trihydrate/clavulanate potassium or a cephalosporin to prevent the development of secondary metritis.

In the later stages of the condition, the number of uterine oxytocin receptors will have decreased and the administration of exogenous oxytocin has little or no clinical effect. In these instances repeated treatment with low-dose prostaglandin may be contemplated; however, it may be difficult to dislodge the placenta, in which case hysterectomy may be indicated. In animals that are not required for breeding, ovariohysterectomy is the treatment of choice.

Inappropriate maternal behaviour

Appropriate maternal behaviour is critical to neonatal survival and includes attentiveness, facilitation of nursing, retrieving neonates, grooming and protecting neonates. Although maternal behaviour is instinctive, it can be influenced negatively by anaesthetic drugs, pain, stress and excessive human interference. Maternal bonding is a pheromone-mediated event initiated at parturition. Whelping and kittening should take place in quiet, familiar surroundings, with minimal human interference yet adequate supervision. Dams with good maternal instincts exhibit caution when entering or moving about the nest box so as not to traumatize neonates by stepping or lying on them. A guard rail along the inside of the whelping box prevents inadvertent smothering of canine neonates.

The neuroendocrine reflex that regulates contraction of mammary gland myoepithelial cells and subsequent let-down of milk is mediated by oxytocin and activated by neonatal sucking. During stress, adrenaline (epinephrine) induces vasoconstriction, blocking the entry of oxytocin into the mammary gland and preventing milk ejection. A nervous, agitated dam is likely to have poor milk availability. Dopamine antagonist tranquillizers (e.g. acepromazine at a rate of 0.01–0.02 mg/kg), administered at the lowest effective dose to minimize neonatal sedation, can improve maternal behaviour and milk ejection in nervous dams. These drugs show minimal interference with prolactin. Placing littermates near the dam facilitates the maintenance of adequate body temperature (neonates cannot thermoregulate or shiver for up to 4 weeks after birth) and ensures the milk is readily available. Normal maternal behaviour includes gentle retrieval of neonates who have become dispersed and isolated across the nest box. Grooming of the neonates immediately following parturition stimulates cardiovascular and pulmonary function and removes amniotic fluids.

Dams that demonstrate little interest in resuscitating the neonates can have poor maternal behaviour throughout the postnatal period; this is a good indication that careful observation is needed to avoid problems later. Maternal grooming stimulates reflex neonatal urination and defecation and maintains the neonatal coat in a clean, dry state. Occasionally, excessive protective behaviour or fear-induced maternal aggression can occur. Mild tranquillization of the dam with an anxiolytic agent can help, but neonatal drug administration via the milk can be problematic. Benzodiazepines, synergists of gamma-aminobutyric acid (GABA), are reportedly superior to phenothiazines for fear-induced aggression (e.g. diazepam at a rate of 0.55–2.2 mg/kg). The effects of newer anxiolytic pharmaceuticals in maternal aggression have not been described in a controlled setting.

Uterine disorders

Metritis

Acute infection of the postpartum endometrium should be suspected if lethargy, anorexia, decreased lactation and poor mothering occur accompanied by fever and a malodorous vaginal discharge. Metritis is a serious condition and is sometimes preceded by dystocia, contaminated obstetrical manipulations, or retained fetuses and/or placentas. Haematological and biochemical evaluations often suggest septicaemia, systemic inflammatory reaction and endotoxaemia. Vaginal cytology shows a haemorrhagic to purulent septic discharge. Ultrasonography of the abdomen allows the evaluation of intrauterine contents and the uterine wall (Figure 14.10). Retained fetuses and placentas can also be identified with ultrasonography. A guarded cranial vaginal culture is likely to be representative of intrauterine bacteria and should be submitted for both aerobic and anaerobic culture and sensitivity testing. This permits retrospective assessment of empirically selected antibiotic therapy. Bacterial ascent from the lower genitourinary tract is more common than haematogenous spread, and *Escherichia coli* is the most common causative organism in both bitches and queens.

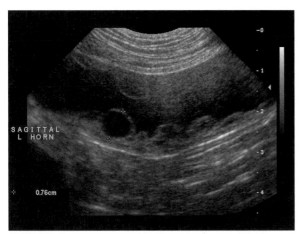

14.10 Postpartum metritis. There is abnormal fluid accumulation within the uterine lumen. Note the endometrium is convoluted and thickened. (Courtesy of T. Baker)

Therapy consists of intravenous fluid and electrolyte support, appropriate bactericidal antibiotic administration and pharmacological uterine evacuation, usually with prostaglandin F2α or its synthetic formulation dinoprost at a rate of 0.10–0.20 mg/kg administered subcutaneously every 12–24 hours for 3–5 days. An ovariohysterectomy may be indicated if the condition of the bitch permits and she is poorly responsive to medical management. Ergometrine (0.02–0.2 mg/kg i.m. given once) is also an effective ecbolic agent, but may cause rupture of a friable uterine wall. Synthetic prostaglandins offer more uterine-specific therapy where available. Oxytocin is unlikely to promote effective uterine evacuation when administered later than 24–48 hours after parturition, because at this time the number of oxytocin receptors is markedly reduced. Neonates should be hand-reared if the dam is seriously sick. Metritis can become chronic and cause infertility, and can be coincident with the development of mastitis.

Subinvolution of placental sites

The persistence of a serosanguineous to haemorrhagic vulval discharge beyond 16 weeks postpartum can indicate subinvolution of one or more of the placental sites of attachment (SIPS) in the bitch. On histological examination, fetal trophoblastic cells appear to persist in the myometrium instead of degenerating, endometrial vessel thrombosis is lacking, and normal involution of the uterus is prevented; however, normal interplacental regions exist. Eosinophilic masses of collagen and dilated endometrial glands protrude into the uterine lumen, oozing blood. The cause is unknown, and fortunately blood loss is usually minimal, intrauterine infection is usually not present and fertility at the subsequent oestrus is unaffected. Given that the condition is usually mild, treatment is generally not necessary, and recovery is spontaneous. In the uncommon situation that vulval blood loss is copious enough to cause serious anaemia, coagulopathies (likely to be defects in the intrinsic pathway or thrombocytopenia/thrombocytopathies), trauma,

neoplasia of the genitourinary tract, metritis and pro-oestrus should be ruled out. Vaginal cytology, vaginoscopy, coagulation testing and abdominal ultrasonography assist in the diagnosis.

In cases of true SIPS where there is significant blood loss, treatment with ergot preparations may also be attempted but is unlikely to be effective. Similarly, the effect of therapeutic prostaglandins and/or oxytocin is questionable and not proven in any controlled study. If the blood loss continues laparotomy and ovariohysterectomy will be necessary and are curative. Histological examination of the uterus is indicated to confirm the diagnosis. Some authors suggest that the administration of oxytocin at the time of parturition has a preventative role in the development of SIPS; however, this remains unproven.

Prolapse

Complete or partial prolapse of the uterus is an uncommon postpartum condition in the bitch, and occurs rarely in the queen. The diagnosis is based on palpation of a firm, tubular mass protruding from the vulva following parturition, and an inability to identify the uterus with abdominal ultrasonography. Vaginal hyperplasia and prolapse, secondary to a hypersensitivity of focal (periurethral) vaginal mucosa to oestrogen, can recur near parturition and should be ruled out by physical examination, vaginoscopy or contrast radiography. The prolapsed uterine tissues are at risk of desiccation and infection from exposure and contamination. The size of most bitches and queens precludes manual replacement; laparotomy and ovariohysterectomy are usually indicated.

Rupture

Rupture of the uterus occurs most commonly with very large litters where there is marked stretching and thinning of the uterine wall, especially in multiparous dams with dystocia. Excessive use of ecbolics can promote uterine rupture (Figure 14.11). Immediate laparotomy for the retrieval of fetuses and repair or removal of the uterus, as well as culture and lavage of the abdominal cavity, is indicated. The uterus should be carefully examined during a Caesarean operation for any areas that appear

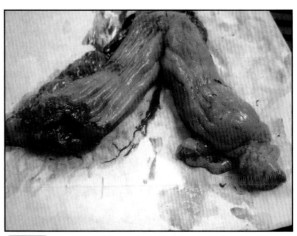

14.11 Rupture of the uterine horn secondary to obstructive dystocia and administration of oxytocin.

prone to rupture. Peritonitis can result from an undetected uterine tear. A unilateral hysterectomy can be considered if the damaged area is limited and the dam is valuable to a breeding programme.

Metabolic disorders

There are critical changes that occur in the physiology of the female at the time of parturition, and these may predispose to the development of a number of metabolic conditions.

Eclampsia

Puerperal tetany or eclampsia is a potentially life-threatening condition that occurs most commonly during the first 4 weeks postpartum, but can also occur in the last few weeks of pregnancy. The condition is seen more frequently in bitches than in queens. Puerperal tetany is caused by a depletion of ionized calcium in the extracellular compartment. There are a number of predisposing factors, including improper perinatal nutrition, inappropriate calcium supplementation and heavy lactational demands. Small dams with large litters are at increased risk. Excessive prenatal calcium supplementation can lead to the development of puerperal tetany by promoting atrophy of the parathyroid gland and inhibiting parathyroid hormone release, thus interfering with the normal physiological mechanisms to mobilize adequate calcium stores and utilize dietary calcium sources. In addition, secretion of thyrocalcitonin is stimulated. The feeding of a commercially available balanced growth (puppy/kitten) feed without additional vitamin or mineral supplementation is optimal during the second half of gestation and throughout lactation. Supplementation with cottage cheese should be avoided because it disrupts the normal calcium–phosphorus–magnesium balance in the diet.

Metabolic conditions that result in protein binding of serum calcium can promote or exacerbate hypocalcaemia, for example, alkalosis resulting from prolonged hyperpnoea during parturition or dystocia. Hypoglycaemia (see below) and hyperthermia can occur concurrently. Therapeutic intervention should be initiated immediately upon recognition of the clinical signs of tetany, without waiting for biochemical confirmation. The signs preceding the development of tonic–clonic muscle contractions (progressing to seizures) include:

- Behavioural changes
- Salivation
- Facial pruritus
- Stiffness/limb pain
- Ataxia
- Hyperthermia
- Tachycardia.

Immediate therapeutic intervention should be instituted with a slow intravenous infusion of 10% calcium gluconate (1–20 ml) given to effect. Cardiac monitoring for bradycardia and arrhythmias should accompany administration; their occurrence warrants temporary discontinuation of the infusion and a slower subsequent rate.

As cerebral oedema can occur as a result of uncontrolled seizures, diazepam (1–5 mg i.v.) or barbiturates can be used to control persistent seizures once normal calcium concentrations are attained. Mannitol may be indicated for cerebral inflammation and swelling. Corticosteroids are undesirable because they promote calciuria, decrease intestinal calcium absorption and impair osteoclasia. Hypoglycaemia should be corrected if present, and exogenous treatment for hyperthermia given if necessary.

Once the immediate neurological signs are under control, a subcutaneous infusion of an equal volume of calcium gluconate, diluted 50% with saline, can be given, repeated every 6–8 hours until the dam is stable and able to take oral supplementation. Oral calcium gluconate or carbonate supplementation (10–30 mg/kg q8h) can be provided using human medicinal products for the treatment of indigestion. Each 500 mg calcium carbonate tablet supplies 200 mg of calcium. Efforts to diminish lactational demands on the dam (by providing supplementary feeding to the neonates) and improve her plane of nutrition are indicated. If the response to therapy is prompt, natural feeding of the neonates by the dam can be reinstituted gradually until the neonates can be safely weaned, usually at a slightly early age (3 weeks). Concurrent supplementation with commercial milk replacement is encouraged. The administration of calcium throughout lactation, but not gestation, may be attempted in dams with a history of recurrent eclampsia (calcium carbonate 500–4000 mg/animal/day in divided doses).

Hypoglycaemia

It is very rare for a bitch or queen to become hypoglycaemic during pregnancy. Normally, progesterone acts as a potent peripheral insulin antagonist and results in hyperglycaemia (see below). Indeed, metoestrus (dioestrus) or gestational diabetes is well recognized, as is the difficulty in stabilizing known diabetics during dioestrus or pregnancy. However, an apparent primary hypoglycaemia of unknown aetiology occurs rarely during pregnancy. The hypoglycaemia may worsen the clinical signs noted with hypocalcaemia. The condition has not been reported in the queen.

Bitches with hypoglycaemia are weak and may become comatose. The clinical features may be mistaken for those of hypocalcaemia, although tetany is not a common clinical finding. The diagnosis is based on measurement of plasma glucose concentration. Investigation may be initiated by a lack of effect of intravenous administration of calcium (in the initial belief that the bitch was hypocalcaemic). The aim of treatment is to restore the plasma glucose concentration and provide frequent intake of glucose until the onset of parturition. The condition disappears following parturition and some authors have suggested that a Caesarean operation may be required in severe cases. Rapid resolution of the clinical signs is achieved following the intravenous administration of glucose to effect. Subsequently, increased frequency of feeding can be used to prevent recurrence until the onset of parturition.

Hyperglycaemia

An increased plasma progesterone concentration has a direct antagonistic effect upon insulin and also stimulates the secretion of growth hormone. Progesterone reduces insulin binding and glucose transportation within tissues. Growth hormone also has an antagonistic effect upon insulin, mediated by a decrease in the number of insulin receptors and inhibition of glucose transport. Development of a diabetic state may therefore occur in the bitch during pregnancy; the condition is rare in the queen.

Mammary disorders

A variety of conditions affect the mammary glands in the postpartum period, including:

- Agalactia (a failure to provide milk to neonates, which can be primary or secondary)
- Galactostasis (engorgement and oedema of the mammary glands)
- Mastitis (septic inflammation of the mammary gland) (Figure 14.12).

These conditions are discussed in further detail in Chapter 17.

References and further reading

Davidson AP (2001) Uterine and fetal monitoring in the bitch. *Veterinary Clinics of North America: Small Animal Practice* **31**(2), 305–313

Davidson AP (2003) Obstetrical monitoring in dogs. *Veterinary Medicine* **98**(6), 508–517

Davidson AP (2009) Dystocia management. In: *Kirk's Veterinary Therapy XIV*, ed. JD Bonagura, pp. 992–998. WB Saunders, Philadelphia

Eilts BE, Davidson AP, Hosgood G, Paccamonti DL and Baker DG (2005) Factors affecting gestation duration in the bitch.

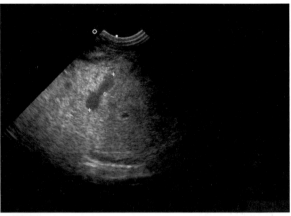

14.12 Acute septic mastitis. Cursors indicate an accumulation of hypoechoic fluid suggesting abscessation. (Courtesy of T. Baker)

Theriogenology **64(2)**, 242–251

Ekstrand C and Linde Forsberg C (2008) Dystocia in the cat: a retrospective study of 155 cases. *Journal of Small Animal Practice* **35**, 459–464

Linde Forsberg C and Eneroth A (1998) Parturition. In: *BSAVA Manual of Small Animal Reproduction and Neonatology*, ed. G England and M Harvey, pp. 127–142. BSAVA Publications, Gloucester

Newman RB, Campbell BA and Stramm SL (1990) Objective tocodynomometry identifies labor onset earlier than subjective maternal perception. *Obstetrics and Gynecology* **76**, 1089–1092

Nyland TG and Mattoon JS (2002) *Small Animal Diagnostic Ultrasound, 2nd edn.* WB Saunders, Philadelphia

Walett Darvelid A and Linde Forsberg C (2008) Dystocia in the bitch: a retrospective study of 182 cases. *Journal of Small Animal Practice* **35**, 402–407

Williams JM and Niles JD (2005) *BSAVA Manual of Canine and Feline Abdominal Surgery.* BSAVA Publications, Gloucester

Useful websites

Veterinary Perinatal Specialties
http://www.whelpwise.com

15

Management and critical care of the neonate

Margret Casal

Introduction

Neonatology is the branch of medicine that deals with the care, development and diseases of newborn infants. In humans, this represents the first 4 weeks of a child's life, at which time many critical events can occur. There is no set definition of neonatology in puppies and kittens. However, most veterinary paediatricians agree that the first 2–3 weeks of life are considered to be the neonatal period in which, just as in humans, the most rapid changes occur. The neonatal period in puppies and kittens is characterized by complete dependence on the mother because of incomplete neurological functions such as auditory and visual abilities and proper spinal reflexes. It is important to consider careful examination of the neonate early in life to aid the rapid detection of abnormalities before disease progression and changing clinical signs have obscured the original cause.

Biology and development

Basic biology

First week of life

During the first week of life newborn kittens and puppies sleep throughout most of the day (80%), and feed vigorously for a short period of time every 2–4 hours. The brain is not completely developed at birth; therefore, some neuromuscular reflexes are absent and the only motor skills present are crawling, sucking and distress vocalization. The neonates only respond to stimuli such as odour, touch and pain. The queen or bitch initiates urination and defecation by licking the urogenital area.

At 3 days of age, kittens and puppies should be able to lift their heads, and by 1 week they can crawl in a coordinated manner. Puppies and kittens are unable to maintain their body temperature during the first few days of life. The body temperature at birth (34.7–37.2°C) is lower than in adults and rises to 36.1–37.8°C during the first week of life. Heart and respiratory rates may be irregular at birth (pulse: 160–200 beats/minute; respiration: 10–20 breaths/minute) and there is no abdominal component to the breathing. During the first week the neonates begin to adjust to the new, extrauterine physiology (pulse: 200–220 beats/minute; respiration: 16–35 breaths/minute). The umbilical cord dries out during the first day of life and should have fallen off by day 3–4. The flexor tone present at birth switches to extensor tone after the fourth day of life (Figure 15.1). While sex determination in normal newborn puppies is unambiguous, it can be challenging in kittens. The sex of kittens can be determined at birth by evaluating the anogenital distance, which is shorter in females (7.6 ± 1 mm) than in males (12.9 ± 1.5 mm). Male kittens are born with descended testicles, which are able to move freely in and out of the scrotum until 5–7 months of age. In dogs, the testicles normally descend at 2 weeks of age, but lack of presence by 8 weeks of age is considered suspicious.

Assessment	Timing	Comments
Reflex		
Sucking	Should be present at birth	Puppy or kitten will try to suck or chew on a finger
Pressing	Should be present at birth	Puppy or kitten will press its head against a bowed hand
Righting	Present at birth	When placed on its back the puppy or kitten should attempt to right itself
Lumbar		Forcefully rubbing a healthy puppy or kitten in the lumbar region will result in vocalization and great activity
Extensor		The patient is placed in dorsal recumbency and the toe of a hindlimb is pinched. If the puppy or kitten is <3 weeks of age, it will adduct the other hindlimb: this is normal
Magnus		The patient is placed in dorsal recumbency and its head bent towards one side. If the puppy or kitten is <3 weeks of age, it will stretch its legs on the side to which the head is bent and bend its legs on the other side

15.1 Neurological examination of the very young paediatric patient. (continues) ▶

Assessment	Timing	Comments
Reflex continued		
Tonic neck	Present until 3 weeks of age	The patient is held by the thorax and its neck is bent towards one side; the puppy or kitten should stretch the limbs on the side to which the neck is turned. If the head is bent dorsally, the forelimbs should be stretched and the hindlimbs adducted.
Hopping	Present at 2–4 days of age	
Anogenital	Present until 3–4 weeks of age	If the anogenital region of the puppy or kitten is stimulated with a moist cloth or cotton wool, it should urinate/defecate
Palpebral and corneal	Should be present as soon as the eyes are open	
Menace	Can be present as early as 2 weeks of age, but usually not until 16 weeks of age	
Tone		
Flexor	Present until 3–4 days of age	When a puppy or kitten is held by the head, it will 'roll up' and adduct its hindlegs
Extensor	After 4 days of age	When a puppy or kitten is held by the head, it will stretch its hindlegs and back

15.1 (continued) Neurological examination of the very young paediatric patient.

Second week of life

During the second week of life, kittens and puppies begin to crawl and their body temperature rises slowly towards normal adult levels. Kittens and puppies will have doubled their birthweight by 7–10 and 10–12 days, respectively. They begin to open their eyes at 10–12 days of age, although in some breeds (e.g. Abyssinians) kittens may open their eyes at birth. Remnants of the hyaloid artery attached to the posterior lens capsule may be seen for a few days after the eyes open.

Third and fourth weeks of life

By the third and fourth weeks of life, the neonate can begin to orientate itself by sound, and its perception of distance becomes clearer. The external ear canals open at 14–16 days of age. The iris is not very well pigmented (it is a blue–grey colour) and the cornea is slightly cloudy owing to its increased water content. Kittens may have divergent strabismus. By the end of the third week, puppies and kittens are able to stand and have good postural reflexes (see Figure 15.1), and by 4 weeks of age they should be able to walk around and explore their surroundings. At this time all sucking, rooting and other reflexes associated with nursing are diminishing or absent. At the end of the fourth week of life neonates are able to regulate their body temperature, which should then be within the normal adult range.

Thermoregulation

When puppies and kittens are born, they have almost no subcutaneous fat and thus little insulation. Initially, body heat is produced by metabolism of brown fat, which is under the control of the sympathetic nervous system (non-shivering thermogenesis). Due to the body size to surface area ratio in neonates, heat loss is much greater than in an older animal. As long as the neonates are close to their dams and the mammary glands, there is not much heat loss so they can maintain thermal balance. However, as the neonate begins to take food, its metabolic rate increases, which in turn elevates its body temperature. Shivering and vasoconstrictive mechanisms may begin around 6–8 days of age, but by about 6 weeks, puppies and kittens are good homeotherms and have a body temperature similar to that of adults.

Hydration and renal function

The neonate is particularly susceptible to dehydration because water makes up 82% of its bodyweight and water turnover is about twice that of an adult. In addition, the skin is more permeable and the body size to surface area ratio is higher than in adults, leading to increased losses. Due to the limited ability of the neonate to conserve fluid and the immaturity of the kidneys, fluid requirements are high at 13–22 ml/100 g bodyweight per day. The nephrons are not completely formed until the third week of life and the glomerular filtration rate increases from 21% at birth to 53% by 8 weeks of age. Tubular secretion is generally thought to be mature by 8 weeks of age. This explains the low urine specific gravity until 8 weeks (1.006–1.017), the increased concentrations of amino acids and proteins, and glucosuria, which is a common finding in neonates up to 2 weeks of age. Given the immaturity of renal function, medications that affect kidney development should be avoided, and dosages of those that are excreted through the kidney must be adjusted according to the patient's age.

Cardiovascular system, acid–base balance and haematology

Before birth, the fetal partial pressure of oxygen (PO_2) is low at only 20 mmHg. During birth, the placenta separates from the uterus, cutting off the maternal oxygen supply. This results in hypoxia and induces gasping respiratory reflexes. As the neonate breathes, the PO_2 rises to 50–60 mmHg. As a result of the increasing oxygen tension, the pulmonary vessels begin to dilate and the ductus arteriosus begins to narrow, closing completely by 1–2 days after birth. The increase in left-sided pressure results in closure of the foramen ovale between the atria. The PO_2 rises even further by day 2, correcting the earlier acidosis that develops in the neonate.

At birth, autonomic innervation of the heart and vasculature is incomplete, the myocardial cells are still dividing and myocardial contractility is decreased. Thus, the ability to react to circulatory distress caused by acid–base shifts, blood loss and hyperthermia is decreased. Neonates may respond to hypoxia with bradycardia or no change in heart rate. The bradycardia is thought to reduce oxygen demand and thus be a protective mechanism. However, by 4 days of age, puppies and kittens have normal blood baroreceptor function, and their heart rates respond to hypoxia in the same manner as adults.

At birth, the right and left ventricles have approximately the same volume, gradually changing to the ultimate adult ratio of 1:2 to 1:3 throughout puberty.

During this time, the canine heart changes from being ellipsoid at birth to more globoid in adulthood. At 1 month of age, puppies and kittens still have lower blood pressures, stroke volumes and resistance in the peripheral vasculature than adults, but they have higher heart rates, cardiac outputs and central venous pressure. Mean arterial pressures in puppies and kittens are 50 ± 10 mmHg immediately after a normal vaginal delivery. Responses to cardiovascular drugs during the neonatal period are less intense than in the adult. It is important to keep these differences in mind when evaluating chest radiographs, electrocardiograms (ECGs) and echocardiograms. Normal haematology profiles for puppies and kittens are given in Figures 15.2 and 15.3.

Parameter	Birth	1 week old	2 weeks old	3 weeks old	4 weeks old	Adults
RBC (10^6/µl)	4.7–5.6 (5.1)	3.6–5.9 (4.6)	3.4–4.4 (3.9)	3.5–4.3 (3.8)	3.6–4.9 (4.1)	5.83–8.87
PCV (%)	45–52 (47.5)	33–52 (40.5)	29–34 (31.8)	27–37 (31.7)	27–34 (29.9)	40.3–60.3
Hb (g/dl)	14.0–17.0 (15.2)	10.4–17.5 (12.9)	9.0–11.0 (10.0)	8.6–11.6 (9.7)	8.5–10.3 (9.5)	13.3–20.5
MCV (fl)	93.0	89.0	81.5	83.0	73.0	62.7–75.5
MCH (pg)	30.0	28.0	25.5	25.0	23.0	22.5–26.9
MCHC (%)	32.0	32.0	31.5	31.0	32.0	32.2–36.3
WBC (10^9/l)	6.8–18.4 (12.0)	9.0–23.0 (14.1)	8.1–15.1 (11.7)	6.7–15.1 (11.2)	8.5–16.4 (12.9)	5.3–19.8
Segmented neutrophils (%)	4.4–15.8 (8.6)	3.8–15.2 (7.4)	3.2–10.4 (5.2)	1.4–9.4 (5.1)	3.7–12.8 (7.2)	3.1–14.4
Band neutrophils (%)	0–1.5 (0.23)	0–4.8 (0.50)	0–1.2 (0.21)	0–0.5 (0.09)	0–0.3 (0.06)	0
Lymphocytes (%)	0.5–4.2 (1.9)	1.3–9.4 (4.3)	1.5–7.4 (3.8)	2.1–10.1 (5.0)	1.0–8.4 (4.5)	0.9–5.5
Monocytes (%)	0.2–2.2 (0.9)	0.3–2.52 (1.1)	0.2–1.4 (0.7)	0.1–1.4 (0.7)	0.3–1.5 (0.8)	0.1–1.4
Eosinophils (%)	0–1.3 (0.4)	0.2–2.8 (0.8)	0.08–1.8 (0.6)	0.07–0.9 (0.3)	0–0.7 (0.25)	0.0–1.6
Basophils (%)	0.0	0–0.2 (0.01)	0.0	0.0	0–0.15 (0.01)	0
Nucleated RBC per 100 WBC	0–13 (2.3)	0–11 (4.0)	0–6 (2.0)	0–9 (1.6)	0–4 (1.2)	0
Reticulocytes (%)	4.5–9.2 (6.5)	3.8–15.2 (6.9)	4.0–8.4 (6.7)	5.0–9.0 (6.9)	4.6–6.6 (5.8)	0

15.2 Haematological profile for neonatal puppies from birth until 4 weeks of age. Hb = haemoglobin; MCH = mean corpuscular haemoglobin; MCHC = mean corpuscular haemoglobin concentration; MCV = mean corpuscular volume; PCV = packed cell volume; RBC = red blood cells; WBC = white blood cells.

Parameter	0–2 weeks old	2–4 weeks old	Adult
RBC (10^6/µl)	5.29 ± 0.24	4.67 ± 0.01	6.56–11.20
PCV (%)	35.3 ± 1.7	26.5 ± 0.8	31.7–48.0
Hb (g/dl)	12.1 ± 0.6	8.7 ± 0.2	10.6–15.6
MCV (fl)	67.4 ± 1.9	53.9 ± 1.2	36.7–53.7
MCH (pg)	23.0 ± 0.6	18.8 ± 0.8	12.3–17.3
MCHC (%)	34.5 ± 0.8	33.0 ± 0.5	30.1–35.6

15.3 Haematological profile for neonatal kittens from birth until 4 weeks of age. Hb = haemoglobin; MCH = mean corpuscular haemoglobin; MCHC = mean corpuscular haemoglobin concentration; MCV = mean corpuscular volume; PCV = packed cell volume; RBC = red blood cells; WBC = white blood cells. (Data from Anderson *et al.*, 1971; Meyers-Wallen *et al.*, 1984; Hoskins, 2001) (continues) ▶

Parameter	0–2 weeks old	2–4 weeks old	Adult
WBC (10⁹/l)	9.67 ± 0.57	15.3 ± 1.2	4.04–18.70
Segmented neutrophils (%)	5.96 ± 0.68	6.92 ± 0.77	2.3–14.0
Band neutrophils (%)	0.06 ± 0.02	0.11 ± 0.04	
Lymphocytes (%)	3.73 ± 0.52	6.56 ± 0.59	0.8–6.1
Monocytes (%)	0.01 ± 0.01	0.02 ± 0.02	0.0–0.7
Eosinophils (%)	0.96 ± 0.43	1.40 ± 0.16	0.0–1.5
Basophils (%)	0.02 ± 0.01	0	0

15.3 (continued) Haematological profile for neonatal kittens from birth until 4 weeks of age. Hb = haemoglobin; MCH = mean corpuscular haemoglobin; MCHC = mean corpuscular haemoglobin concentration; MCV = mean corpuscular volume; PCV = packed cell volume; RBC = red blood cells; WBC = white blood cells. (Data from Anderson *et al.*, 1971; Meyers-Wallen *et al.*, 1984; Hoskins, 2001)

Glucose metabolism, hepatic function and serum biochemistry

In the neonate the risk of hypoglycaemia is great because there are limited glycogen stores and the liver shows a poor gluconeogenic response. As long as the neonate is healthy it can maintain normal blood glucose concentrations for up to 24 hours without feeding. However, failure to suck will result in hypoglycaemia after 24–36 hours owing to depletion of hepatic stores. Many of the microsomal enzymes that are involved in drug metabolism are not functional during the neonatal phase, and it is not until about 8 weeks of age that the liver has functional capacity similar to that of an adult liver. Thus, drugs that require hepatic metabolism should only be administered cautiously to neonatal patients.

Neonatal concentrations of albumin and plasma protein are also significantly lower than in adults. Therefore, dosages of drugs that are bound to albumin or plasma proteins must be adjusted accordingly. Extramedullary haemopoiesis may be noted in the liver of neonatal puppies and kittens. The liver is also the site of production of most coagulation factors. Due to the immaturity of the liver, many coagulopathies may be exacerbated during the prepubertal period. As growing requires rapid bone turnover, the serum concentration of alkaline phosphatase (ALP) is often elevated, but in a healthy, growing animal it should never be increased to more than 2–3 times that of an adult. Serum ALP and gamma glutamyltransferase (GGT) are not reliable indicators of liver disease during the first 2 weeks of life, because both are present in colostrum and are absorbed through the gut, increasing the levels in the neonate. The lack of ALP and GGT in a puppy <2 weeks old can be used as an indicator for failure to receive colostrum. In neonatal kittens the elevations in ALP and GGT are not as dramatic as those seen in puppies. Differences in ALP are only seen during the first 2 days of life. A better indicator of colostral absorption in the kitten is an elevation in serum immunoglobulin G (IgG) during the first 2 weeks of life. Normal serum biochemistry profiles for puppies and kittens are given in Figures 15.4 and 15.5.

Parameter	1–3 days old	2 weeks old	4 weeks old	Adult
Bile acids (µM/l)	<10	<10	<15	
Total bilirubin (mg/dl)	0.5 (0.2–1.0)	0.3 (0.1–0.5)	0 (0–0.1)	(0.3–0.9)
ALT (IU/l)	69 (17–337)	15 (10–21)	21 (20–22)	(16–91)
AST (IU/l)	108 (45–194)	20 (10–40)	18 (14–23)	(23–65)
ALP (IU/l)	3845 (618–8760)	236 (176–541)	144 (135–201)	(20–155)
GGT (IU/l)	1111 (163–3558)	24 (4–77)	3 (2–7)	(7–24)
Total protein (g/dl)	4.1 (3.4–5.2)	3.9 (3.6–4.4)	4.1 (3.9–4.2)	(5.4–7.1)
Albumin (g/dl)	2.1 (1.5–2.8)	1.8 (1.7–2.0)	1.8 (1.0–2.0)	(2.5–3.7)
Cholesterol (g/dl)	136 (112–204)	238 (223–344)	328 (266–352)	(128–317)
Glucose (mg/dl)	88 (52–127)	129 (11–146)	109 (86–115)	(65–112)

15.4 Normal serum biochemical values in young dogs (median and range). ALP = alkaline phosphatase; ALT = alanine aminotransferase; AST = aspartate aminotransferase; GGT = gamma glutamyltransferase. (Data from Hoskins, 2001)

Parameter	Birth	1 week old	2 weeks old	4 weeks old	Adult
Glucose (mg/dl)	55–290	105–145	107–158	117–152	70–150
Blood urea nitrogen (mg/dl)	26–45	16–36	11–30	10–22	17–35
Creatinine (mg/dl)	1.2–3.1	0.3–0.7	0.4–0.6	0.4–0.7	0–2.3
Phosphorus (mg/dl)	5.9–11.2	6.7–11.0	7.2–11.2	6.7–9.0	3.3–7.5
Calcium (mg/dl)	9.4–13.9	10.0–13.7	9.9–13.0	10.0–12.2	7.5–10.8
Total protein (g/dl)	3.8–5.2	3.5–4.8	3.7–5.0	4.5–5.6	5.4–8.1
Albumin (g/dl)	2.5–3.0	2.0–2.5 (2.1–2.6)	2.1–2.6 (2.2–2.8)	2.4–2.9	2.4–4.1
IgG (mg/dl)	0–0	350–1500 (0–64)	250–1146 (0–100)	161–648 (88–565)	703–2481
ALT (IU/l)	7–42	11–76	10–21	14–55	5–130
AST (IU/l)	21–126	15–45	14–23 (14–27)	15–31	5–55
ALP (IU/l)	184–538	126–363	116–306	97–274	10–80
GGT (IU/l)	0–2	0–5	0–4	0–1	1–7
Total bilirubin (mg/dl)	0.1–1.1	0.0–0.6	0.0–0.2	0.0–0.3	0.0–0.4
Cholesterol (mg/dl)	65–141	119–213 (148–277)	137–223	173–253	42–170
Triglycerides (mg/dl)	23–132	129–963	38–475	43–721	20–90
Amylase (IU/l)	310–837	187–438	170–611	275–677	500–1500
Creatine kinase (IU/l)	91–2300	107–445	99–394	125–592	88–382
LDH (IU/l)	176–1525	117–513	107–388	98–410	80–345
Lipase	12–43	8–46	5–56	4–86	10–195

15.5 Normal serum biochemical ranges in young kittens (numbers in parentheses indicate concentrations in kittens that did not receive colostrum). ALP = alkaline phosphatase; ALT = alanine aminotransferase; AST = aspartate aminotransferase; GGT = gamma glutamyltransferase; IgG = immunoglobulin G; LDH = lactate dehydrogenase. (Data from Levy *et al.*, 2006)

Immunological development

Virtually no antibodies are transferred to canine and feline fetuses *in utero* because of the type of placentation (endotheliochorial), and they are therefore born immunologically immature. Consequently, puppies and kittens are dependent upon colostral transfer of antibodies (passive immunity) for postnatal protection against infectious diseases. In puppies and kittens, colostrum needs to be ingested within the first 24 and 16 hours of life, respectively, following which mainly IgG and IgA are absorbed. Thereafter, the gut seems to be closed to further absorption, most likely as a result of changes in intestinal pH, proteolytic enzymes and the loss of specific receptors. At 24 hours of life, the peak serum IgG concentration is 17.2 ± 8.6 mg/ml, the IgA concentration is 5.2 mg/ml, and IgM is virtually undetectable in puppies. Peak serum concentrations of IgG (19.15 ± 8.51 mg/ml) and IgA (0.32 ± 0.31 mg/ml) have been measured at 48 hours following birth in kittens. However, the half-life of IgG in kittens appears to be 4 5 days, compared with approximately 8 days in puppies. Similarly, the half-life of IgA was determined to be much shorter in kittens than in puppies. This highlights the importance of vaccination protocols in kittens that include a 6 weeks time point for the first injectable vaccination or the intraocular vaccination at 2 weeks of age against

feline herpesvirus infections in countries where this vaccine is available.

Colostrum is rich in IgG and IgA in both queens and bitches, but IgM can be produced by the neonates around the time of birth. Depending on the type of maternally derived antibodies, these may last from 6–16 weeks after birth. The longer lasting antibodies, such as those against canine parvovirus, are important because they may interfere with the efficacy of vaccines. In addition, puppies are born with an immature thymus and reduced T cell reactivity, and they have inadequate neutrophil (phagocytic) responses because of incomplete development of the complement system. Interestingly, neonatal kittens do not appear to be dependent on colostrum for proper neutrophil phagocytic and oxidative burst function and plasma opsonic capacity, which suggests that an alternative pathway exists for defence against bacterial infections. Puppies and kittens are born with low lymphocyte counts, which increase during the first weeks of life. While B cell populations are relatively high at birth and decrease over time in puppies, they begin low in kittens and steadily increase. The concentration of CD8+ cells begins low in both species, but CD4+ cells increase during the first weeks of life in kittens and remain constant in puppies. Therefore, the CD4+ to CD8+ ratio decreases over time in both species.

Thyroid gland development

Plasma concentrations of tri-iodothyronine (T3) and free T3 (fT3) differ between kittens and adult cats during the first 5 weeks of life. The concentrations of all thyroid hormones change significantly with time during the first 12 weeks of life in both kittens and puppies (Figures 15.6 and 15.7). In addition, the concentration of thyroxine (T4) does not increase outside the normal adult range in kittens as it does in puppies and other neonates. However, the low concentrations of T3 and fT3 in both puppies and kittens during the first weeks of life suggest that, although the thyroid is able to produce these hormones, the process of peripheral conversion of T4 into T3 may not be completely developed at this age. Due to these differences, it is critical to know the exact age of the puppies or kittens when measuring thyroid hormone concentrations and not to use the standard reference ranges for normal adult dogs or cats.

Adrenal gland development

Normally, corticotrophin-releasing hormone (CRH) is produced in the hypothalamus and acts on the pituitary gland to produce adrenocorticotrophic hormone (ACTH), which in turn acts on the adrenal glands to secrete cortisol. From the time of birth this is a well regulated system with positive and negative feedback mechanisms. However, there are differences that are noteworthy: in a study undertaken in young dogs, lower serum concentrations of cortisol were demonstrated in puppies 1–6 weeks old (0.57 ± 0.04 µg/dl) than in the 6–12 week age group (1.04 ± 0.06 µg/dl). Interestingly, breed-related differences were noted in serum cortisol concentrations in adult dogs but the study did not investigate breed differences in puppies.

Gastrointestinal tract development

The neonate is born with a sterile gastrointestinal system, which develops its own microbiological flora to aid in digestion during the first few days of life. Gastrointestinal peristalsis is weaker (slower) than in the adult, there is lower intestinal blood flow and the gastric pH is higher. It is clear that medication, changes in the environment or disease will cause upset to this fragile system, which is most commonly apparent in the form of diarrhoea. One of the most common causes for diarrhoea in the orphaned neonate is overfeeding and inappropriate dilution of milk replacer (see below).

Supplementary nutrition

Supplementary nutrition becomes necessary when one or more of the neonates is not gaining weight or is losing weight. In general, puppies and kittens should be able to take up their daily requirements in 4–5 meals per day during the first few weeks of life. The caloric requirements in the early weeks of life are given in Figure 15.8. As a rule of thumb, puppies should gain 2.2 g/kg of expected adult weight per day. Kittens are born weighing 80–120 g and should gain between 70 and 100 grams weekly. During the first week of life kittens will ingest only 10–15% of their bodyweight in milk. Thereafter, this volume increases to 20–25% of their bodyweight (weeks 1–4). Puppies get their energy from fat during the first weeks of life, whilst kittens get their energy from protein. Therefore, the ingredients of the milk replacer should be reviewed carefully before use. A comparison of the composition of milk from different species is given in Figure 15.9.

Age (weeks)	T3 ± SE (nmol/l) (n = 60)	fT3 (± SE) (pmol/l) (n = 60)	rT3 (± SE) (nmol/l) (n = 60)	T4 ± SE (nmol/l) (n = 60)	fT4 (± SE) (pmol/l) (n = 60)
Birth	0.14 ± 0.05 [a]	0.99 ± 0.18 [a]	1.422 ± 0.07 [a]	34.0 ± 1.5	18.5 ± 0.6
1	0.30 ± 0.06 [a]	1.89 ± 0.16 [a]	1.491 ± 0.06 [a]	74.3 ± 1.5 [a]	20.9 ± 0.6
2	0.62 ± 0.05 [a]	2.64 ± 0.17 [a]	1.680 ± 0.06 [a]	98.0 ± 1.5 [a]	24.0 ± 0.6
3	0.86 ± 0.06 [a]	3.40 ± 0.16	1.887 ± 0.06 [a]	101.2 ± 1.5 [a]	36.4 ± 0.6 [a]
4	1.10 ± 0.06	4.14 ± 0.16	1.490 ± 0.06 [a]	87.0 ± 1.5 [a]	25.8 ± 0.6
Normal adult range	1.0–2.5	2.8–6.5	0.4–0.9	20–50	12–33

15.6 Serum concentrations of T3, fT3, rT3, T4 and fT4 in neonatal puppies. [a] Outside normal adult range.

Age (weeks)	T3 ± SE (nmol/l) (n = 60)	fT3 (± SE) (pmol/l) (n = 85)	rT3 (± SE) (nmol/l) (n = 85)	T4 ± SE (nmol/l) (n = 60)	fT4 (± SE) (pmol/l) (n = 85)
Birth	0.14 (± 0.04) [a] (65)	1.36 (± 0.43) [a] (18)	0.4 (± 0.16) (4)	15.2 (± 1.3) (78)	10.0 (± 1.2) [a] (42)
1	0.21 (± 0.04) [a] (76)	0.90 (± 0.28) [a] (43)	0.80 (± 0.06) [a] (30)	20.6 (± 1.3) (80)	15.3 (± 1.0) (56)
2	0.34 (± 0.04) [a] (78)	0.91 (± 0.23) [a] (59)	1.22 (± 0.05) [a] (40)	29.9 (± 1.3) (83)	23.5 (± 0.9) (70)
3	0.46 (± 0.04) [a] (78)	1.04 (± 0.26) [a] (48)	1.53 (± 0.05) [a] (43)	36.5 (± 1.3) (80)	28.2 (± 0.9) (67)
4	0.57 (± 0.04) [a] (78)	1.40 (± 0.23) [a] (63)	1.43 (± 0.05) [a] (44)	34.5 (± 1.3) (79)	26.0 (± 0.9) (65)
Normal adult range	0.6–1.9	4.2–7.2	0.28–0.62	12–49	12–36

15.7 Serum concentrations of T3, fT3, rT3, T4 and fT4 in neonatal kittens. [a] Outside normal adult range.

Age (weeks)	Puppies: calories (kcal/100 g/day)	Kittens: calories (kcal/100 g/day)	Puppies and kittens: fluids (ml/100 g)
1	13–15	<38 at birth; 28 thereafter	18 (average)
2	15–20	28	13–22
3	20	27	13–22
4	≥20	25	13–22

15.8 Caloric and fluid requirements in puppies and kittens during the neonatal period.

Parameter	Bitch	Queen	Cow [a]	Goat [a]
Fluid content	77	79	88	87
Fat (%)	9.5	8.5	3.5	4.1
Protein (%)	7.5	7.5	3.3	3.6
Lactose (%)	3.4	4.0	5.0	4.7
Calcium (g/100 ml)	0.24	0.18	0.12	0.13
Phosphorus (g/100 ml)	0.18	0.16	0.10	0.16
Metabolizable energy (kcal/100 ml milk)	146	121	70	69

15.9 Comparison of milk composition in different species. [a] When using cow or goat milk as the base of the replacer for neonates, be aware that in order to provide the fat/protein requirements, the lactose concentration will be far too high and cause diarrhoea. Commercial milk replacers are a better choice.

Complications: Common pitfalls in supplementary nutrition are over- and underfeeding. Overfeeding with milk replacers often results in diarrhoea whilst underfeeding results in dehydration and lack of weight gain. Homemade formulas are often deficient in growth factors, amino acids and other nutrients essential for growth. Many of the commercial milk replacers are made using cow's milk as a base and are therefore not always complete either. The energy density of the formula may be too high and this results in fluid requirements not being fulfilled. Fluid deficits may also occur when the energy density of the formula is too low, or the stomach capacity of the neonate is too small to hold the required amount of formula.

Many milk replacers can easily cover the daily caloric requirements of both puppies and kittens; however, the fluid requirements are not easily met. Therefore, it may be necessary to dilute the formula to provide the appropriate amount of fluids. Regular feeding is important to maintain good hydration in the neonate. If only three feedings per day can be provided, then one of the higher density milk replacers should be used, and the extra fluids provided subcutaneously. Overfeeding or a high lactose content in the milk replacer often causes diarrhoea. Nothing is better than mother's milk: it also contains bile salt-activated lipase, which is necessary for proper digestion. After each meal the neonate should be encouraged to urinate and defecate by stimulating the anogenital region with a moistened cotton ball. Lack of defecation may indicate constipation or other problems, such as malformations or hypothyroidism in kittens. Neonates should be weighed daily until 3 weeks of age to ensure proper weight gain.

Before supplementing the neonate, body temperature should be assessed. If the body temperature is too low, gut motility is reduced and may cease, the abdomen becomes distended and finally the neonate may regurgitate, resulting in aspiration pneumonia. The body temperature should be at least 35.6°C, or intestinal sounds should be assessed prior to commencing supplemental feeding. If intestinal sounds are present at a lower body temperature, feeding can be initiated.

Feeding methods: Supplemental feeding can be provided by:

- Placing the puppy or kitten with a surrogate
- Bottle feeding
- Tube feeding.

Surrogacy: When placing the neonate with a surrogate mother, the neonate should be rubbed with the other puppies or kittens in the litter to entice the mother to take care of the orphan. It may help to rub faeces from the bitch's or queen's own litter on the orphan to make it acceptable to the mother.

Bottle feeding: Bottle feeding can be performed if there is only one neonate that needs to be supplemented. The neonate should be held upright, allowing it to place the front paws on the bottle or the person's hand. Ample time should be given for the neonate to swallow and breathe in between sucks. If puppies or kittens feed too quickly from the bottle, they tend to swallow air or aspirate. Feeding straight from a syringe is not recommended because aspiration is a common problem.

Tube feeding: Tube feeding is an easy and convenient method to provide the neonate with the supplemental nutrition it needs. Tube feeding allows for exact amounts to be fed, is less time-consuming when multiple animals need to be supplemented, and is appropriate if the neonate is not nursing from a bottle or malformations are present. A 5 or 8 Fr rubber or plastic feeding tube is used depending on the size of the neonate. The feeding tube is measured alongside the neonate from the tip of the nose to the end of the ribs. To indicate the portion of the feeding tube that will be placed, a mark is made at three-quarters of the measured length from the tip of the feeding tube. Criteria for the correct placement of the tube include negative pressure when pulling back on the plunger of the syringe, vocalization by the neonate, or swallowing of the tube by the neonate (Figure 15.10).

During the first week of life, 12–13 ml/100 g per day should be given divided every 2 hours. During the second week of life, this dose should be increased to 14 ml/100 g and in the third week to 18 ml/100 g. Meals at this time can be given every 2–4 hours. Longer feeding intervals will lead to larger volumes of milk replacer being given per feed, which in turn may cause diarrhoea, stomach distension, abdominal discomfort and possibly aspiration (as noted above). If tube feeding is the only method used for replacement (rather than supplementation), the maximum amount of milk that can be given at a single feeding is 4 ml/100 g bodyweight.

By approximately 3–4 weeks of age, a soft gruel can usually be offered by mixing kitten or puppy food with warm water at a ratio of 50:50. Care should be taken to offer the food warm to increase palatability. The food can be offered by hand or using a tongue depressor. Metal spoons should be avoided so as not to accidentally damage the newly erupting teeth. It is also possible to try placing some of the gruel on a front paw, which the neonate will attempt to clean, thus tasting the food; this works particularly well with kittens.

Critical care of the neonate

A variety of common clinical problems occur in the neonate because of their particular physiology. It is important that these are identified and treated rapidly.

Hypothermia

Hypothermia in the neonate is a serious problem. Gut motility slows with decreasing body temperature, ultimately causing ileus. When hypothermic neonates are tube-fed, the milk replacer is either regurgitated and aspirated resulting in pneumonia, or the ingesta begin to ferment leading to a bloated neonate. This causes increased pressure on the thorax and thus laboured breathing. Most neonates in pain or respiratory distress swallow air, worsening the bloat. A vicious cycle begins, leading to circulatory collapse and death. Hypothermia also inhibits cellular immune functions, which may lead to increased susceptibility to bacterial infections. A neonate is considered hypothermic if its body temperature decreases below 34.4°C at birth, below 35.6°C at 1–3 days of age, or below 37.2°C at 1 week of age.

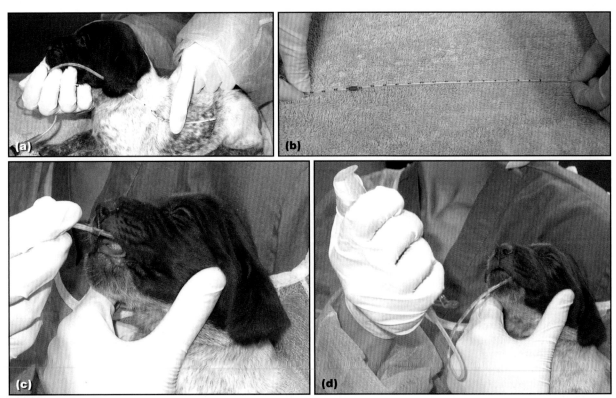

15.10 Principles of tube feeding. **(a)** The tube is used to measure the distance from the tip of the nose to the last rib. **(b)** A mark is made on the tube at three-quarters of the length measured in (a), indicated here by the red tape. **(c)** With the puppy upright on the table or in the hand, the tube is gently fed into the mouth. Most puppies and kittens will swallow the feeding tube. **(d)** Check for negative pressure when drawing back on the plunger of the syringe.

Clinical signs

Clinical signs in a chilled neonate with a body temperature above 31.1°C include restlessness, loss of appetite, continuous crying, red mucous membranes and skin that is cool to the touch. However, muscle tone is still good; the respiratory rate is >40 breaths/minute and the heart rate >200 beats/minute. When the body temperature falls into the range of 28–29.4°C, the neonate appears lethargic and uncoordinated but responsive. Moisture is seen around the corners of the lips, the heart rate drops below 50 beats/minute and the respiratory rate is between 20 and 25 breaths/minute. No abdominal sounds are heard and metabolism is impaired, resulting in hypoglycaemia (see below). Below 21°C, the neonate appears to be dead. If extreme measures of arousal result in a response, treatment may be attempted.

Treatment

Treatment consists of slowly reheating the neonate (not more than 1°C/hour) whilst keeping it dry, because rapid reheating may cause delayed organ failure. Reheating increases caloric demand because of the increase in the metabolic rate. Thus, whilst reheating the neonate, dextrose should be added to the fluid supplementation to assist in stabilizing the patient. The required ambient temperature is dependent on the neonate's age: until 1 week of age the temperature should be kept at 29.5–32.2°C; from 1–4 weeks of age at 26.7°C; and thereafter at 21–24°C. Humidity levels should be kept at 55–65% to prevent the skin from drying out. Higher humidity levels are conducive to infections. However, the humidity may be raised to 85–90% and the temperature to 29.5–32.2°C if the neonate is premature and has a low birthweight. The higher temperature and humidity help to maintain the patient's core temperature, hydration and metabolism.

To provide external heat to the neonate, heating pads, heat lamps and warm water bottles/gloves can be used cautiously. Owing to lethargy or the young age (no thermoregulation or reflexes), the patient cannot crawl away from excessive heat. Thus, the ambient temperature should be checked often to avoid overheating. The neonate's body position should be shifted frequently, which appears to reduce the tendency to vomit. Oxygen and warm air, as present in oxygen cages or human paediatric incubators, can rewarm the hypothermic neonate quickly and safely.

If a neonate is severely hypothermic, it can be given warm fluids by intravenous, intraperitoneal or intraosseous routes. The temperature of the administered fluids should never be more than 1°C higher than the patient's body temperature. It had been previously thought that hypothermic patients could be rewarmed by giving warm liquids orally. However, for the reasons mentioned above, a hypothermic neonate should never be fed if the body temperature is <34.4°C and/or if no bowel sounds are heard. Rapid rewarming results in heat prostration, which is characterized by increased respiratory rate and effort, cyanosis, diarrhoea and seizures. If the body temperature is raised by ≥2°C within an hour, life-threatening conditions usually result. Treatment of overheating consists of cool air and tepid water baths. Even if the neonate survives initially, such animals often die because of delayed organ failure (heart and kidneys). Thermal burns may occur when the neonate cannot escape excessive heat and these need to be treated locally, and systemically with fluids. Antibiotic ointments are not recommended because they are easily absorbed and can be harmful. If the burns are severe, crystalloids are administered at shock doses, bactericidal antibiotics such as ampicillin or amoxicillin/clavulanate are given systemically, and caloric supplementation at 2–4 times maintenance doses is recommended.

Hypoglycaemia

Hypoglycaemia can be serious in the neonate. Two studies have shown that there is little or no response to glucose administration in the youngest of neonates. However, normal serum glucose concentrations are far lower in neonates than in adult animals, and neonates can tolerate lower serum concentrations better than adults.

Clinical signs

A variety of clinical signs may occur in the hypoglycaemic neonatal patient (serum glucose <30 mg/dl) including tremors, crying, irritability, increased appetite, dullness, lethargy, coma, stupor and seizures. Besides starvation and hypoxia, other common causes for hypoglycaemia in the neonatal patient are sepsis, poor environmental conditions, congenital metabolic defects such as glycogen storage disease, portosystemic shunts, 'toy breed hypoglycaemia' and hypopituitary dwarfism. Placental insufficiency and prematurity have been cited as causes of hypoglycaemia in the neonate immediately after birth.

Treatment

The treatment consists of giving dextrose slowly intravenously at a rate of 0.5–1 g/kg as part of a 5–10% solution of dextrose in lactated Ringer's or normal saline. Higher concentrations of intravenous dextrose should be avoided because of its irritant nature (phlebitis). Dextrose can be given at higher concentrations directly into the mucous membranes of the mouth. After treatment the blood glucose concentration should be monitored because of the risk of hyperglycaemia attributable to the poor regulatory mechanisms in the neonate.

Dehydration

Dehydration occurs very quickly in neonatal patients because of their high body water content, the inability to concentrate urine and the impaired ability to autoregulate renal blood flow, which results in a decreased glomerular filtration rate. In addition, other compensatory mechanisms such as increasing the heart rate and cardiac contractility are not well developed in neonates.

Clinical signs

The mucous membranes are best used to assess hydration, because skin turgor is different in neonates, making it not ideal to check for dehydration. At 5–7% dehydration the mucous membranes become tacky or dry. At 10% they will be very dry and there will be a noticeable decrease in skin elasticity, and at >12% dehydration circulatory collapse occurs. The most common causes of dehydration are diarrhoea, vomiting, pneumonia, decreased milk intake and excessive ambient temperatures. Even mildly dehydrated neonates are severely compromised and need to be treated immediately.

Treatment

Treatment consists of providing fluids. In very mild cases, fluids may be given orally provided that gut sounds are heard (i.e. gut motility is intact). In mild cases, fluids may be administered subcutaneously. The fluids will be absorbed slowly without overloading the homeostatic system. However, ideally, fluids should be given intravenously via the jugular vein (Figure 15.11) or the cephalic vein using a 26 gauge three-quarter inch (19 mm) intravenous catheter. Aseptic technique must be used, because the immature immune system may be even more compromised through illness.

If intravenous access is not available or possible, intraosseous catheters may be placed in the femur (Figure 15.12) or humerus using an 18–22 gauge injection needle or a 20 gauge x 3.75 cm spinal needle. For placement in the femur access to the marrow space is gained through the trochanteric fossa, and through the greater tubercle for humerus placement.

- The hair at the placement site must be clipped and the area aseptically prepared.
- Less than 0.5 ml of lidocaine is injected to infiltrate the cutaneous and subcutaneous tissues and the periosteum.
- With one hand, the bone is stabilized and rotated outwards so that the injection site is easily accessible.
- The leg is held so that the long axis can be felt to facilitate visualization of the bone.
- The catheter (injection or spinal needle) is held in the other hand and, with a rotating motion, inserted into the bone, ensuring that it is being pushed into the marrow space.
- Once the needle feels solidly in place, an injection cap is placed on top, and bone marrow aspirated to ensure correct placement and to dislodge any small particles of bone that may have become stuck when using an injection needle as an intraosseous catheter.

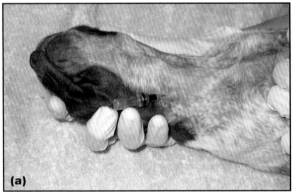
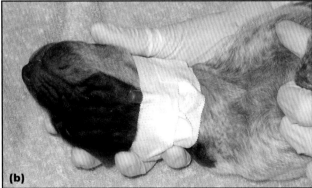

(a) (b)

15.11 **(a)** Correct placement of a jugular catheter in a newborn puppy. The catheter can be taped carefully to the fur using cloth tape. **(b)** Care must be taken not to block the airways by keeping the neck somewhat stretched.

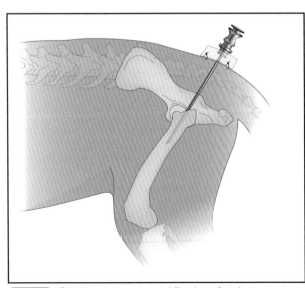
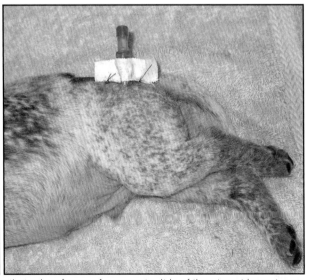

15.12 Correct placement and fixation of an intraosseous catheter in a femur of a neonate. It is of the utmost importance that sterile techniques are used and that the bone is not punctured more than once. Fixation of the catheter is achieved by attaching tape to the top of the catheter and suturing the tape to the skin.

- Two pieces of tape are placed at the top of the catheter in a butterfly shape and sutures are placed through the tape and into the skin to prevent the catheter from moving or slipping out.

The catheter must be bandaged to prevent infection. If after puncturing the bone the catheter is incorrectly placed, attempts *should not* be made to place the catheter in the same bone again, because the administered fluids will leak out into the subcutaneous tissues. In the author's opinion every attempt should be made to place intravenous catheters rather than intraosseous because of the risk of damage to the bone and subsequent infection with the latter.

The maintenance fluid requirements of the neonate are high at 80–100 ml/kg/day because of increased total body water, increased body surface area, higher metabolic rate, reduced renal concentrating ability and increased insensible losses (evaporation, respiration). However, the total volumes given are small because volume overload can quickly ensue. Fluids should therefore be given at a rate of 3–4 ml/kg/h. In cases of shock or moderate to severe dehydration 30–40 ml/kg of isotonic crystalloids can be given as a bolus over 5–10 minutes. The fluids should be given warm but not more than 1°C above body temperature. It is difficult to keep the fluids from cooling down to room temperature when administering them through a fluid pump. One method to keep the fluids warm is to run the fluid line through a warm water bath or under a heating blanket shortly before the line reaches the patient. After stabilizing the patient, the maintenance dose is 6 ml/kg/h. To this, 50% of the deficit is added over 6 hours (deficit = bodyweight x % dehydrated). Signs of overhydration and resulting disease include laboured breathing and froth around the lips due to pulmonary oedema, peripheral oedema and increased heart rate due to cardiovascular overload, and seizures or coma due to intracranial haemorrhage. The extent of rehydration should be monitored carefully by assessing the mucous membranes and by weighing the neonate frequently.

Hypoxia

Hypoxia is not easily recognized clinically because puppies and kittens tend not to hyperventilate until they are several days old. During the first 1–2 hours after birth, respiratory and metabolic acidosis has been shown to be normal. In fact, most neonates recover within <45 minutes after birth without intervention. However, treatment should be considered for those patients with acidosis that lasts longer than 3 hours. Hypoxia causes serious stress that can lead to complications such as respiratory depression, translocation of intestinal bacteria and chilling, resulting in decreased resistance against bacterial infections. Studies have also shown that neonates that lack colostrum may develop necrotizing enterocolitis when they become hypoxic. Aspiration or infections, coughing and nasal discharge are considered to be causes of hypoxia.

Clinical signs

The clinical signs include increased respiratory effort and rate and a distended abdomen caused by aerophagia. Hypoxia, like hypothermia, leads eventually to decreased or absent gut motility, which in turn leads to a vicious cycle resulting in circulatory collapse and death (see above). Decreased heart rate and blood pressure are common findings in hypoxic neonates. Immediate care is warranted in patients with these signs of generalized hypoperfusion.

Treatment

The treatment consists of oxygen supplementation, provided either by using a commercially available incubator or creating a makeshift oxygen cage. An oxygen line can be fed into a box covered with a blanket, which will also help to retain warmth and prevent drafts. Care must be taken to prevent 100% oxygen from coming in direct contact with the eyes if they are open, because this may cause retinal detachment.

Preventive strategies for neonatal health

Healthy breeding animals

Ensuring healthy neonates begins even before breeding. The breeding animals should be checked for prevalent diseases, e.g. brucellosis in dogs (in countries where this is present), and feline leukaemia virus (FeLV) and feline immunodeficiency virus (FIV) in cats. At that time, a physical and faecal examination should be performed. Any parasites that may be present should be treated appropriately and vaccinations updated as needed. In bitches, the first dose of the vaccine against canine herpesvirus is given at the time of breeding and repeated later during pregnancy. Modified live vaccines should not be given during pregnancy because they may cause disease or death in the fetus. Dogs can be safely dewormed with fenbendazole during the last week of pregnancy to minimize transplacental transmission of *Toxocara canis*, and cats can be given pyrantel to minimize shedding of *Toxascaris leonina*. The animals to be bred should have all of the required breed-specific testing done, such as hip, elbow, eye and von Willebrand's disease screening.

Diet

The diet should be complete and ensure that the bitch or queen gains weight at an appropriate rate. Over- or underweight dams tend to have higher neonatal mortality rates. Commercial canine or feline diets are generally nutritionally complete and can be supplemented with puppy or kitten food, respectively, during the last third of pregnancy. Raw diets should be avoided because they may contain parasites, *Salmonella* or other infectious agents. In addition, they may not be well balanced, e.g. they may contain too much phosphorus leading to a decrease in available calcium and potentially stunted bone growth in the fetuses. Supplements and vitamins are not needed if a balanced diet is being fed. Many supplements, and especially fat-soluble vitamins, can be detrimental if given at too high a dose.

Parturition

Equipment

A whelping or kittening room should be made available to minimize stress to the expectant mother and to provide a quiet, clean and dry area with minimal animal and human traffic. Drafts, overcrowding, poor sanitation and noise should be avoided. Food and water should be readily accessible because some queens and bitches are unwilling to leave their litters even for a moment to eat.

Close to the time of parturition, basic supplies should be present in sufficient quantities. Plastic or metal whelping or kittening bins are ideal because they can be easily cleaned. If wooden whelping boxes are used, they should be bleached before use. Layers of newspaper or disposable nappies (diapers) can be used and may be covered with blankets. Once the neonates have been born, the blankets should be washed daily or more often if required. The newspaper or nappies should be replaced as often as needed. Milk replacer appropriate to the species should be available, as should nursing bottles or neonatal feeding tubes.

Weighing scales calibrated in grams, such as kitchen or baby scales, allow for accurate daily weight measurements. Digital thermometers that are able to measure as low as 32°C are useful, because neonates cannot regulate their body temperature, which can be quite low during the first weeks of life. In addition, if the neonate needs supplementary feeding, the body temperature has to be determined because at temperatures <36°C the gut may shut down. Heating or warming devices such as warm water blankets, heat lamps, warm water gloves or other forms of heating pads should be available but used with caution.

Ambient temperature

Once the puppies or kittens are born, the ambient temperature should be 29.5–32.2°C during the first day of life. The temperature can then be decreased slowly during the first week of life to 24–26.7°C. In the case of orphaned puppies or kittens, or hairless breeds such as the Mexican Hairless dog or Devon Rex cat, the ambient temperature may be maintained somewhat higher during the first few weeks of life. It is important to remember that the cellular immune system is not functional in neonates with low body temperatures; therefore, maintaining proper environmental temperatures aids greatly in decreasing morbidity.

Reduced exposure to fomites

It is essential to reduce the risk of infection by reducing exposure to fomites. Shoes are the most important fomite and should be left outside the whelping/kittening room, and hands should be washed before and after handling the neonates to minimize transmission of disease. If multiple animals are kept in the household, the youngest ones (those most susceptible to disease) should be taken care of first before tending to the older ones.

Examination

At birth the neonate should be examined for a cleft palate or other malformations. If desired the umbilicus can be dabbed with a disinfectant (such as diluted chlorhexidine solution) but care should be taken to use a solution that is non-toxic because the mother is likely to lick it off. Puppies and kittens should be observed and weighed twice during the first 24 and 16 hours of life, respectively, to determine whether colostrum has been ingested.

The neonates should be weighed daily for the first 3 weeks of life and the weights recorded, because the first sign of illness is often a lack of weight gain or weight loss. In fact, various studies have shown that the lack of weight gain, or the presence of weight loss, is directly correlated with a decreased chance of survival. Daily weighing also allows overall assessment of the neonate's vitality (muscle tone, responses to stimuli). During this time it is equally important to assess the dam's body temperature daily, examine the quality and amount of vulvar discharge (to check for endometritis) and examine the mammary glands (to check for mastitis). After 3 weeks, if the neonates are doing well and have been gaining weight consistently, they can be weighed 2–3 times weekly until weaning. If there is a question about milk uptake, neonates can be weighed before and after nursing using an accurate gram scale.

References and further reading

Anderson L, Wilson R and Hay D (1971) Haematological values in normal cats from four weeks to one year of age. *Research in Veterinary Science* **12**, 579–583

Bebiak DM, Lawler DF and Reutzel LF (1987) Nutrition and management of the dog. *Veterinary Clinics of North America: Small Animal Practice* **17**, 505–533

Blunden TS (1998) The neonate: congenital defects and fading puppies. In: *BSAVA Manual of Small Animal Reproduction and Neonatology*, ed. GCW England and MJ Harvey, pp.143–152. BSAVA Publications, Gloucester

Casal ML (1995) Feline pediatrics. *Veterinary Annual* **35**, 210–235

Casal ML, Zerbe CA, Jezyk PF, Refsal KR and Nachreiner RF (1994) Thyroid profiles in healthy puppies from birth to 12 weeks of age. *Journal of Veterinary Internal Medicine* **8**, 158

Hoskins JD (2001) *Veterinary Pediatrics, 3rd edn.* WB Saunders, Philadelphia

Hotston Moore P and Sturgess K (1998) Care of neonates and young animals. In: *BSAVA Manual of Small Animal Reproduction and Neonatology*, ed. GCW England and MJ Harvey, pp.153–169. BSAVA Publications, Gloucester

Johnston S, Root Kustritz M and Olson P (2001) The neonate – from birth to weaning. In: *Canine and Feline Theriogenology*, ed. S Johnston *et al.*, pp.66–104. WB Saunders, Philadelphia

Lawler DF (2008) Neonatal and pediatric care of the puppy and kitten. *Theriogenology* **70**, 384–392

Levy JK, Crawford PC and Werner LL (2006) Effect of age on reference intervals of serum biochemical values in kittens. *Journal of the American Veterinary Medical Association* **228**, 1033–1037

Macintire DK (1999) Pediatric intensive care. *Veterinary Clinics of North America: Small Animal Practice* **29**, 971–988

Meyers-Wallen VN, Haskins ME and Patterson DF (1984) Hematologic values in healthy neonatal, weanling, and juvenile kittens. *American Journal of Veterinary Research* **45**(7), 1322–1327

Monson WJ (1987) Orphan rearing of puppies and kittens. *Veterinary Clinics of North America: Small Animal Practice* **17**, 567–576

16

Clinical approach to neonatal conditions

Margret Casal

Introduction

Neonatal mortality in dogs and cats is lamentably high with, in some circumstances, up to 40% of off-spring dying before they reach 12 weeks of age. Good husbandry can reduce this problem significantly, particularly as infectious disease accounts for a relatively small proportion of deaths, although it is a major cause of morbidity. Importantly, the physiological status of the neonate may influence the manner in which it responds to various disease conditions, which results in variation in the clinical signs seen when compared with the adult. Furthermore, the inability of the neonate to deal with various pathological processes results in a more rapid demise once disease is established. The rapid detection of clinical disease is therefore essential to ensure prompt treatment and give the greatest chance of survival.

Diagnostic investigation

An approach to the investigation of disease in neonates is described in this Chapter. Common methods for examination of the critically ill neonate are described in Chapter 15, and these will facilitate recognition of the severity of the clinical disease and implementation of appropriate stabilization techniques.

Clinical history

The fact that neonates may have unusual or non-specific clinical signs highlights the importance of collecting a comprehensive clinical history, not only of the patient, but also of the littermates, parents and other relatives. Breed, sex and age as well as previous observation of the disease in these breeding animals or relatives may give insight into possible underlying genetic diseases. The relevant history relating to the queen or bitch should include nutritional status, vaccination dates, oestrous cycle information (intervals and duration), breeding practice, medications or supplements given during pregnancy, possible exposure to toxins, problems during pregnancy or birth, and finally mothering abilities. In cats, the blood types of the parent animals may suggest neonatal isoerythrolysis as a differential diagnosis for the presenting illness in a kitten.

Information relating to the neonates that should be collected includes the record of daily weight gain or loss since birth because this will provide information on the duration of illness. A lack of colostrum ingestion may point to specific infectious diseases or overall poor immunity in the neonate. The history should also include the number of animals affected, the method by which they were raised, their normal environment, the behaviour, size and weight of each puppy or kitten within the litter, duration and type of clinical signs, milk replacers and medications given.

Physical examination

A request to investigate neonatal disease should include a full clinical examination of the affected animal, the dam and preferably one healthy littermate (if available) for comparison. Neonates are often immunologically incompetent and may not have been vaccinated, so the examination environment should be cleaned carefully before use, and the patient should not come into contact with other animals in the clinic.

A physical examination can be challenging because sick neonates may be completely unresponsive, whilst those that are moderately unwell may be markedly distracted by the new environment and may not demonstrate normal responses to stimuli. The following section attempts to characterize some of the differences between adults and neonates in terms of clinical examination findings. The list is not complete, but comparison with some of the developmental landmarks described below may be helpful in detecting other abnormalities.

For the physical examination of a neonate, a paediatric stethoscope with a 2 cm bell is helpful. A digital thermometer allows rapid measurement of the body temperature without causing discomfort, and this should be able to record as low as 32°C because neonates can have a body temperature lower than 34°C. Neonates cannot regulate their body temperature during the first few weeks of life, therefore they should be examined on a warm, clean surface rather than a cold metal table. Checking the oral mucous membranes allows assessment of hydration in the neonate, because the skin turgor is not as well developed as in adults. The mucous membranes are hyperaemic during the first 2–4 days of life, after which they become pink. Cyanotic or white mucous membranes are a sign of distress. Neonates in an adequate state of hydration have moist mucous membranes, but tacky to dry membranes indicate 5–7% dehydration. At 10% dehydration, the mucous membranes are very dry and there is a noticeable decrease in skin elasticity.

The neonate is born with hair that covers most of the body, except the ventral abdomen. Lack of hair or a sparse haircoat may indicate either a genetic abnormality of the skin or premature birth. The ventral abdominal skin of the neonate is normally hairless and dark pink. Bluish or dark red discoloration of the skin is indicative of a neonate in distress (cyanosis or sepsis, respectively). Other than urine and faeces, discharge from any orifice is abnormal in the neonate.

The neonate's head, body, limbs and tail should be examined for symmetry and normal conformation. The head should be examined specifically for open fontanelles, cleft palate, bulging from behind closed eyelids (ophthalmia neonatorum) and for normal formation of the nose and external ears. The presence of flattening or malformations of the chest should be noted (e.g. swimmer syndrome, pectus excavatum), as should bulges in the neck area (e.g. gas in the oesophagus, ectopic heart, goitre). The gross appearance of normal neonatal puppies and kittens shows a mildly rounded abdomen but this should never be bloated, which may be a sign of respiratory distress with aerophagia, or of urinary or faecal retention. Aerophagia is common in neonates in pain.

The abdomen and urachus should be examined for defects of the abdominal wall and ventral urine scalding, which may be caused by cannibalism by an overzealous mother, ventral closure defects or persistent urachus. The genitals and the anus should be checked for patency by stimulating urination and defecation using a moistened cotton ball. The presence of haircoat abnormalities over the dorsum may indicate the presence of spina bifida. The tail should be examined for muscle tone, length, curliness and kinks. Abnormalities in tone may be an indication of associated defects or problems (e.g. abnormal innervation of the distal pelvis).

Diagnostic techniques

Blood

Blood can be obtained easily from the jugular vein in neonates. However, no more than 10% of the circulating volume should be drawn over the course of a week (e.g. if a neonate weighs 250 g, no more than 2.5 ml of blood should be drawn in 1 week). If a neonate remains in the hospital for several days, it is essential to record each time blood is drawn to avoid causing the patient to become anaemic (Figure 16.1). Such monitoring is essential, especially in neonates that cannot maintain normal blood glucose concentrations and are subjected to multiple blood samples (one drop of blood equals about 0.1 ml). Due to the small amount of blood drawn at any one time, an appropriately sized collection tube should be used to ensure that any anticoagulant does not dilute the sample and result in false laboratory data.

Urine

To obtain a urine sample, the neonate can be stimulated to urinate simply by gently rubbing the genital area with a moistened cotton ball. Alternatively, the bladder can be expressed carefully. It is rarely necessary to obtain sterile urine samples by cystocentesis, which should be avoided because of the fragility of the skin and organs of the neonate. In the normal neonate, urine is almost colourless; the presence of significant colour to the urine usually indicates dehydration.

Imaging

Diagnostic imaging techniques, including radiography and ultrasonography, can be extremely useful. An ultrasound examination is best performed using a 7.5 MHz or higher frequency transducer, and neonates generally tolerate this imaging technique better than radiography. Radiography of neonates requires the use of a high detail intensifying screen with a single emulsion film to ensure a radiograph of diagnostic quality. For optimum contrast, whole body radiographs should not be taken and the kilovoltage should be reduced to half that used for an adult, because there is little body fat and poor mineralization at this age. If possible, radiographs of healthy littermates should be obtained to allow comparison with a normal age-matched neonate.

Patient	Record number	Bodyweight (BW) (g)	Blood for week (ml) A = BW x 0.01	Blood drawn (ml) B	Comments
Date	Weekday		A–B = remaining for week		
	Monday				
	Tuesday				
	Wednesday				
	Thursday				
	Friday				
	Saturday				
	Sunday				

16.1 Form to assess the amount of blood that can be drawn from a sick neonate to prevent excessive blood loss through phlebotomy. No more than 10% of the total blood volume should be drawn in the course of 1 week. Total blood volume is approximately 10% of the bodyweight. The shaded areas should be completed.

Scoring system

An objective method of scoring neonatal health, adapted from the Apgar score used in humans, may be useful in the assessment of neonates during the first few minutes of life. The parameters that comprise the Apgar score include respiratory effort, muscle tone, heart rate, response to stimulation and colour of the mucous membranes (Figure 16.2).

The Apgar score provides information about the overall condition of the neonate, and the values are highly correlated with the chances of survival and viability. Studies have also shown that a decreased rooting reflex, decreased sucking and decreased swallowing response are highly correlated with decreased Apgar scores. These values are not correlated with blood or serum biochemistry

---

Canine Neonatal Status and Treatment Form
Matthew J Ryan Veterinary Hospital – University of Pennsylvania

Dam's Name:	Owner's Name:	Breed:	Medical Record#:

WEIGHT: ____ grams | **Puppy #**(use birth order if possible): ____ of ____ total pups. | **Birth Date:** | **Birth TIME:** ____ AM PM

Markings (*complete only when puppy or kitten is stable*): (B=Black, R=Red, W=White, G=grey, Y=Yellow, T=Ticked)

Sex: | **Temp:**

Birth Type (circle): Vaginal unassisted   Vaginal assisted   C-section

If birth was by C-section: Was an Opioid used on the Dam prior to birth? ☐ Y ☐ N List anesthetic regimen:

Deformities present on exam: ☐ Y ☐ N Type:

Exam comments:

Top   Bottom

Canine Neonatal Vitality Score: (Circle condition in the first minute of birth prior to resuscitation efforts)

Factor	0 points	1 point	2 points
Heart rate	Absent	Present, less than 180	180 or greater
Respiratory efforts	Absent	Gasping	Normal
Mucus membrane color	White or Cyanotic	Pale Pink	Normal Pink
Spontaneous activity & muscle tone	Absent	Tone - no movement	Movement
Suckle Reflex	Absent	Weak	Strong
Lumbo-sacral stimulation (vigorously rub fur back or scratch region)	No response	Movement – no crying	Movement and Crying
Canine Neonatal Vitality Score (out of 12): (add points from all circled values above)			

Puppy management and interventions:

Type of external heat source provided:	Suction of airways? ☐Y ☐N With:
Rubbed puppy to dry and wake it up? ☐Y ☐N Length of time performed:	____ min.

Naloxone	**0.25 mL/kg** (0.4mg/mL) **IV** or under tongue	Time given:	Route:
Oxygen	Record % oxygen, method and times given	Time given:	Route:
Ventilatory Assistance	30 breaths / min (record time and length)	Time given:	Route:
Acupuncture	Nasal philtrum; 25g needle	Time given:	Route:
Chest compressions	1-2 per sec. (record time and length)	Time given:	Route:
Doxapram	**0.1 mL** / pup or kitten **IV, IO** or under tongue	Time given:	Route:
Dextrose and Fluids	10% 2-4 mL/kg (0.8 mL of 50% dextrose + 3.2 mL saline) **IV or IO**	Time given:	Route:

Other treatments (Such as Epinephrine or Bicarb:

Time when puppy management and interventions were completed (not including providing warmth) _____ AM PM

☐ Dead ☐ Alive   Signature: _____

Canine Neonatal Status and Treatment Form Version 4b   Last Revised: 3/11/09 by Dr. Anne Traas, DVM DACT

16.2 Neonatal resuscitation form, including vitality (Apgar) scoring table, used at the Veterinary Hospital of the University of Pennsylvania.

abnormalities; however, a poor Apgar score should be used as an indication for increased resuscitation effort, because this has been shown to increase the chance of neonatal survival.

Common presentations

The most common reasons for neonatal cats and dogs to be presented to veterinary surgeons include:

- 'Fading'
- Lack of weight gain, or weight loss
- Trauma
- Diarrhoea
- Respiratory disease
- Stunted growth
- Seizures.

Paediatric patients are not just small versions of the adult animals. The immune system and many metabolic processes are not fully developed. Therefore, paediatric patients are more susceptible to infectious diseases, and the illness will be more pronounced owing to limited reserves and decreased ability to respond to dehydration, hypothermia and hypoglycaemia. The most common clinical signs associated with neonatal infectious diseases are of gastrointestinal and respiratory origin. For the reasons stated above, clinical signs need to be recognized early and a working diagnosis made as soon as possible, so that treatment can be initiated to reduce both the severity and the duration of the illness.

Fading puppies and kittens

Clinical signs
Fading puppies and kittens show two different types of initial presentation, but ultimately the condition is characterized by anorexia, weight loss, lethargy, emaciation and death. With the first type of presentation the neonates are sick at birth; they are either born small or weak or with birth defects. The neonates are unable to or fail to nurse, resulting in dehydration, hypothermia, hypoglycaemia and death within the first few days of life. In these cases, perinatal bacterial infections and respiratory distress have been described as the most common reasons for organ failure and death. With the second type of presentation the neonates initially appear healthy, but during the first week of life they become weak, depressed and anorexic, and enter the fatal cycle of dehydration and hypothermia (described above). In some instances, particularly in kittens, death may occur later, at around the time of weaning.

Diagnosis
The diagnosis of the cause of fading in a neonate can be challenging.

History: Attempts to make a diagnosis should start with a very thorough history. Some of the information required has been mentioned above, but in the context of the fading puppy or kitten the following are particularly important:

- Vaccination status (especially herpesvirus vaccinations in dogs)
- Timing of vaccination
- Type of vaccines given
- Brucellosis testing in dogs (in the countries where this is present)
- Supplements or medications given during pregnancy
- Information about trauma and activities (such as cat or dog shows) attended by the dam during pregnancy
- Information about the environment.

Dams that go outside are potentially exposed not only to infectious diseases but also to toxins (e.g. rodenticides and antifreeze). Information about previous litters and current littermates includes: the number of neonates affected; the exact clinical signs; age of onset; the sexes of the affected neonates; and the number and sex of any stillborn littermates. Furthermore, enquiries as to the type of milk replacer or medications given, and the housing, environment and bedding of the neonates may also indicate a potential cause of disease. For example, wood chips may be contaminated with *Klebsiella* spp. and lead to umbilical infections if not changed frequently; paint chips that contain lead may be ingested and result in lead poisoning; phenols in shampoos may cause liver disease with pulmonary congestion and oedema in kittens because of an inherent deficiency in glucuronyl transferase. Inappropriate milk replacer or the use of antibiotics may be the cause of severe diarrhoea in a neonate.

Clinical examination: The second part of information gathering lies in a physical examination that should be thorough yet rapid and provide a minimum database of information. Immediate, stabilizing care should be provided, and then once the neonate is stable, the rest of the physical examination and any further diagnostic tests can be pursued. The results of complete blood cell counts, serum biochemistry screens and urinalyses must be compared with those from age-matched normal control animals. If radiographs are taken, they should be compared with those from a healthy littermate, because neonatal radiographs can be quite difficult to interpret. Specialized tests include screening for inherited or congenital metabolic defects, and biochemical analyses that detect specific enzyme defects or other protein deficiencies. If multiple neonates in a litter are affected, it may be necessary to elect euthanasia for the most severely ill individual and submit the fresh carcass for post-mortem examination.

Infectious diseases: These tend to progress quite rapidly in neonates, and whilst it may be too late to save the dying animal, microbial culture and sensitivity testing from samples of its tissues may provide valuable information for the rest of the litter. Besides testing the neonate, bitches should be tested for canine brucellosis (in countries where it is present) and for the presence of antibodies against canine herpesvirus. Queens should be tested for feline leukaemia virus, feline immunodeficiency virus, and antibodies against feline coronavirus. Other infectious agents to consider are roundworms, *Neospora caninum*, *Toxoplasma gondii*, canine and feline parvovirus, fleas and ticks. Septicaemia can be caused by group B streptococci, *Escherichia coli* and perhaps other bacteria. Clinical signs in neonates include a swollen, red and/or purulent umbilical stump and/or bright red to purple toes, especially around the area of the nail bed.

Congenital and genetic diseases: Many neonates with congenital or genetic diseases present as fading puppies or kittens or poor doers. Pedigree information and previous post-mortem reports should be reviewed carefully. Biochemical tests can be performed in specialized laboratories to elucidate the underlying cause. Metabolic screening of urine samples allows for the detection of abnormal metabolites, which may be formed in the absence of specific enzymes or protein transport mechanisms. The results obtained from the screening inform the selection of direct enzyme assays, Western blots to detect the absence of specific proteins, and DNA and RNA analyses.

There are several laboratories throughout Europe and the USA that specialize in DNA testing for specific genetic diseases. If the fading neonate should die or be euthanased, any available urine and samples of both the kidney and liver should be frozen immediately. The urine can be submitted to a metabolic screening laboratory. The liver sample should be stored either for further biochemical testing by the metabolic laboratory, or for submission to a toxicology laboratory if there is suspicion of toxicity either through ingestion or absorption. The kidney sample should be stored to enable DNA extraction for analysis of genetic diseases. Karyotyping may also be performed because abnormal numbers and shapes of chromosomes are known to cause neonatal morbidity and mortality.

Sepsis and neonatal death

Neonatal deaths are not uncommon, but they should certainly not be considered normal. Figure 16.3 details the neonatal losses observed in a research colony, averaged over a 10-year period. The most important observations from this data are that the majority of puppy losses are seen during the first few days of life, whilst kitten losses tend to occur around the time of birth and then again around the time of weaning. This underlines the importance of daily weight measurements, especially in kittens.

Clinical signs

Clinical signs in neonates with sepsis consist of hyperaemic (dark red to purplish) mucous membranes and abdominal skin, restlessness or lethargy,

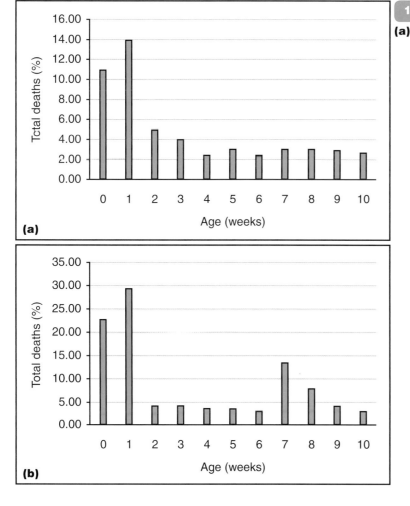

16.3   Neonatal deaths in a research colony, averaged over 10 years. **(a)** Puppies. **(b)** Kittens.

reluctance to suck, aerophagia and dehydration. Often, in the author's experience, the first sign noted is a dark red discoloration around the nail beds. In addition to causing sepsis, bacterial infection may also cause arthritis and meningitis.

Aetiology

Typical organisms implicated include *E. coli*, *Streptococcus* spp. and *Actinomyces pyogenes*, which generally invade through the umbilicus. Other causes of neonatal disease and death in puppies are canine herpesvirus, canine distemper virus, canine parvovirus, *Brucella canis*, *Leptospira* spp. and, more rarely, canine coronavirus. In kittens, feline herpesvirus, feline leukaemia virus, feline coronavirus and *Trichomonas* may be implicated. Coccidia may cause disease, but rarely cause death without secondary infection. Early neonatal infections with canine herpesvirus cause death in puppies born to bitches without colostral antibodies against the virus. On rare occasions, a puppy may survive only to succumb later due to delayed organ failure. Animals that survive may suffer from neurological or cardiac disease. Both brucellosis and leptospirosis are more likely to cause neonatal death in puppies than in livestock.

Diagnosis

Obtaining a diagnosis ante-mortem is difficult and often impossible. Complete blood cell counts and serum biochemistry findings that may indicate impending organ failure and death include increases in haematocrit, decreases in serum glucose and calcium, and increases in blood urea nitrogen (BUN), creatinine, bilirubin, aspartate aminotransferase (AST), alanine aminotransferase (ALT), phosphorus, magnesium and potassium. Any puppies or kittens that have died should be sent for post-mortem examination, and the remaining neonates should be treated symptomatically for the most likely cause of disease until a diagnosis is made.

Treatment

Fluids, dextrose and oxygen should be administered (see Chapter 15). In addition, broad-spectrum antibiotics such as penicillins or cephalosporins should be given systemically if possible. Enrofloxacin should be used with caution as it may cause cartilage damage in puppies. A combination of penicillin G and amikacin is used frequently in neonatal puppies with no apparent side effects. However, the dose and interval must be adjusted to accommodate the immature kidneys. The kidney is surprisingly resistant to the nephrotoxic effects of aminoglycosides in puppies <2 weeks of age.

Diarrhoea

Diarrhoea is among the most common presenting clinical signs observed in neonatal and other paediatric patients. The younger the patient, the more quickly diarrhoea will lead to dehydration and hypoglycaemia. Given that compensatory mechanisms are not fully developed in paediatric patients, the clinical signs need to be dealt with as quickly as possible.

Aetiology

Diarrhoea is generally a multifactorial condition caused by infectious agents, the lack of a mature immune system and often poor management practices. In the youngest of patients, bacteria very quickly colonize the naïve intestinal tract. However, disease can be prevented by colostral IgA, which enhances local immunity; clearly this emphasizes the importance of ingesting colostrum soon after birth.

The most common infectious agents are coronavirus, rotavirus and *E. coli*. These viruses are rarely fatal in the absence of secondary infections. Anecdotally, infection with *Salmonella* spp. is more common in litters born to bitches or queens fed with raw food diets. Other infectious agents include *Clostridium perfringens*, *Toxoplasma gondii*, *Cryptococcus*, *Neospora caninum*, coccidian and roundworms. Less common causes of diarrhoea are metabolic in origin, and of these the most common include portosystemic shunts, vitamin B12 deficiency and exocrine pancreatic insufficiency.

Diagnosis

A thorough history will reveal nutritional deficits or dietary indiscretion. Careful abdominal palpation may reveal the presence of foreign bodies, inflamed intestines, pancreatitis or other abnormalities. A faecal sample should be examined for quality (consistency, colour and smell), viruses (e.g. parvovirus), parasites (e.g. *Toxocara*, *Toxascaris*, *Trichuris vulpis*, *Ancylostoma* spp., *Giardia* spp., *Isospora* spp.) and bacteria (e.g. *Campylobacter*, *Salmonella*, *Clostridia*, bacterial overgrowth). Diagnosis of metabolic and genetic disorders takes longer and requires complete blood cell counts, serum biochemistry screens, urinalyses, specialized assays (such as serum ammonia and bile acids), biopsy samples of relevant organs, ultrasonography, radiography and biochemical enzyme assays.

Treatment

While the specific treatment depends on the inciting cause, dehydration, hypoglycaemia and other clinical signs must be addressed immediately. Fluid replacement, oral or intravenous dextrose, anti-diarrhoeal products and/or antiemetics may need to be administered.

Respiratory distress

Respiratory disease is an important reason for a neonate to be presented to the veterinary surgeon. In kittens, upper respiratory diseases are more common, whilst lower respiratory diseases predominate in puppies.

Clinical signs

Clinical signs of respiratory disease include nasal and ocular discharge, sneezing, coughing, lethargy, weakness and anorexia.

Diagnosis

The vaccination history and origin of the paediatric patient may suggest specific causes for the respiratory disease. A physical examination should include careful auscultation of the lung fields, trachea and heart to allow localization of the most prominent clinical signs. An ocular examination should also be performed if appropriate. In kittens, upper respiratory infections are usually diagnosed on the basis of the presenting clinical signs. For example, copious ocular discharge and conjunctivitis accompanied by a mild nasal discharge suggests feline herpesvirus infection, whereas oral lesions with few or no ocular lesions are more common in calicivirus infections in the absence of secondary infections. If an accurate diagnosis is needed, samples can be submitted for specialized viral cultures (e.g. for distemper in puppies, rhinotracheitis in kittens). Microbiological cultures and antibacterial sensitivity assays should be performed either from nasal swabs, tracheal washes or ocular swabs if primary or secondary bacterial infections are suspected.

Treatment

The treatment may involve antibiotics, fluids and high quality nutrition. In kittens with an upper respiratory tract infection, it may be necessary to suggest creative ways of enticing them to eat (e.g. warming the food, adding fish sauce). After the puppy or kitten has improved clinically, an appropriate vaccination and health programme should be discussed and recommended to the owner.

Stunted growth

A variety of environmental, congenital and genetic disorders may result in stunted growth. A detailed history may reveal inappropriate nutrition, poor environmental conditions or inadequate preventative healthcare. The history should also include that of the parents, the littermates and related animals, and it should be determined whether any animals of the same breed have shown any of the same clinical signs.

A careful physical examination may reveal facial dysmorphia (e.g. growth hormone deficiency, mucopolysaccharidosis, hypothyroidism), short limbs (e.g. achondroplasia) or cataracts (e.g. diabetes mellitus). Comparison with littermates and review of weight charts may reveal the extent and onset of growth retardation. Diagnostic tests may need to be extensive in cases of congenital or genetic diseases. Blood smears may show inclusions in the white blood cells that suggest a lysosomal storage disease. Elevations in serum bile acids suggest a portosystemic shunt, while decreased trypsin-like immunoreactivity may indicate pancreatic insufficiency or vitamin B12 deficiency. A diagnosis is essential to determine the type of treatment and whether therapy is even possible. Congenital diseases such as hypothyroidism need to be treated as soon as possible in neonates to prevent the long-term effects of hormone deprivation.

Hypothyroidism

The lack of thyroid hormones in the neonate leads to much more serious disease than in the adult, because of the involvement of the thyroid gland in development. Clinical signs in affected puppies and kittens may be as mild as apathy and failure to thrive, or as severe as joint and bone abnormalities, complete apathy, extremely stunted growth and constipation (kittens). Serum concentrations of thyroid hormones should be determined at the slightest suspicion of hypothyroidism, because the earlier treatment is initiated the better the outcome. Therapy is performed as in adults, but thyroid hormone concentrations should be checked frequently in young patients and the results compared with normal values for the corresponding age group.

Neonatal isoerythrolysis

Neonatal isoerythrolysis can be an emergency in certain breeds of cats. Three blood groups have been described in the cat: A, B and AB. Cats with blood type A have low titres of naturally occurring antibodies against blood type B red blood cells. Therefore, type B kittens born to type A queens do not show any clinical signs of incompatibility reactions after ingestion of colostrum containing alloantibodies. However, all blood type B cats have high titres of naturally occurring antibodies against type A erythrocytes. This may lead to incompatibility reactions when type A kittens receive colostral antibodies from a type B queen.

Clinical signs of neonatal isoerythrolysis may vary from sudden death to haemoglobinuria, icterus, reluctance to suck or slight depression with no other signs. In the latter case, a positive Coombs' test reveals that an incompatibility reaction has taken place, but with no obvious effect on the viability of the kitten. From studies, it is clear that kittens at risk of neonatal isoerythrolysis must be removed from their dams during the first day of life. These neonatal kittens can be fostered to a type A queen with colostrum or can be hand-reared and given subcutaneous injections of immune serum from a well vaccinated type A cat.

General treatment considerations

Generic treatments that may be employed to support neonates are described in Chapter 15. The purpose of this section is to detail the principles of treatment that influence the therapeutic choices available to the clinician. These principles are important because there are basic differences in absorption, distribution, metabolism and elimination of drugs between neonatal and adult patients.

Drug absorption

Oral absorption of drugs in the neonate is often different from that in the adult due to:

- Slower transit time as a result of delayed/slowed gastric emptying
- Increased volume of mucus within the stomach
- Higher stomach pH.

The overall effect can be complex because the slow transit time may result simply in a delay to reach peak plasma concentration, or to a reduction in peak concentrations. In addition, the mucus can decrease absorption, and the high stomach pH may destroy some drugs, decrease the absorption of others (e.g. acidic drugs such as chloramphenicol) and increase the absorption of others (e.g. basic drugs such as penicillin and ampicillin). There may also be decreased bile production, which in turn reduces the uptake of lipid-soluble drugs.

Drug distribution

The distribution of drugs is quantified by the volume of distribution. For water-soluble drugs, the volume of distribution is increased in the neonate because it has a greater body water content than an adult. For lipid-soluble drugs the volume of distribution is decreased owing to the lower body fat content. There is also decreased binding to plasma proteins and a lower concentration of albumin in neonates. The blood–brain barrier is not complete in very young neonates; therefore, there is increased permeability to lipid-soluble drugs and higher concentrations of drugs that are normally removed by *p*-glycoprotein. This may lead to increased effects of anaesthetics and raised concentrations of avermectins and digoxin in the brain. Thus, these classes of drugs should be avoided.

Drug metabolism and excretion

Metabolism and excretion account for most of the differences in neonatal physiology and thus greatly influence drug disposition. Decreased hepatic and renal function leads to increased or poorly metabolized drugs and may result in toxicity. Examples of drugs that should be used with great care because of incomplete renal function include digoxin, aminoglycosides, angiotensin-converting enzyme (ACE) inhibitors and non-steroidal anti-inflammatory drugs (NSAIDs). Drugs that are metabolized by the liver and should be used with caution in neonates include phenobarbital, phenytoin, theophylline, caffeine and ciclosporin. However, caffeine has been used successfully as a respiratory stimulant in newborns with no apparent side effects.

Antibiotics that are safe in neonates include penicillins and cephalosporins. Tetracyclines should be avoided because of enamel hypoplasia and tooth discoloration, and chloramphenicol because of possible bone marrow toxicity. Interestingly, puppies <2 weeks of age are very resistant to the nephrotoxic effects of aminoglycosides. Due to the increased volume of distribution in neonates, the dose of this class of drugs needs to be increased two-fold to achieve the same concentrations;

however, the drugs are administered only once daily. Macrolides appear to be safe because no adverse effects have been documented. Sulphonamides have not been linked directly to any side effects, but thyroid function may be decreased in some animals.

Phenothiazine tranquillizers should be used with caution because they induce hypothermia and hypotension as a result of vasodilation which, in the neonate, is not compensated for by an increase in heart rate. Midazolam is probably a better choice of drug for this purpose. For analgesia, opiates are very safe in neonates when given at the correct dosage. Buprenorphine is administered at a rate of 0.005– 0.01 mg/kg orally up to four times daily; the injectable form can be given to neonatal kittens orally. Butorphanol is administered at a rate of 0.1– 0.5 mg/kg s.c., i.m. or i.v. q1–4h as needed. NSAIDs can be used, although advice on dosage regimes may be best sought from the manufacturer because in most cases these products are not authorized for this purpose.

It is clear that these principles of treatment are complex and that the 'best choice' agent and the dose to be administered will vary according to the age of the neonate, its physiological state and the particular properties of the therapeutic agent. Most importantly, the clinician must not assume a simple transposition of drug and dosage regime from knowledge of these in the adult.

References and further reading

Boothe DM and Bücheler J (2001) Drug and blood component therapy and neonatal isoerythrolysis. In: *Veterinary Pediatrics: Dogs and Cats from Birth to Six Months*, ed. JD Hoskins, pp. 35–56. WB Saunders, Philadelphia

Casal ML (1995) Feline pediatrics. *Veterinary Annual* **35**, 210–235

Lavely JA (2006) Pediatric neurology of the dog and cat. *Veterinary Clinics of North America: Small Animal Practice* **36**, 475–501

Lawler DF (1995) The role of perinatal care in development. *Seminars in Veterinary Medicine and Surgery (Small Animal)* **10**, 59–67

Macintire DK (1999) Pediatric intensive care. *Veterinary Clinics of North America: Small Animal Practice* **29**, 971–88, vii–viii

Monson WJ (1987) Orphan rearing of puppies and kittens. *Veterinary Clinics of North America: Small Animal Practice* **17**, 567–576

Moon PF, Erb HN, Ludders JW, Gleed RD and Pascoe PJ (2000) Perioperative risk factors for puppies delivered by cesarean section in the United States and Canada. *Journal of the American Animal Hospital Association* **36**, 359–368

Moon PF, Massat BJ and Pascoe PJ (2001) Clinical theriogenology: neonatal critical care. *Veterinary Clinics of North America: Small Animal Practice* **31**, 343–367

Traas AM (2008) Resuscitation of canine and feline neonates. *Theriogenology* **70**, 343–348

Traas AM, Abbott BL, French A and Giger U (2008) Congenital thyroid hypoplasia and seizures in two littermate kittens. *Journal of Veterinary Internal Medicine* **22**, 1427–1431

Veronesi MC, Panzani S, Faustini M and Rota A (2009) An Apgar scoring system for routine assessment of newborn puppy viability and short-term survival prognosis. *Theriogenology* **72**, 401–407

17

Clinical approach to mammary gland disease

Josep Arus Marti and Sonia Fernandez

Introduction

The mammary glands are modified skin glands, which are commonly affected by a variety of pathological conditions. The mammary glands are functionally related to hormone secretion from the ovaries, such that there may be an effect upon the mammary gland when there is ovarian pathology.

Normal anatomy and physiology

- The bitch usually has five pairs of mammary glands on each side of the ventral thorax and abdomen, although it is not unusual to see four or six pairs.
- The queen usually has four pairs of mammary glands.
- By convention the glands are numbered from cranial to caudal (M1 to M4/M5).

The normal function of the mammary glands is to provide nutrition for the neonates, and although they are mainly exocrine glands, they also have endocrine and paracrine functions. Each gland drains to the outside through a nipple. The parenchyma of the gland is formed at the base by numerous glandular acini or alveoli grouped within a fibrous capsule. The stroma is composed of adipose tissue, muscle cells, blood vessels, lymph vessels and nerve endings, whilst the acini themselves are composed of secretory cells grouped in clusters and surrounded by myoepithelial cells. Contraction of the myoepithelial cells causes the expulsion of milk into the collecting system. From there the milk flows into the lactiferous duct (which can increase in size to accommodate the milk that is secreted) and then into the gland sinus, which terminates at the nipple orifice.

The vascular supply to the mammary glands is complex. The arterial supply to glands M1 and M2 is provided by the internal thoracic arteries and lateral cutaneous branches of the intercostal arteries. Gland M3 is supplied by the cranial epigastric artery, and glands M4 and M5 by the caudal epigastric artery and the external pudendal artery. Venous return parallels the arterial supply and is also symmetrical. Vascular patterns may differ according to the size of the animal, any

mammary disease and species. The lymphatic drainage of M1, M2 and M3 is provided by the ipsilateral axillary lymph nodes, whereas M4 and M5 drain to the ipsilateral superficial inguinal lymph nodes. There is some lymphatic communication between M1–M2 and M3–M4, which varies among individuals depending on size, and also between dogs and cats.

Mammary gland function can be divided into three distinct physiological phases:

- Mammary gland development
- Lactation
- Galactopoiesis.

Mammary gland development

Mammary gland development begins with the growth of epithelial tissue from the embryonic mammary ridge. The gland continues its development until puberty, when the first hormonal stimulus occurs.

Lactation

Lactation is defined as the onset of the synthesis and secretion of milk. The onset of lactation coincides with the hormonal changes that occur during pregnancy and parturition. However, this is promoted initially by oestrogen at the first and subsequent oestrous cycles, which stimulates canicular development. This is followed by a subsequent maturation of the alveolar–lobe system, which is stimulated by progesterone secreted during the luteal phase. The decrease in progesterone that occurs towards the end of pregnancy has a negative effect on the production of prolactin inhibitory factors (PIF) from the pituitary gland, such that as progesterone declines there is a rise in the plasma concentration of prolactin. Prolactin, a peptide hormone produced in the anterior pituitary gland, along with many other factors (defined as the 'lactogenic complex') is then responsible for the initiation of lactation (Figure 17.1). The regulation of prolactin depends primarily on the PIF (dopamine) and the prolactin-releasing factors (PRF), which include serotonin, melatonin, histamine, thyrotrophin-releasing hormone (TRH) and endorphins (Figure 17.2). At physiological concentrations oestrogens stimulate the secretion of prolactin, and at therapeutic concentrations they inhibit lactation.

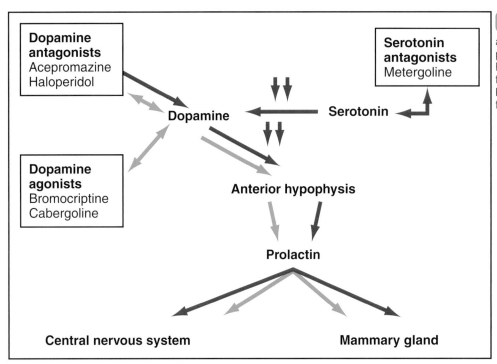

17.1 Prolactin regulation and drugs that affect prolactin production. Prolactin inhibitory factors = red arrows; Prolactin-releasing factors = blue arrows.

Factor	Type of action	Releasing factors	Inhibitory factors
Central	Direct	Thyrotrophin-releasing hormone Gamma-aminobutyric acid Oxytocin Others	Dopamine Somatostatin Others
	Indirect	Serotonin	Opioids
Peripheral	Direct	Oestrogens	Progestogens
	Indirect		Thyroxine

17.2 Prolactin inhibitory and releasing factors.

Galactopoiesis

Galactopoiesis is defined as the phase during which the mammary glands maintain lactation. Milk production and the ejection reflex are controlled by a neuroendocrine pathway, in which the afferent nerve (activated by nipple stimulation or conditioned stimuli such as the smells and sounds of neonates) triggers the release of oxytocin from the posterior lobe of the pituitary gland. Oxytocin, itself the efferent pathway, activates the excretion of milk by stimulating the contraction of the myoepithelial cells. This stimulation also promotes the neuroendocrine release of other hormones that are part of the 'lactogenic complex', such as thyroid-stimulating hormone (TSH), growth hormone (GH), adrenocorticotrophic hormone (ACTH) and prolactin, which are produced by the anterior pituitary gland and hypothalamus. The pulsatile release of oxytocin is rapid (30–40 seconds), although its effect is longer but still brief (approximately 5 minutes). Any physical or psychological stress can inhibit the neuroendocrine reflex by releasing adrenaline, thus blocking the release of oxytocin. This is an important consideration in bitches that are stressed (e.g. primiparous) or in pain (e.g. following a Caesarean operation).

Clinical approach

History and physical examination

Along with a general physical examination and history (Figure 17.3), inspection of the mammary glands is an important aspect of a clinical reproductive examination. Early detection of any pathology, such as small nodules, inflammation and abnormal secretions, increases the chances of successful

When did the problem start?
What has been the progression of the problem?
What is the reproductive history of the bitch/queen?
Oestrous period duration Pseudocyesis: frequency and duration Previous parturitions Lactation and milk production Neuter status
What hormonal treatments have been used?
Type of product How often have they been applied/administered?

17.3 Relevant clinical history relating to mammary gland disease.

treatment. Palpation of the regional lymph nodes and the area surrounding the mammary glands should also be performed as part of the examination. If any pathology is suspected further diagnostic tests should be undertaken.

Cytology

Examination of cells from any secretion of the gland or a fine-needle aspirate of the nipple or any nodules/tumours is possible. This may reveal bacteria, neoplastic cells or inflammatory cells. Due to the compartmentalized nature of the mammary gland, the absence of these cells does not automatically exclude the suspected diagnosis and further tests may be necessary.

Bacteriological culture

Material may be collected by stripping the mammary gland or, in the case of mastitis, by needle aspiration. Performing an antibiogram is always important in cases of mastitis. In addition, the veterinary surgeon should always consider the potential effects of the antibiotic on any puppies or kittens that are still nursing.

Biopsy

Surgical biopsy (often complete excision) is without doubt the diagnostic technique of choice in cases of suspected neoplasia, although aspiration cytology may provide limited information. It is the normal practice of the authors to remove local lymph nodes if possible and submit these for histopathological examination.

Imaging

Thoracic radiography

Radiographs of the left and right lung fields are mandatory when there is suspicion of mammary gland neoplasia. It is recommended that three radiographic views are taken (left lateral, right lateral and ventrodorsal). It is important to recognize the potential differences between the bitch and queen in terms of the radiographic signs of thoracic metastases (Figure 17.4).

Ultrasonography

Ultrasound examination of mammary gland disease may be useful because it can demonstrate the extent and margin of any abnormality. Some authors believe that on the basis of the ultrasonographic appearance, it may be possible to differentiate benign from malignant disease, although there is only anecdotal information at present.

Other tests

Other diagnostic tests such as thermography, tumour markers, determination of tumour hormonal receptors for progesterone and oestrogens and possibly genetic analysis may become useful in the future as further investigations in these areas are undertaken in animal species.

Mammary gland diseases

Lactation problems

Agalactia

The term agalactia refers to the absence of milk in a female that should be lactating. Agalactia can be classified as either true agalactia or temporary agalactia. True agalactia is defined as the persistent and almost total absence of milk production. It is rare and its aetiology is poorly understood. The suspected aetiologies and predisposing factors include physical malformations, an unfavourable environment and malnutrition; these effects may be related to neurogenic inhibition of prolactin secretion. Agalactia has also been associated with the presence of uterine infections and fetal retention, and these again may possibly be related to a general malaise of the female.

Temporary agalactia usually occurs in primiparous females, or those who have undergone a Caesarean operation perhaps slightly prematurely such that there is a lack of synchronization between the appearance of milk and the birth. This necessitates artificial feeding of neonates for the first few days postpartum. The treatment of both types of agalactia is difficult and not always effective. In the authors' clinics, whatever the cause of agalactia,

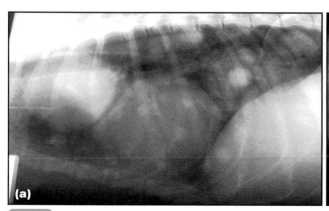
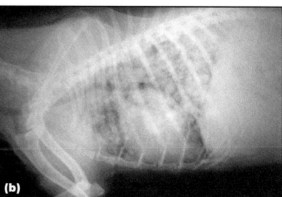

17.4 Radiographic appearance of metastatic pulmonary disease in **(a)** a bitch and **(b)** a queen.

attempts are made to stimulate prolactin secretion indirectly using drugs (Figure 17.5). At the same time it is important to encourage the suckling reflex and the release of oxytocin by placing the puppies or kittens with the mother, even if they are fed artificially.

Drug	Dose	Comments
Metoclopramide	0.1–0.2 mg/ kg s.c. q6–8h	Recommended therapy; initially s.c. or orally until milk production

17.5 Drug treatment for agalactia.

Absence of milk ejection

An absence of milk ejection must be differentiated from agalactia. In the former, there is milk production but a failure to eject it from the mammary glands. This may occur because the mammary glands are congested (preventing milk from passing through the glands), or because of an inhibition of milk letdown caused by adrenaline produced in response to pain or stress. Treatment consists of the administration of oxytocin (2–5 IU every 6 hours or immediately before feeding the litter). Sedatives (diazepam, anxiolytics) may be of help for very nervous bitches, and analgesia is an important consideration in bitches that have recently had a Caesarean operation. Careful examination of these cases is important because inadequate treatment may lead to mastitis.

Excessive milk production

Excessive milk production occurs in a small number of bitches at the peak of lactation, and there may be an excessive fat content within the milk. The common sequel to this is an osmotic diarrhoea in the puppies. The condition can be diagnosed by staining the faeces of the puppies with Sudan III, which demonstrates the excessive fat content. Treatment is by artificial rearing or weaning of the puppies, depending upon their age.

Mastitis

Mastitis (mammary inflammation or mammitis) is defined as inflammation with or without infection of one or more mammary glands, and usually occurs during the postpartum period or in a pseudopregnant female. There are several forms:

- Acute mastitis
- Gangrenous mastitis
- Chronic mastitis
- Subclinical mastitis.

The most frequent cause is ascending bacterial infection from the nipple, although haematogenous spread has been described more rarely. The main pathogens isolated in these cases are *Escherichia coli*, staphylococci and streptococci. Suction trauma of the nipple by the puppies or kittens and poor hygiene can contribute to the aetiology.

Acute mastitis: In acute mastitis, one of the caudal mammary glands is usually hot and painful on palpation, and swollen and reddened in appearance (Figure 17.6a). The female may have a fever and be lethargic. Death of one of the puppies or kittens ('toxic milk syndrome') may also occur. The secretion from the gland can be brownish, purulent or haemorrhagic, but in some rare cases the secretion may appear normal. It is not uncommon for acute mastitis to be associated with uterine infection, although the reason for this is unclear.

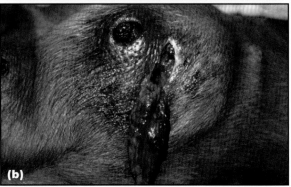

17.6 **(a)** Acute mastitis. The mammary gland is reddish and swollen; the secretion is haemorrhagic. **(b)** Gangrenous mastitis. The abscess has ruptured and necrosis is evident.

Gangrenous mastitis: As a consequence and often as a progression of severe acute mastitis, there may be production of abscesses and subsequent necrosis within one or more mammary glands (Figure 17.6b). The affected glands often appear darkened, cold and/or ulcerated. The female usually has systemic signs of septicaemia.

Chronic and subclinical mastitis: There has been limited investigation of both the incidence and the significance of chronic mastitis in the bitch and queen. Chronic infections may be an incidental finding in older queens. The affected glands often show minimal inflammatory changes, or may appear swollen with palpable nodules that resemble a neoplasm. Subclinical mastitis should be suspected when there is an increase in neonatal mortality or when the litter does not gain weight properly.

Diagnosis: The diagnosis of inflammatory disease of the mammary gland may be supported by haematological changes, cytological investigation of fluid from the affected gland, and changes in the pH of the milk.

- Haematology tends to reveal a neutrophilic leucocytosis in the case of acute mastitis.

- Cytology of the milk reveals a large number of degenerate neutrophils (>3000/ml) with ingested bacteria and macrophages.
- Evaluation of the milk pH will help in selecting the most appropriate antibiotic.

Treatment: The treatment of mastitis is based on the use of broad-spectrum antibiotics (as determined by the antibiogram) and should always take into consideration the pH of the milk.

- If the milk has a lower pH than the plasma (bitch <7.3; queen <7.2), antibiotics such as trimethoprim/sulfadiazine (15–30 mg/kg orally q12h for 21 days), erythromycin (10 mg/kg orally q8h for 21 days) or lincomycin (15 mg/kg orally q8h for 21 days) should be used.
- If the pH of the milk is >7.4, antibiotics such as ampicillin (20 mg/kg i.m. q8h for 21 days) or cefalexin (30 mg/kg orally q12h for 21 days) should be used.
- Some antibiotics, such as chloramphenicol, doxycycline and tetracycline, reach the milk in reasonable concentrations regardless of pH. However, their use is not recommended because of the side effects on the puppies or kittens (e.g. staining the tooth enamel).

- In cases of acute mastitis, the choice of antibiotic is not so important because the barrier that separates the plasma from the milk has already been broken.
- Adjuvant therapies include manual emptying of the milk to avoid accumulation within the mammary gland.
- Whether or not to remove the neonates from the dam is a controversial issue. It is generally recommended to continue natural feeding, except in the presence of abscesses or gangrenous mastitis because the ingestion of infected milk or milk containing antibiotics creates problems in the litter. However, milk composition is always poorer in patients with mastitis and this affects the weight gain of the offspring.
- In cases of gangrenous mastitis, effective surgical drainage is required in addition to antibiotic therapy.
- Mastectomy is a last choice treatment necessary in some cases.

Galactostasis

Galactostasis is defined as an excessive accumulation of milk along with a lack of milk let-down. In the early phase of the disease the animal is systemically well; however, the condition may progress and result in a sterile or septic mastitis. Affected females are often those with significant milk production during false pregnancy, or those that are subjected to rapid weaning, particularly of a large litter. Physical examination usually demonstrates swollen glands, which are inflamed and oedematous. In queens it appears that the cranial glands are more commonly affected than the caudal glands.

Cytology of the milk usually reveals the presence of an increased number of non-degenerated neutrophils, macrophages and polymorphonuclear leucocyte elastase (PMNE). Macrophages are often observed phagocytozing fat droplets. The treatment is aimed at reducing the production of milk in the affected glands by using cold compresses for 10–15 minutes several times per day. Administration of diuretics and gluco-corticoids for up to 2–5 days may also be useful for symptomatic treatment. The use of prolactin-inhibitor drugs should be reserved for females in pseudopregnancy or in those cases where the young have already been weaned. Massage of the glands is not recommended because emptying of the mammary gland can stimulate milk production.

Galactorrhoea

Galactorrhoea is defined as the development of the mammary gland and the production and excretion of excessive and inappropriate milk. The most common cause is a sudden decrease in the plasma concentration of progesterone, and the condition often occurs in cases of pseudopregnancy (pseudocyesis). This condition may also be observed in females that are surgically neutered during dioestrus, where prostaglandin is used to cause regression of the corpora lutea, and in cases from which progesterone therapy is withdrawn suddenly. The condition is usually diagnosed via a physical examination and medical history. Treatment is usually unnecessary because within a few days the condition resolves. Symptomatic treatments include the withdrawal of food for 24–48 hours. Specific treatment using prolactin inhibitor drugs is normally curative in long-standing cases.

Problems not related to lactation

Pseudopregnancy

Pseudopregnancy (pseudocyesis/false pregnancy) is a normal physiological event during which there may be some uncomfortable physical, psychological or behavioural manifestations in a proportion of bitches. Importantly, there is also some evidence that repetitive pseudopregnancy may increase the risk of subsequent mammary gland neoplasia.

The events of pseudopregnancy are interesting because the bitch is unique physiologically among mammalian species. All bitches have a luteal phase (dioestrus), in which progesterone is dominant, for 2 months following ovulation regardless of whether or not they are pregnant. In general, there are no differences between pregnant and non-pregnant bitches regarding plasma progesterone concentrations. For this reason all non-pregnant bitches can be considered to have pseudopregnancy; although, clearly there are significant differences among bitches in the manifestation of any clinical signs. Some authors refer to two groups that show 'overt' or 'covert' pseudopregnancy. It appears from a small number of studies that the difference between these two groups relates to the actual plasma concentrations of prolactin; bitches with pseudopregnancy appear to have prolactin concentrations higher than bitches at the same stage of the cycle that do not have pseudopregnancy (Gobello *et al.*, 2001).

Pseudopregnancy is rare in queens because they are induced ovulators and the non-pregnant luteal phase only follows sterile mating or, rarely, spontaneous ovulation. Most commonly, the clinical signs of pseudopregnancy in the queen include only absence of oestrus, rather than significant mammary gland enlargement or galactorrhoea.

Clinical signs: Clinical signs are most commonly observed in the bitch 6–8 weeks after the end of oestrus, when there is enlargement of the mammary glands, milk production, changes in behaviour (lethargy, aggressiveness) and occasionally signs typical of parturition and lactation, including anorexia and nursing of inanimate objects. Where there is lactation the milk secreted may be normal, or it may be more watery or brownish in colour. In some cases a secondary mastitis develops as a result of the false pregnancy. The intensity of these clinical signs varies greatly among animals and may depend upon environmental factors, the relationship of the animal with its owners, the level of physical activity and food intake. The most important differential diagnosis is, of course, a real pregnancy.

Treatment: This will only be necessary in those females that have marked behavioural changes or excessive secretion of milk, or where mastitis develops. In the past, common treatments have included oestrogen (diethylstilboestrol) and progestogens (proligestone, medroxyprogesterone acetate), but because of the risk of adverse effects and the development of prolactin inhibitor drugs, these protocols are now not commonly used. Currently, drugs derived from ergotamine (anti-prolactinics) are the products of choice for use in females with pseudopregnancy.

- Cabergoline is widely used in the UK. It is a dopamine agonist and as such inhibits the release of prolactin. The molecule is the latest generation drug that has come to market. Cabergoline is normally used at a dose of 5 µg/kg/day for 6 days. Side effects are rare and consist of vomiting and anorexia.
- Bromocriptine is also a dopamine agonist that is available as a human medicinal product. It may be used in dogs at a dose of 20 µg/kg/day for 10 days. Bromocriptine has significant adverse effects including nausea, vomiting and sometimes diarrhoea. Often the drug can only be administered in conjunction with antiemetics and commonly the dose is increased gradually from 5 µg/kg/day to 20 µg/kg/day to ensure tolerance of the adverse effects. Administration of the oral tablet to small dogs can be difficult because the tablets cannot be broken accurately into small enough pieces.
- Metergoline is available to veterinary surgeons in some countries. It is an anti-serotonin compound, which is administered at a dose of 0.2 mg/kg/day for 8 days. Adverse effects include excitement, aggressiveness and vocalization.

Other adjuvant therapies that can help in cases of pseudopregnancy include withdrawal of food for 24–48 hours, increasing the level of exercise, and the application of local cold compresses. The use of anxiolytics (e.g. selegiline) or tranquillizers (diazepam) can sometimes be useful. The phenothiazine tranquillizers (e.g. acepromazine) or butyrophenones (e.g. haloperidol) are generally considered to be contraindicated because of their anti-dopaminergic activity. The best method for control of pseudopregnancy and to reduce the risk of mammary gland neoplasia is to undertake ovariectomy or ovariohysterectomy. This is normally performed after the signs of pseudopregnancy have ceased.

Physical abnormalities

In non-breeding animals abnormalities of the mammary gland may not be significant or may go unnoticed unless there is inappropriate milk production which cannot escape the gland, resulting in mastitis. Physical or anatomical abnormalities of the mammary gland are usually most apparent in the breeding animal. Careful examination prior to breeding to establish the number of nipples, or to detect abnormalities of the mammary gland and nipple, is important to ensure that the resultant litter can be supported; otherwise supplementary feeding of the neonates will be required. In addition, some abnormalities may also produce dermatological signs or aesthetic conditions that will result in a request for cosmetic surgery.

It is noteworthy that sometimes small nipples are found more cranially or caudally than normal. Cranially positioned nipples usually become non-functional and cause no problems to the female. However, caudally positioned nipples may become functional if the litter sucks from them; in this case, it is important to keep the nipple clean and to ensure that it is not damaged during sucking.

Inverted nipples: Inverted nipples tend to be pruritic and frequently there is a waxy discharge. Affected nipples should be carefully everted to be cleaned, and if necessary a topical ointment containing an antibiotic and/or anti-inflammatory agent should be applied if pruritus or dermatitis is intense.

Nipple hyperplasia: Nipple hyperplasia (Figure 17.7) may be congenital, or acquired as a result of chronic irritation of the nipple. In congenital cases the functionality of the nipple should be examined carefully and a decision made about whether to remove it surgically if the animal is going to be used for breeding. In cases of acquired hyperplasia some response may occur to application of topical creams to protect the nipple and prevent irritation. The extent of hyperplasia will determine whether the nipple remains a suitable size to allow neonates to feed. Nipples that are very large, or where there is a significant cosmetic effect, may be surgically removed with the underlying gland.

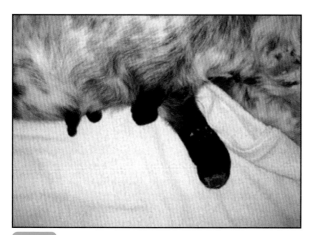

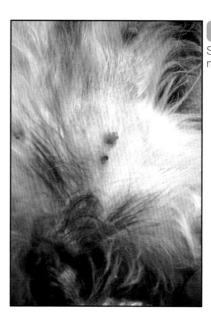

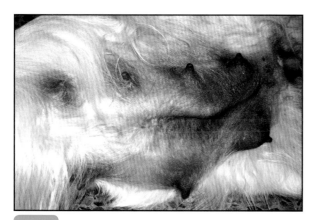

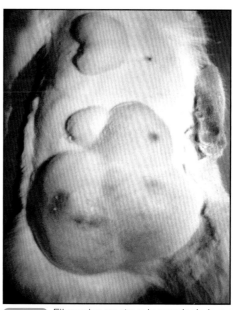

17.7 Extensive nipple hyperplasia in a bitch.

Mammary gland asymmetry: This is not uncommon and produces no adverse effects unless the size of the litter is greater than the number of mammary glands (Figure 17.8). In these cases supplementary feeding of the litter will be necessary, or those puppies or kittens that are not sucking should be identified and fed artificially. (NB A hierarchy of puppies that suck from each nipple is often established; if there are not enough nipples some puppies will be excluded.)

17.8 Canine mammary gland asymmetry.

Supernumerary nipples: Supernumerary nipples are also known as accessory nipples (Figure 17.9). They are minor malformations that consist of nipples in addition to the ones corresponding to the normal mammary glands. They may be complete and functional, partially functional or non-functional. They tend to be of smaller size than the 'original' nipple.

Other nipple abnormalities: A few other nipple abnormalities are seen uncommonly in both dogs and cats. These include nipple hypoplasia, unusually shaped nipples and trauma to the nipple. The management of these cases should follow the broad guidelines above and the particular clinical circumstances of the individual.

Proliferative conditions

Proliferative disease of the mammary gland may be either neoplastic or non-neoplastic in nature. The

17.9 Supernumerary nipple.

following section deals with common conditions arising from both groups.

Mammary fibroadenomatous hyperplasia

Mammary fibroadenomatous hyperplasia (also known as mammary hypertrophy, juvenile mammary hypertrophy, fibroepithelial hyperplasia) is an important non-neoplastic proliferative disorder of the mammary gland (Figure 17.10).

17.10 Fibroadenomatous hyperplasia in a queen after progestogen therapy.

The condition is characterized by a rapid, benign and uniform growth of the mammary stroma and duct epithelium, affecting one or more glands. It mainly affects young queens and bitches after their first oestrus, but can affect all ages and both sexes. It may also be observed after gestation or pseudopregnancy, following ovariohysterectomy (up to 1 year after surgery), in young cycling cats or in the adult queen or tom cat receiving progesterone treatment (to control oestrus or to prevent urine spraying in the case of the male), and it has been described even in neutered adult cats.

Mammary hyperplasia appears to be related to progesterone; some authors suggest that the condition is a result of a hypersensitivity to physiological concentrations of endogenous progesterone or administration of exogenous progestogen or progesterone (Burstyn, 2010). Interestingly, the affected tissue has a high number of progesterone receptors and a low number of oestrogen receptors. This observation fits with the most common choice of treatment, which is depot administration of the progesterone receptor antagonist aglepristone.

Grossly, lesions of fibroadenomatous hyperplasia often contain multiple fluid-filled cysts of different sizes that can be easily palpated. If drained by needle aspiration these cysts refill rapidly. Histologically, two types of condition have been identified within the classification of fibroadenomatous hyperplasia:

- Lobular hyperplasia (also called intraductal papillar hyperplasia) appears to be a result of proliferation of the mammary gland duct epithelium, and is typically observed in animals given exogenous progestogen therapy
- Fibroepithelial hyperplasia (also called diffuse fibroepithelial hyperplasia) is manifest as general enlargement of the mammary glands, and is typically seen in queens after oestrus.

Clinical signs: The clinical signs at the onset of the disease may vary from just one simple cyst in one mammary gland to multiple enlarged, uniform, firm, non-painful masses, with the animal showing significant discomfort. The condition may evolve quickly, resulting in a very painful, reddish lesion, sometimes with skin ulceration with or without discharge, and the animal then becomes anorexic and lethargic and may be pyrexic. Generally, the condition in the bitch is not as common nor as severe as it is in the queen. In the bitch the condition may resolve spontaneously following cessation of progestogen administration. In the cat it is often a severe condition that may end in death as a result of acute ischaemic necrosis or thrombosis.

Treatment: The recommended treatment protocol includes:

- Withdrawal of progestogen/progesterone treatment (if any)
- Administration of aglepristone at a rate of 10–15 mg/kg s.c. on days 1, 2 and 8
- Administration of cabergoline at a rate of 5 µg/kg orally for 5–7 days or bromocriptine at a rate of 0.25 mg/cat orally for 5–7 days. (NB Aglepristone and cabergoline should be used at the same time)
- Administration of broad-spectrum antibiotics if ulceration or infection present
- Administration of non-steroidal anti-inflammatory drugs (NSAIDs) to help diminish inflammation and fever, and to control pain

- When the animal is stable ovariohysterectomy should be performed. This may be especially important in queens that are severely affected. In these cases, it is recommended to perform a lateral incision to avoid damage to the swollen tissue
- A mastectomy should be undertaken in those cases where medical treatment has failed or when the size of the mammary gland is large or the degree of ulceration is significant
- The administration of diuretics or corticosteroids, or the application of cold towels, should be considered to help manage the clinical signs.

Mammary gland neoplasia

Mammary gland neoplasia is the second most frequent type of tumour seen in bitches, after skin tumours, with an incidence amounting to 25–50% of all tumours. Approximately 50% of mammary gland tumours are benign and easily treated by surgery alone. In the queen, mammary gland neoplasia represents the third most frequent cancer after lymphosarcoma and skin tumours. Special attention must be paid to those queens with very painful glands, as they may develop serious tachycardia or cardiac arrhythmias. Although unlicensed, the authors have successfully used the beta receptor antagonist, atenolol, in the management of these cases. The incidence in the queen is almost half that seen in the bitch; however, a large proportion (almost 80%) are adenocarcinomas.

Although there is no known genetic predisposition, there is a higher incidence of mammary gland neoplasia in poodles, terriers and some spaniel breeds, as well as in the Siamese cat. The average age at presentation is 10–12 years, with a noticeable increase in incidence from 6 years of age onwards. Males can also be affected, with very low incidence, and are usually affected by tumours with malignant characteristics. Interestingly, the caudal mammary glands are more frequently affected, probably owing to their large size. It is important to remember that there are lymphatic connections between *all* the mammary glands, either in the form of lymphatic vessels or a plexus of vessels. These connections involve the right and left glands as well as adjacent glands on the same side.

In the bitch there are a number of factors that predispose to mammary gland neoplasia:

- Previous treatments for oestrus control: there is a strong relationship between the use of progestogens and mammary gland tumours, although not uncommonly the lesions induced are benign
- Ovariohysterectomy/ovariectomy: early removal of the ovaries plays an important protective role against mammary gland disease. Intact females or females that are neutered later than 2.5 years of age have a higher incidence of disease, whereas females neutered prior to their first oestrus have an incidence of neoplasia of approximately 0.05%. The incidence appears to increase to 8% if neutered after the first oestrus

and to 24% if neutered after two or more oestrous cycles

- Hormonal receptors: both oestrogen and progesterone receptors have been identified in canine mammary gland tumours. Benign tumours generally have a high number of receptors, whereas malignant tumours, especially those that are poorly differentiated, have a low number of progesterone receptors. It has also been demonstrated that there is a higher concentration of aromatase in mammary gland neoplasia than in normal mammary gland tissue. These facts may play an important role in the future management of this cancer
- Obesity: early obesity is a clear risk factor for mammary gland tumours and may also influence survival time after surgery
- Pseudopregnancy: it has been suggested that pseudopregnancy may predispose to mammary gland neoplasia, although this relationship has not been confirmed in all epidemiological studies.

One of the key factors in the investigation of mammary gland neoplasia is the likely prognosis. In general, the clinical signs that indicate a poor prognosis (i.e. either recurrence of the tumour or malignant behaviour of the tumour) include:

- Rapid growth of the neoplasm
- Infiltration of the nearby tissues or poor delineation of the tumour (Figure 17.11a)
- Inflammatory changes (lymphoedema of the limbs or vulva) or ulceration of the tissues
- Metastasis to regional lymph nodes or to target organs such as lungs (Figure 17.11b), liver or bones
- Size of tumour: in dogs, a tumour <3 cm in diameter is associated with a better prognosis. In cats, a tumour with a diameter >0.75 cm and involvement of a large number of mammary glands indicates a high likelihood of local metastasis
- The earlier the detection and management of the tumour, the better the prognosis.

Diagnosis: The first and most important step when investigating suspected mammary gland neoplasia is palpation, not only of the lump or nodule, but of all the glands and the regional lymph nodes. To make a differential diagnosis it is necessary to gather information about previous progestogen treatments, episodes of oestrus and pseudopregnancy, and the age at which neutering was undertaken. The differential diagnoses include mammary gland hyperplasia, mammary gland cysts, galactostasis, mastitis and extramammary pathology, such as mast cell tumours, lipomas, hernias and foreign body reactions.

The recommended diagnostic steps used to investigate a mass in a mammary gland are as follows:

- Complete physical examination, palpation and collection of a clinical history
- Thoracic radiography, preferably using three radiographic views. It should be noted that a normal radiological appearance does not exclude the presence of micrometastases

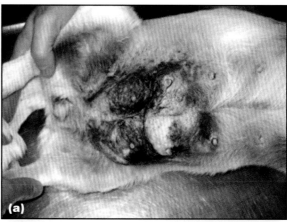
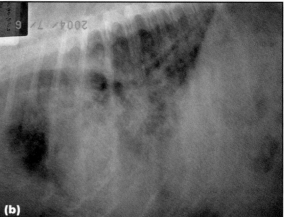

17.11 (a) Clinical appearance of an inflammatory carcinoma. (b) Radiographic findings of pulmonary metastasis.

- Ultrasound examination of the mass to help eliminate other possible diagnoses, such as hernias or cysts, and to evaluate metastases in target organs or enlargement of iliac lymph nodes
- Blood analysis to evaluate liver and kidney function, blood calcium concentration and to count white and red blood cells
- Fine-needle aspiration may diagnose the nature of the neoplasia or indicate the presence of inflammatory cells. It should be noted that the absence of malignant cells does not exclude the possibility of neoplasia
- Biopsy may be indicated if there is doubt over the nature of the mass and there is particular interest in preserving all the glands/nipples for breeding purposes. However, in most cases a mastectomy is performed
- Complementary studies such as electrocardiography may be used, especially prior to chemotherapy treatments and surgery.

All the tissues removed (including suspected nodules or complete mammary glands, together with the regional lymph nodes) must be sent for histopathological examination.

Treatment: There are a number of treatment options in cases of suspected or confirmed mammary gland neoplasia but the basic treatment is always surgical excision.

Surgical treatment: Before embarking upon surgical treatment it is imperative that the surgeon has a good understanding of the normal anatomy of the mammary glands, lymph nodes, blood vessels and also the number of glands (usually five in each chain in the bitch, four in the queen). The aggressiveness of the planned surgery may vary in relation to the type of lesion and the number of mammary glands affected. Interestingly, however, no scientific study supports the theory that radical surgery will produce significantly longer survival times than more simple approaches. The surgical options include:

- Nodulectomy: this is best planned only for small, encapsulated, non-invasive nodules
- Simple mastectomy: excision of the entire gland (gland, nipple and skin)
- Block mastectomy: excision of the whole gland and the regional lymphatic node. It should be noted that the axillary lymph node may be difficult to excise, but removal of the inguinal lymph node is essential if a caudal mammary gland is affected. If the tumour is in M4 or M5, ultrasound examination of the iliac lymph node should be undertaken and, if enlarged, this should be subjected to cytological examination or excision
- Medium chain mastectomy: excision of the whole affected gland and the glands related to it, together with the regional lymph nodes
- Total or radical mastectomy: this may be unilateral or bilateral. If bilateral surgery is planned it is recommended that two separate surgeries are performed (the first to undertake a unilateral mastectomy and ovariohysterectomy and, once this is healed, a second surgery for mastectomy of the other chain).

It is usually recommended that an ovariohysterectomy is performed simultaneously with any mastectomy because, even though its protective role against mammary gland neoplasia may be limited (if performed after the second oestrus), ovariohysterectomy will help to prevent future mammary gland dysplasia and benign tumours; it will also reduce the progesterone concentration. At the time of surgery, ovariohysterectomy is normally performed prior to mastectomy to avoid spreading neoplastic cells into the abdomen. It is also best to use a separate incision for the ovariohysterectomy and the mastectomy. The use of cabergoline, a prolactin inhibitor, prior to mammary gland surgery will help to eliminate any existing milk and diminish the size of the mammary gland tissue. It will therefore allow better definition of the affected tissue and decrease the chances of contamination of the surgical area by milk.

Chemotherapeutic treatment: When malignancy of the tumour or metastatic potential is suspected, or with tumours that are not resectable (e.g. inflammatory carcinoma), chemotherapy may be contemplated (Figure 17.12). Combined therapy with more than one chemotherapeutic agent is more efficient than single agent therapy.

Agent	Dose
Doxorubicin	Dogs (<15 kg) or cats: 25 mg/m^2 or 1 mg/kg i.v. Dogs (>15 kg): 30 mg/m^2 i.v. Every 3 weeks, 4–6 cycles
FAC protocol (only for dogs)	Doxorubicin (same dose as above) day 1 Cyclophosphamide (100 mg/m^2 i.v. or orally) day 1 5-Fluorouracil (150 mg/m^2 i.m.) days 1 and 15 Every 3 weeks, maximum of 6 cycles
Mitoxantrone	5.5 mg/m^2 i.v. Every 3 weeks, 4–6 cycles
Doxorubicin + Cyclophosphamide	Doxorubicin (same dose as above) Cyclophosphamide (200 mg/m^2 i.v.) divided on days 4 to 7 of the first week of each treatment cycle Cyclophosphamide (cats: 25 mg/m^2 orally) divided on days 2 and 4 of the first week
Other chemotherapeutics accepted	Carboplatin Cisplatin Paclitaxel

17.12 Possible chemotherapy protocols for mammary gland tumours.

Chemotherapeutic agents

The basic properties and methods of use of some of the different chemotherapeutic agents are given below.

Doxorubicin

- The most commonly used agent.
- May be nephrotoxic (cats) and cardiotoxic (dogs); an effect that is dose-limiting.
- Renal and cardiac function must be evaluated prior to every dose.
- If cardiomyopathy is present, epirubicin or mitoxantrone should be used.
- It should be injected slowly; if extravasation occurs, cold should be applied to the area.
- Metoclopramide, maropitant or butorphanol should be administered to prevent vomiting.
- Collies, Bobtails, Boxers and West Highland White Terriers may develop haemorrhagic diarrhoea for 3–7 days after administration; this usually resolves within 24–48 hours without specific treatment.

FAC treatment

FAC is the acronym for the trade names of three drugs administered in combination. The actual pharmaceutical preparations are fluorouracil, doxorubicin and cyclophosphamide.

- Most commonly used when the surgical margins are not clear.
- Fluorouracil is **not** to be used in cats.
- Cyclophosphamide produces a risk of haemorrhagic cystitis.
- The protocol is best administered in the morning.
- The animal should be encouraged to drink water or cortisone should be administered during the treatment.

Hormonal treatment: Hormone therapy is not widely used because there have been only limited studies on the types of receptor within canine and feline mammary gland tumours. However, hormone therapy may be of special interest in benign tumours, in which there tends to be a high concentration of hormonal receptors.

The most popular treatment is tamoxifen (0.4–0.8 mg/kg/day orally over 4–8 weeks). Tamoxifen is an oestrogen antagonist, which unfortunately has some residual oestrogen-like effects itself. This may result in adverse effects, including ovarian disease and pyometra in intact females, urinary incontinence, and some oestrogenic signs including an increase in the size of the vulva and alopecia.

Aglepristone may also be used (20 mg/kg s.c. single dose 1 week prior to surgery or at 10 mg/kg s.c. on days 1, 2, 7 and 8 prior to surgery; Figure 17.13). Some authors describe its use for reduction in tumour size, whilst others describe its use only for reduction of the Ki67 cell proliferation index and progesterone receptors. Aglepristone interferes with the progesterone receptors in the mammary gland tumour and prevents the binding of progesterone.

Active agent	Dose	Disorders
Aglepristone	20 mg/kg s.c. once weekly until normal recovery (5–11 weeks) [a]	Mammary fibroadenomatous hyperplasia (mainly cats) Prior to mammary surgery (tumour)
Cabergoline	5 mg/kg orally q24h for 5–7 days [b]	Mastitis Mammary tumours Pseudopregnancy Galactostasis Mammary fibroadenomatous hyperplasia (if aglepristone not available)
NSAIDs (firocoxib, piroxicam)	Firocoxib: 5 mg/kg/day orally	Mastitis Mammary tumours (before and after surgery)
Antibiotics (ampicillin, erythromycin, lincomycin, clindamycin, tetracycline; cephalosporins are generally the treatment of choice if the female is nursing a litter)		Mastitis (avoid all except ampicillin if pregnant or puppies still sucking) Ulcerated tumours

17.13 Useful drugs for treatment of mammary gland disorders. [a] Aglepristone can result in abortion or resorption in pregnant females. [b] Cabergoline can result in abortion or resorption in pregnant females. If given for more than 8–10 days the return to oestrus may be more rapid than expected.

Aromatase inhibitors: These have been quite widely used in human medicine in the treatment of breast and ovarian cancers. Aromatase inhibitors work by inhibiting the action of the enzyme aromatase, which converts androgens into oestrogens, and they thereby slow the growth of a number of oestrogen-responsive tumours. Some of the aromatase inhibitors used in human medicine include exemestane, anastrozole, letrozole and resveratrol. In the dog, studies have shown a significantly higher expression of aromatase in mammary gland tumours in comparison with normal tissue; therefore, these products may have a future role in treatment of some of these cancers.

NSAIDs: These drugs, including cyclo-oxygenase 2 (COX-2) inhibitors, may be useful in two ways: controlling inflammation and pain before and after surgery; and potentially as anti-neoplastic agents. The COX-2 enzyme is responsible for the formation of prostaglandins. It is undetectable in normal tissues, but inflammation and mitosis induce COX-2 expression; in fact, malignant tumours show an overexpression of COX-2. In both cats and dogs, increased levels of COX-2 have been correlated with poorer prognosis (metastasis, recurrence of tumour and reduced survival time). NSAIDs bind to and block the action of COX. Firocoxib is a selective COX-2 inhibitor, and can be used at a rate of 5 mg/kg/day orally in the management of mammary gland tumours to help reduce the size of the tumour and decrease the risk of metastasis, but further studies are needed (see Figure 17.13). Other COX-2 inhibitors include meloxicam, piroxicam and carprofen but no detailed information is available about their role in the management of mammary gland neoplasia. In human medicine the combination of surgery, chemotherapy and COX-2 inhibition has been shown to be more effective than surgery and chemotherapy alone.

Radiotherapy: In small animal practice, radiotherapy is rarely used for the treatment of mammary gland tumours, probably because of a lack of expertise and equipment, the potential for high cost, and the need to anaesthetize the animal for each treatment. However, radiotherapy may be of value for control of well localized tumours.

Antibiotics: In many cases of neoplasia systemic antibiotics are useful, especially when ulceration or secondary infection is present.

References and further reading

Burstyn U (2010) Management of mastitis and abscessation of mammary glands secondary to fibroadenomatous hyperplasia in a primiparturient cat. *Journal of the American Veterinary Medical Association* **236 (3)**, 326–329

Fontbonne A, Levy X, Fontaine E and Gilson C (2007) *Guide practique de reproduction clinique canine et feline.* Med'Com, Paris

Gobello C, Concannon PW and Verstegen J (2001) Canine Pseudopregnancy: a review. www.ivis.org

Görlinger S, Kooistra HS, van den Broek A and Okkens AC (2002) Treatment of fibroadenomatous hyperplasia in cats with aglepristone. *Journal of Veterinary Internal Medicine* **16**, 710–713

Martín de las Mulas J, Millan Y, Bautista J, Pérez J and Carrasco L (2000) Oestrogen and progesterone receptors in feline fibroadenomatous change: an immunohistochemical study. *Research in Veterinary Science* **68**, 15–21

Sorenmo K (2003) Canine mammary gland tumors. *Veterinary Clinics of North America: Small Animal Practice* **33**, 573–596

18

Clinical approach to conditions of the non-pregnant and neutered bitch

Alain Fontbonne

Vulvar, vestibular and vaginal abnormalities

Embryologically, the vagina differs from the vestibule–vulva entity. This explains why the pathology of the distal part of the female genital tract is different from that of the cranial vagina. The paired paramesonephric ducts (Müllerian ducts) fuse to form the uterus, cervix and vagina of the bitch. The reunion of the tips of the Müllerian ducts and the urogenital sinus gives rise to the paramesonephric tubercule, which is canalized and fuses with the genital folds to form the vestibule. The hymen forms at the junction of the vagina and vestibule and is usually opened soon after birth. The genital swellings are transformed into the vulvar lips.

Pathology of the vulva and vestibule

Congenital abnormalities
Congenital abnormalities of the vulva are more commonly detected in breeding bitches, because most of the time they do not cause any obvious clinical signs.

Persistence of the hymen: This is rare in the bitch. It leads to an apparent lack of puberty, because no discharge is visible during oestrus (although the vulva is swollen). Digital palpation of the vagina is virtually impossible because the hymen blocks the cranial part. A swab or catheter also cannot be passed.

Incomplete perforation of the hymen: This leads to a circumferential vaginovestibular stricture and is seen more frequently than persistence of the hymen. A slight circumferential narrowing at the junction between the vagina and the vestibule, called the cingulum, is normal in adult bitches and should not be misinterpreted as a vaginal anomaly. Incomplete perforation of the hymen can sometimes be corrected manually through digital dilation. More often surgery, which involves an episiotomy, or vaginal endoscopy are necessary.

Vestibular–vulval hypoplasia and stenosis: Incomplete joining of the genital folds to the vestibule may lead to this condition, which is easily diagnosed by digital exploration of the vestibule. In small breeds of dog, difficulty passing a swab into the vestibular lumen is indicative of this condition. It may affect any breed, but the author's data suggest a predisposition in collies and Bergers Picards.

Dorsal occlusion of the vulva: The occlusion (Figure 18.1) is caused by a dorsal perineal fold that covers almost the entire vulva, making natural mating impossible. In addition these bitches are more likely to develop vaginitis and/or cystitis. Bitches that suffer from this vulvar abnormality may show no or very subtle clinical signs such as a slight creamy vulvar discharge, attracting male interest outside oestrus, failure to mate and possibly dystocia, if a pregnancy has been achieved. Obesity is a predisposing factor, especially in small breeds. Surgical vulvoplasty (episioplasty) is the treatment of choice.

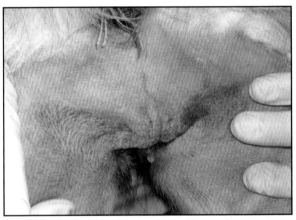

18.1 Occluded vulva in a bitch. The perineum covers the vulvar lips completely, predisposing the bitch to developing vaginitis. (Courtesy of X. Lévy)

Rectovestibular fistulae: Fistulae between the rectum and vestibule have been described. The true incidence is unknown because most affected neonates are euthanased. In affected bitches faeces pass through the vulvar lips. Contrast vaginography or vaginoscopy helps to make the diagnosis. The condition has to be treated surgically. Prevention can be achieved by detecting and surgically treating imperforate anus in newborn puppies.

Breeding: Technically, bitches with anatomical abnormalities of the vulva, vestibule and vagina can be bred by artificial insemination with a subsequent planned Caesarean operation. This will depend very much on the ethical dimension of such a decision and this may vary from person to person and country to country. If a litter is produced the female neonates should not be kept for breeding purposes owing to the potential genetic predisposition.

Acquired abnormalities

Vestibulitis: Inflammation of the vestibule, without signs of vaginitis (see below), is sometimes observed. The origin is often unknown, but it may be caused by alkaline urine (the urine pH should be checked), as a consequence of certain diets, immune-mediated cause, or following treatment with methionine. Bacteriological culture is nearly always unrewarding. Affected bitches may show signs of discomfort, excessive licking, slight vulvar discharge and a strong odour. Endoscopy can be used to rule out problems in the more caudal part of the vagina, such as follicular hyperplasia (Figure 18.2) and local hyperaemia. Local application of antiseptic creams and/or treatment with glucocorticoids may resolve the clinical signs. It is advisable to use an Elizabethan collar to prevent licking.

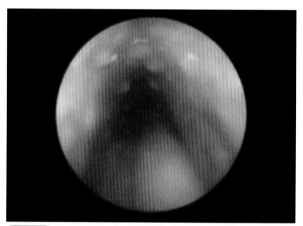

18.2 Vaginoscopic view of mild vestibulitis in a bitch. The follicular reaction can be clearly seen. (© A. Fontbonne)

Vulvar tumours: Tumours of the vulva represent around 30% of all genital tumours. The tumour types are identical to vaginal tumours (see below). Owing to the external position of the vulva, visual diagnosis and palpation is easy (Figure 18.3).

18.3 Malignant tumour of the vulva in a bitch. (Courtesy of X. Lévy)

Pathology of the clitoris

The clitoris of the bitch lies within the ventral clitoral fossa. Its development is dependent on hormonal control.

Clitoral hypertrophy

Clitoral hypertrophy may occur in:

- Cases of irritation caused by excessive licking of the area (perivulvar dermatitis)
- Masculinized bitches, attributable to treatment of the dam with progestogens or androgens during pregnancy
- Intersex animals. Approximately 50% of male pseudohermaphrodites (Figure 18.4) and 100% of true hermaphrodites show clitoral hypertrophy
- Hyperadrenocorticism: 20–30% of affected bitches show clitoral hypertrophy to some extent
- Treatment of bitches with androgens (to prevent oestrus in racing dogs).

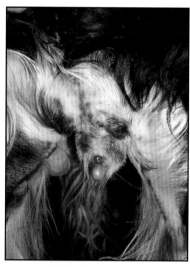

18.4 Enlarged clitoris in a Cocker Spaniel bitch, which was in fact a male pseudo-hermaphrodite. (© A. Fontbonne)

Castration and/or removal of the stimulus may not be sufficient to reduce the size of the clitoris, and in many cases surgical clitoridectomy must be performed.

Clitoral tumours

Clitoral tumours (Figure 18.5) occur rarely and are of the same types as tumours of the penis in male dogs: squamous cell carcinomas and malignant mast cell tumours. Chondrosarcomas of the clitoris bone (if present) may be found.

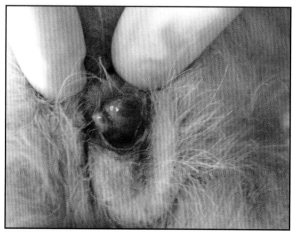

18.5 Haemangiosarcoma of the clitoris in a Lhasa Apso bitch. (© A. Fontbonne)

Pathology of the vagina

Congenital abnormalities

Vaginal septum: Incomplete or abnormal fusion of the Müllerian ducts can lead to the formation of a vertical septum within the vagina (Figure 18.6). Its length can vary from a few millimetres to the complete formation of a double vagina. Clinical signs are often absent but this anomaly may lead to chronic vaginitis, inability to breed naturally, chronic cystitis and infertility. Vaginal endoscopy is the best method to diagnose the presence and extent of the septum, although the distal part of the septum is often easily explored by digital palpation. When endoscopy is not available, contrast radiography of the vagina may be useful. The vaginal septa are visible as dark bands within the opaque contrast medium (Johnston *et al.*, 2001). Very small vaginal septa may be broken down manually. Elongated septa require episiotomy and resection.

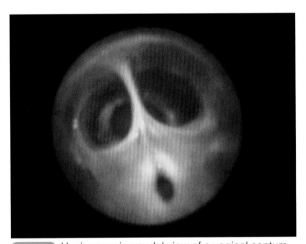

18.6 Vaginoscopic caudal view of a vaginal septum in a bitch. Note that the vestibule (located caudal to the septum) is not involved as it is not derived embryologically from the Müller's canals. (© A. Fontbonne)

Segmental aplasia of the vagina: This anomaly has been described in bitches. It leads to an apparent lack of puberty, because affected bitches do not show any discharge at the time of oestrus. Vaginal endoscopy can be used to diagnose the problem, but simply assessing the length of the vagina by inserting a swab can also be useful.

Vaginal hyperplasia/prolapse: This disease is caused by excessive oedema and subsequent prolapse of the vagina within the vaginal lumen, and in more severe cases prolapse through the vulvar lips may occur under normal oestrogen stimulation. It is an excessive form of the normal physiological hyperplasia of the vaginal epithelium that occurs during oestrus. Vaginal hyperplasia may sometimes (<10% of cases) appear during the second month of pregnancy, when plasma concentrations of oestradiol begin to rise again. Young bitches around the time of their second or third oestrus seem to be most frequently affected. Brachycephalic breeds, including Pugs, French and English Bulldogs,

Boxers and Bull Mastiffs, and other breeds, such as the Dogo Argentino, Great Dane and Labrador Retriever, are more commonly affected. The heritability of this disease is unknown but a genetic trait is strongly suspected.

Clinically, three different stages may be observed (Figure 18.7):

- Stage 1 is characterized by a slight protrusion of the vaginal wall within the vestibule lumen, without any protrusion outside the vulvar lips
- Stage 2 is a moderate protrusion of the vaginal wall and parts of the lateral vagina, forming a pear-shaped mass between the vulvar lips
- Stage 3 is a protrusion of the wall, the lateral and the dorsal part of the vagina, forming a voluminous oedematous mass at the vulvar opening.

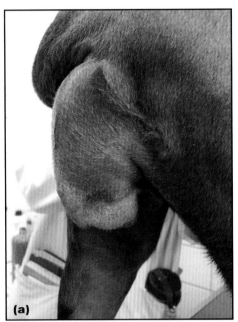

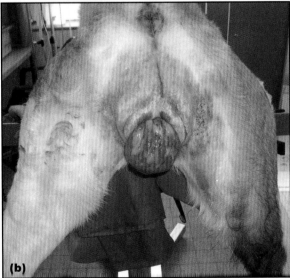

18.7 **(a)** Stage 1 vaginal prolapse in a Great Dane bitch in oestrus. Note the deformation of the perineum, although no mucosa could be seen through the vulvar lips when the bitch was standing. **(b)** Stage 2 vaginal prolapse in a bitch just before corrective surgery. (a, © A. Fontbonne; b, Courtesy of X. Lévy) (continues) ▶

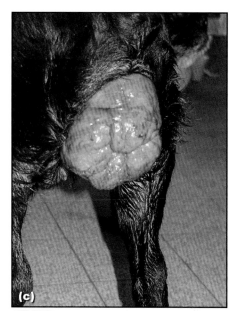

18.7 (continued) **(c)** Stage 3 vaginal prolapse in a Brabançon bitch in oestrus. Note the voluminous ulcerated oedematous ring at the vulvar opening. (c, © A. Fontbonne)

Vaginal prolapse always originates from the floor of the vagina, cranial to the urethral opening, and the external urethral orifice can be found on the ventral surface of the prolapsed mass. Whatever the stage, vaginal hyperplasia may lead to the development of vaginitis, cystitis and mating failure, owing to the inability of the male penis to penetrate inside the vagina. Clinical diagnosis is easy. The condition must be differentiated from true vaginal prolapse (which is not oestrogen dependent) and vaginal neoplasia (see below). Vaginal cytology may help to identify a pro-oestrous or oestrous stage of the cycle.

Treatment: The treatment involves removing the oestrogen stimulation and/or removing the protruding mass. In bitches that are not intended for breeding, ovariectomy during oestrus (or just after) is a surgical option, although this may not always lead to complete shrinkage of the protruding vagina in stages 2 and 3. Stage 1 often resolves spontaneously at the end of the oestrous period, when the oestrogen concentration declines.

In stage 2, some authors recommend treatment with proligestone, although this, in combination with the endogenous oestrogen produced by the bitch, carries the risk of inducing pyometra. Another approach is to maintain the prolapsed tissue in a clean and intact state (which involves cleaning it with saline solution several times daily and/or applying a local protective cream such as petroleum jelly or malic, benzoic and salicylic acid) and remove it surgically at the beginning of dioestrus. An Elizabethan collar should be used to prevent licking and biting.

Stage 3 almost always requires surgical treatment (circumferential amputation). Care must be taken to catheterize the urethra to avoid damage during the removal of the protruding tissue. Even after successful surgery, the risk of recurrence at the following oestrus is high (65–100%). Manual

reduction of the prolapse and placement of purse-string sutures is not recommended owing to the increased risk of vaginitis and perivulvar trauma and pain. The potential genetic predisposition and possible problems towards the end of pregnancy (when oestrogen rises) make the use of artificial insemination questionable from an ethical point of view.

Acquired abnormalities

Juvenile vaginitis: Juvenile vaginitis is a disease that affects young prepubescent bitches and may be observed from 8 weeks of age onwards. Clinically, a mucoid to mucopurulent vulvar discharge, sometimes in copious amounts, may be observed (Figure 18.8). Apart from frequent licking of the vulva the animals are in good general health.

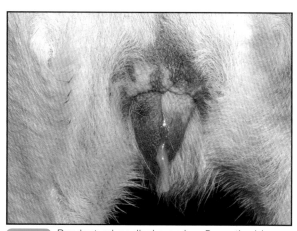

18.8 Purulent vulvar discharge in a 5-month-old Beagle bitch suffering from prepubertal vaginitis. Note the enlarged aspect of the vulva caused by frenetic licking. (Courtesy of G. Casseleux)

The aetiology of the condition has not been identified. It appears mainly to be an inflammatory process. Bacteriological swabbing of the vagina often leads to no significant bacterial growth. Spontaneous resolution occurs in >80% of cases after the first oestrus, probably due to the regeneration of the vaginal epithelium induced by oestrogen. Some authors consider early neutering to be contraindicated because it may induce a chronic form of vaginitis, which is very difficult to treat (see below). Treatment of juvenile vaginitis with antibiotics tends to give short-term relief, but the vaginitis recurs shortly after antibiotic treatment is discontinued. Small doses of oestrogen may help. In most cases it is best to wait until the first oestrus, which will resolve the condition.

Vaginitis in the intact adult bitch: This disease may be described as an inflammatory state of the adult vagina. It is often underdiagnosed by veterinary surgeons.

Aetiology: A distinction is made between *primary* and *secondary* vaginitis.

Primary vaginitis is rare and may be viral (caused by canine herpesvirus) but is most often of bacteriological origin. Unlike in women, fungal causes of vaginitis are almost unknown in bitches. The normal canine vaginal flora contains many different bacteriological strains, including *Escherichia coli*, *Pasteurella* spp., *Streptococcus* spp., *Staphylococcus* spp. and other aerobic bacteria, anaerobic bacteria and mycoplasmas. These bacteria may proliferate for unknown reasons and become pathogenic.

Secondary vaginitis is the most common form and may develop as a result of underlying vaginal problems including:

- Hermaphroditism or pseudohermaphroditism
- Anatomical problems such as vulvar atresia, stenosis (vulvar, vestibular or vaginal), persistence of the hymen and a vaginal septum
- Urinary problems, especially an ectopic ureter or cystitis
- Trauma or foreign bodies (e.g. grass seeds)
- Uterine diseases (see below) such as endometritis and pyometra
- Vaginal neoplasia leading to local irritation of the epithelium
- Endocrine diseases such as diabetes and hyperadrenocorticism.

Clinical signs: The clinical signs can be very mild, and vaginitis in the adult bitch is often underdiagnosed. Slight vulvar discharge (mucoid to purulent), recurrent cystitis, perivulvar pruritus and attention from males (outside of oestrus) are the most common clinical signs. More unusually, a heavy purulent vulvar discharge (which may be mistaken for pyometra), pain, discomfort and excessive licking of the vulva may be seen.

Diagnosis: Other sources of vulvar discharge such as oestrus, early dioestrus (when discharge is normally dark and odorous) and pyometra must be excluded. Vaginal cytology will show a large number of polymorphonuclear cells, suggesting inflammation. Inspection of the vulva and vestibule with a speculum allows detection of anatomical abnormalities. Vestibulitis may occur, with numerous reactive lymphoid nodules (non-specific follicular lesions) caused by local irritation. These can be visualized on the surface of the vestibule mucosa (see Figure 18.2).

Vaginal endoscopy, when available, will confirm the diagnosis and help identify the underlying problem. It will show local or diffuse areas of hyperaemia of the vestibular and vaginal mucosa, and possibly anatomical abnormalities (Figure 18.9). In cases of primary vaginitis a swab for bacteriology should be taken from as cranial a point as possible (Figure 18.10). For unknown reasons one or two strains of bacteria may be present in large numbers and become pathogenic, causing inflammation of the vagina. However, these dominant bacteria may only be considered to be pathogenic if accompanied by clinical signs. More than 15% of normal bitches show a heavy growth of one or more bacteria in the cranial vagina without any problems (Johnston *et al.*, 2001) and therefore this in itself does not justify the diagnosis of primary vaginitis. All bitches with vulvar discharge should undergo a serological test to detect *Brucella canis* in countries where this may be a possibility.

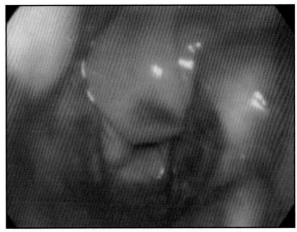

18.9 Vaginoscopic appearance of the vagina of a bitch with secondary vaginitis. Note the inflammation of the vaginal mucosa and the circular stenosis in the centre, probably caused by a congenital abnormality. (Courtesy of X. Lévy)

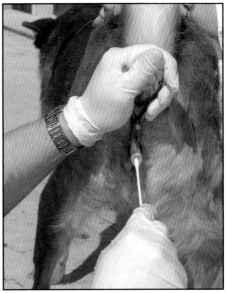

18.10 Technique for swabbing the vagina in order to perform bacterial culture and identification. A tubular speculum is inserted into the vestibular lumen in order to be able to pass the cotton part of the sterile swab into the anterior vagina. (© A. Fontbonne)

Treatment: In cases of secondary vaginitis, the underlying cause must be treated first. Primary vaginitis caused by bacterial infection may be treated with antibiotics, but no definite protocol has been agreed. Some authors recommend treating with antibiotics only if a culture or a swab collected from the anterior vagina reveals a heavy growth of a single organism and the bitch shows clinical signs of the condition. Antibiograms should be used to identify the correct antibiotic for treatment. In a recent study in breeding bitches suffering from reproductive problems (Lévy *et al.*, 2004), bacteria found within the vagina were sensitive to cefquinome in 94.7% of cases, sensitive to marbofloxacin or enrofloxacin in 90.3%, sensitive to cefalexin in 74%, sensitive to trimethoprim/sulphonamide in 60% and sensitive to amoxicillin in only 48% of cases.

The duration of treatment is still under discussion. Most authors recommend treatment for 2–4 weeks. A longer period may lead to the development of bacterial resistance and/or a rise in the occurrence of mycoplasmas, which may become pathogenic in the absence of the normal vaginal flora.

Vaginitis in the neutered adult bitch: Vaginitis may develop in a neutered adult bitch. This can become chronic and is often difficult to cure. The bitch presents with the same clinical signs as an intact female, but licking of the vulva may become excessive. The initial cause is unknown, but the condition may be due to the lack of oestrogen and the subsequent changes in the normal vaginal flora. Oral administration of oestrogens (estriol: 0.5–2.0 mg q24h for 1 month) may be helpful. In severe cases, corticosteroids are necessary to reduce the pain and excessive licking of the vulva.

Vaginal tumours: Vaginal tumours are the second most common reproductive tumour of the bitch (after mammary gland tumours). They are usually seen in bitches >10 years of age. All breeds may be affected but Boxers have a higher incidence. In nearly 90% of cases the tumours are benign. Leiomyomas (most common), fibromas, fibroleiomyomas and fibropapillomas (vaginal polyps) can be found and are often hormone dependent. Therefore, the vast majority of vaginal tumours are found in intact bitches. Malignant tumours found in the vagina are leiomyosarcomas, adenocarcinomas, epidermoid carcinomas, transitional carcinomas originating from the urethral epithelium, haemangiosarcomas (Figure 18.11) and mast cell tumours. The most common malignant vaginal tumour is the leiomyosarcoma, which may metastasize.

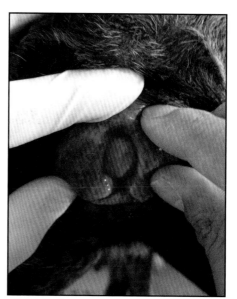

18.12 Slight protrusion of a vaginal leiomyoma in a bitch in oestrus. The differential diagnosis includes stage 1 vaginal prolapse. (Courtesy of Alfort Veterinary College)

analysis using a fine-needle aspirate will help to differentiate between tumour types. Vaginal prolapses are sometimes mistaken for tumours, but have a much softer consistency on palpation.

Treatment: Surgical removal (episiotomy) is the treatment of choice. Concurrent ovariectomy is recommended to prevent recurrence, owing to the hormonal background of these tumours. Sometimes a complete surgical vulvectomy and/or vaginectomy may be required in cases of highly infiltrative malignant tumours. Chemotherapy may be given as an adjuvant treatment (Figure 18.13).

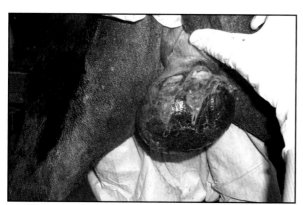

18.11 Malignant haemangiosarcoma of the vagina. (Courtesy of X. Lévy)

Clinical signs: Clinical signs of vaginal tumours include an intravaginal mass modifying the shape of the perineum, protrusion of a mass through the vulvar lips (Figure 18.12), vulvar discharge, urinary problems and faecal tenesmus.

Diagnosis: The clinical diagnostic investigation begins with digital palpation of the rectum and vagina. Vaginal endoscopy will help to diagnose the site of implantation, the number of tumours and the degree of infiltration into the vaginal wall. Cytological

Tumour type	Chemotherapeutic agent	Comments
Leiomyosarcoma Transitional adenocarcinoma	Carboplatin: 250–300 mg/m², one slow i.v. administration q3wks	In association with radiotherapy when possible
Mastocytoma	Prednisolone Vinblastine Lomustine	Radiotherapy preferred. Use chemotherapy only with elevated neoplastic grade tumours (grade 2 with proliferative index Ki-67 >10% and grade 3)
Transmissible venereal tumour	Vincristine: 0.5–0.75 mg/m², two to six injections at 1-week intervals Adriamycin (use in cases of failure with vincristine): 30 mg/m², one injection q3wks	Treatment has to be continued for one or two treatments beyond complete remission of the tumour. Radiotherapy (10–30 Gy) also shows very good efficacy

18.13 Suggested chemotherapy protocols for malignant vaginal and vulval tumours. (McEntee, 2002; Tierny *et al.*, 2008)

Transmissible venereal tumour: Also known as TVT or Sticker's sarcoma, this highly contagious tumour exists worldwide, especially in hot countries, but its appearance has decreased to almost zero in developed countries owing to control of the stray dog population. It is not endemic in the UK, but may be brought in quite easily from other European countries through the pet travel scheme.

The tumour spreads when neoplastic cells are passed from the vaginal mucosa of an infected bitch to the penis (or *vice versa*) during coitus. Clinical signs of TVT in the bitch are swelling of the perineum, serosanguineous vulvar discharge and/or protrusion through the vulvar lips of single to multiple, grey to red, cauliflower-like masses. Occasionally, the tumours may also be seen in the nasal or oral mucosa. The diagnosis, prognosis and treatment are identical to those in male dogs (see Chapter 20).

Miscellaneous conditions

Trauma: Vaginal wounds, tearing of the vaginal wall or presence of foreign bodies may occur on rare occasions after trauma (e.g. natural mating or catheterization with a rigid endoscope, especially during the anoestrous phase of the oestrous cycle when the vaginal wall is thin). In the case of a natural mating tearing the vagina, the bitch may still become pregnant. The lesion may remain unnoticed and the vaginal wall may heal spontaneously. A slight pinkish vulvar discharge may be noticed throughout dioestrus and there is a risk of peritonitis. Vaginal endoscopy and contrast radiography of the vagina are the best diagnostic methods.

True vaginal prolapse: This may occur following parturition, owing to excessive force during labour, or after dystocia. It can also been seen after a natural mating when the dogs have been separated forcibly by their owners. The bitch is presented with the vaginal mucosa everted in a conical structure outside the vulvar lips (Figure 18.14). In mild cases, the vagina may be cleaned and replaced using lubricant. In more severe cases or with traumatized vaginal tissue, vaginal pexia under laparotomy and/or surgical amputation may be required.

18.14
True vaginal prolapse in a bitch. Note the tubular appearance of the protruded mucosa. (Courtesy of JP. Mialot)

Vulvar discharge
In anatomical terms, vulvar discharge can originate from the uterus, the vagina/vestibule/vulva area or the urinary tract.

Discharge from the uterus: It is necessary to distinguish between oestrogen-induced discharge and that not induced by oestrogen. Ultrasonography can often be helpful in the diagnosis.

- Oestrogen-induced discharge from the endometrium:
  o Normal pro-oestrous/oestrous
  o Ovarian disease (follicular cyst, granulosa cell tumour)
  o Ovarian remnant syndrome in neutered bitches
  o Iatrogenic origin.
- Non-oestrogen induced discharge from the endometrium:
  o Blood:
    – Postpartum: uterine haemorrhage, subinvolution of placental zones
    – Abortion during pregnancy: dark greenish vulval discharge due to uteroverdin from the placenta
    – Haematometra: a severe form of pyometra
    – Uterine tumours
    – Intoxication with anticoagulants.
  o Mucus:
    – Cystic endometrial hyperplasia
    – Low-grade endometritis.
  o Pus:
    – Pyometra
    – Postpartum acute metritis.

Discharge from the vagina/vestibule/vulva area:

- Vaginitis: mucoid to heavy purulent discharge.
- Vaginal tumours: mucoid, bloody or slightly purulent discharge, due to irritation of the mucosa.
- Intravaginal foreign bodies or vaginal trauma: often pinkish mucopurulent discharge.

Discharge from the urinary tract: Urinary contamination may have a characteristic odour of ammonia and is transparent in colour. Urine is an irritant to the vaginal mucosa and may induce vaginitis.

- Urinary incontinence (often combined with perivulval dermatitis).
- Ectopic ureter (predisposed breeds include Golden Retriever, Labrador Retriever, Siberian Husky, English Bulldog and Poodle).

Diagnosis: The first step is to take a sample for vaginal cytology, to differentiate between oestrogen and non-oestrogen induced forms of vaginal discharge. This should be followed by a gynaecological examination, including rectal and abdominal palpation to detect any mass or anatomical abnormality. An algorithm detailing the diagnostic procedure is given in Figure 18.15.

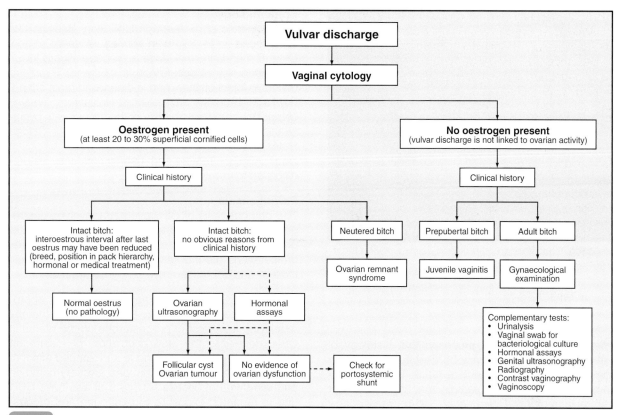

18.15 Diagnostic approach to bitches with vulvar discharge.

Pathology of the uterus

The endometrium and myometrium undergo normal physiological cyclical changes that must be differentiated from uterine pathology. The luteal phase in the bitch, regardless of whether or not she is pregnant, is very long compared with that in other species. During this phase the uterine horns will increase in diameter and the lumen will assume a cork like appearance. However, the uterus will atrophy markedly with age or following ovariectomy. It is also useful to note that, after several litters, changes may occur in the uterine wall. Interstitial fibrosis and embedding of glands and the blood vessels often occurs. Following severe pyometra, individual endometrial glands may be lost and replaced with diffuse endometrial fibrosis (Schlafer, 2003).

Endometritis

Endometritis is inflammation of the endometrium. This disease mainly affects breeding bitches and may cause infertility, as in many other species.

Aetiology

Bacterial infection may play a role in the development of endometritis. Controversy still exists about the role of mycoplasmas. Endometrial lesions of viral aetiology are not well recognized. Many bacteria are commonly isolated during oestrus from the uterus of normal fertile bitches. Bacteria found in the uterus are usually also found in the cranial vagina during pro-oestrus and oestrus, when the cervix is open (Watts and Wright, 1995). Numerous bacteria are found in bitches with no reproductive problems, including:

- Aerobic bacteria – *Streptococcus* spp., *E. coli*, *Pasteurella multocida*, *Staphylococcus* spp., *Proteus* spp. and *Corynebacterium* spp.
- Anaerobic bacteria *Lactobacillus* spp., *Bifidobacterium* spp., *Clostridium* spp. and *Corynebacterium* spp.
- Bacteroidaceae have been isolated from 55% of vaginal swabs in normal bitches.

Diagnosis

Culture: Uterine cultures may be performed following a cytological assessment of the endometrium by laparotomy or transcervical endoscopy (see below). However, swabbing the anterior vagina during pro-oestrus is much easier and less invasive and is therefore the preferred method (Root-Kustritz, 2006).

Cytology: Samples for endometrial cytology have been collected via surgery (Groppetti *et al.*, 2007). The most common technique involves passing a 4–7 Fr urethral catheter through the cervix, guided by a rigid endoscope, followed by infusion and aspiration of a sterile saline solution (Root-Kustritz, 2006; Fontaine *et al.*, 2009; Groppetti *et al.*, 2010) (Figure 18.16).

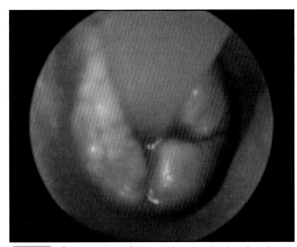

18.16 Caudal view of the uterine cervix showing the technique used to collect endometrial samples for cytological and bacteriological identification. A sterile human ureteral catheter is guided through the cervical ostium in order to inject and aspirate a small quantity of sterile saline solution. (Courtesy of E. Fontaine)

Histology: It has been proposed that histology of uterine biopsy samples may be useful, especially for the prognostic assessment of future fertility. A 2–4 mm skin punch biopsy sample may be taken after ventral laparotomy (Downs *et al.*, 1994). Transcervical passage of the biopsy forceps (as guided by endoscopy) may be achieved, but biopsy samples are difficult to obtain and are not always well correlated with the true lesions (Günzel-Apel *et al.*, 2001).

Endoscopy: Hysteroscopy has been used in an attempt to assess the endometrium, but it can sometimes lead to severe vaginal and uterine inflammation, with petechiae and ecchymosis and a potential adverse effect on future fertility.

Cystic endometrial hyperplasia and pyometra

These two pathological conditions are often presented together, although some authors believe that they are not connected (De Bosschere *et al.*, 2001).

Definitions

Cystic endometrial hyperplasia (CEH) is hyperplasia of the endometrium and abnormal dilatation and proliferation of the endometrial glands. Pyometra is the presence of purulent contents within the uterus. Four histological stages of canine CEH/pyometra have been described:

- Stage I: uncomplicated CEH
- Stage II: CEH with plasma cell infiltration. No tissue destruction is visible histologically
- Stage III: CEH with acute endometritis. Endometrial ulceration and small areas of haemorrhage may be visible. Myometrial inflammation is present in 40% of cases
- Stage IV: CEH with chronic endometritis. The endometrium is atrophied, with infiltration of plasma cells and lymphocytes. The myometrium is hypertrophied and fibrotic and in other places atrophic.

Epidemiology

The incidence of pyometra in laboratory Beagles has been found to be 15.2% in bitches >4 years of age. Over all breeds the incidence is probably slightly lower. The mean age of occurrence is 9.5 years. Pyometra occurs in intact bitches, and is diagnosed most often 2 weeks to 4 months after oestrus. Nulliparous females comprise approximately 75% of the bitches with pyometra (Pretzer, 2008). Some breeds, such as Golden Retrievers, collies, Rottweilers, Bernese Mountain Dogs, Airedale Terriers and Irish Terriers, may be predisposed to develop the condition.

Pathogenesis

The pathogenesis of CEH/pyometra involves oestrogen stimulation of the uterus, followed by periods of progesterone influence, which results in endometrial proliferation, uterine glandular secretions, cervical closure and decreased myometrial contractions.

Hormones: Oestrogen increases the number of progesterone receptors in the endometrium, thereby amplifying the effect of progesterone on the uterus. If CEH develops, the downregulation of oestrogen receptors is hindered, leading to a prolonged effect of oestrogen. These uterine changes are very common and two-thirds of bitches >9 years of age have CEH due to repeated exposure to oestrogen and progesterone, without any signs (Johnston *et al.*, 2001).

Synthetic progestogens, which may be administered over long periods of time to prevent oestrus in bitches, play a similar role in promoting CEH and pyometra. The effects of these hormones on the uterus are often cumulative over several oestrous cycles.

Mechanical effects: These also play a role because the endometrium is hyperactive during the luteal phase, and any injury or scarring during this phase may induce a proliferative reaction. Endometrial trauma can modify the structure of the endometrium and the characteristics of the progesterone receptors.

Infection: This follows as the second stage. Progesterone has a negative effect on the immune protection of the uterus by leucocytes. Lymphopenia and a marked suppression of the immune system are found in 35% of cases of pyometra. *E. coli* is often found in cases of pyometra (70% of cases) and causes endotoxin release, which may result in septic shock. Many strains also release cytotoxic and necrotizing factors, which are often haemolytic. These strains are identical to the ones identified in urine and faeces. Furthermore, subclinical urinary tract infection may increase the risk of uterine infection. Other bacteria may be found in cases of pyometra, with a much lower prevalence.

There are some theories that the primary cause of CEH is a subclinical uterine infection, which provides the stimulus leading to endometrial proliferation and the development of CEH (Schlafer, 2003). Studies of uterine receptors have shown that the expression of oestrogen and progesterone receptors is different in cases of CEH and pyometra (De Bosschere *et al.*, 2002). This may mean that the pathogenesis of CEH is mediated by hormones and that of pyometra by infection. It could also mean that pyometra may occur *de novo*, without preliminary CEH.

Clinical signs
Cases of CEH often go unnoticed clinically, except in breeding bitches where CEH may lead to infertility. Sometimes, a slight mucous vulvar discharge can be seen.

Two forms of pyometra are differentiated: open-cervix pyometra and closed-cervix pyometra. Hysterectomized bitches may develop stump pyometra of the remnant, which is always associated with failure to remove all the ovarian tissue and subsequent stimulation with oestrogen and progesterone. Both have to be removed surgically to resolve the problem.

Non-specific signs: The body temperature is rarely increased. The non-specific clinical signs noted may include anorexia, depression, lethargy and abdominal distension. Haematology is valuable, with most bitches having elevated polymorphonuclear leucocyte counts (>15 x 10^9/l) with a marked left shift. Elevated blood urea concentrations commonly accompany dehydration (with hyperproteinaemia and hyperglobulinaemia). Endotoxaemia may also occur owing to a decreased glomerular filtration rate and a decreased ability of the renal tubules to concentrate urine. A secondary membranous glomerulonephritis, caused by the deposition of antigen–antibody complexes against *E. coli* in the glomeruli, can be a complicating factor. However, renal dysfunction is not always found in cases of pyometra.

In cases of open-cervix pyometra, vulvar discharge will be seen, which may vary from purulent to sanguineous or mucoid. Closed pyometras are more difficult to diagnose because the bitch can show more or less severe signs of systemic disease without any indication of the reproductive system being involved. When haematometra develops, vulvar bleeding with blood clots occurs and a reduced red blood cell count may be identified.

Diagnosis
Very often diagnosis occurs late in the process, owing to a lack of early clinical signs. In older bitches, a recent oestrus may not have been noticed by the owners.

Clinical examination: Bitches with open-cervix pyometra are generally less systemically ill than those with closed-cervix pyometra. The clinical signs may include lethargy, depression, anorexia, polyuria, polydipsia, vomiting and diarrhoea. Bitches with closed-cervix pyometra may have marked abdominal distension because the uterus can weigh as much as 10 kg in large breeds. Bitches are often dehydrated, septicaemic or even in endotoxic shock. Fever is rarely present. Even bitches affected by endotoxaemia are often hypothermic. Abdominal palpation should be avoided owing to the risk of uterine rupture.

Clinical pathology: In cases of open-cervix pyometra, cytological examination of the vulvar discharge will reveal large numbers of degenerating polymorphonuclear leucocytes. Haematology often shows an elevated leucocyte count with a clear left shift. Monocytosis may be encountered. A normocytic, normochromic anaemia may be seen (packed cell volume (PCV) 30–35%). This is believed to reflect the chronic nature of the disease and the toxic suppression of the bone marrow. Evaluation of anaemia is often complicated by concomitant dehydration. Metabolic acidosis is frequent, and abnormalities in serum biochemistry may sometimes include elevated serum alkaline phosphatase (in 50–75% of cases), hyperproteinaemia (as a consequence of dehydration), hypergammaglobulinaemia and hypoalbuminaemia. Serum blood urea nitrogen and creatinine are not usually elevated, unless renal complications develop as a consequence of dehydration.

Imaging: Ultrasonography, if available, is the best way to diagnose pyometra. The ultrasound images show an anechoic or hypoechoic fluid-filled organ with variable wall thickness and proliferative changes (Figure 18.17). Diagnosis of pyometra on lateral radiographs is more difficult. New generation ultrasound machines also allow the early detection of CEH and can help to differentiate between the various stages of the CEH/pyometra complex (Bigliardi *et al.*, 2004) (Figure 18.18).

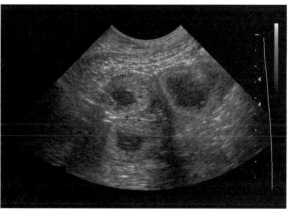

18.17 Ultrasonographic appearance of the uterine horns of a bitch suffering from a moderately advanced pyometra. Note the thick endometrial walls and the hypoechoic content (pus). (Courtesy of Alfort Veterinary College)

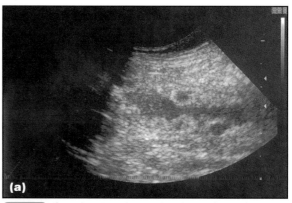

18.18 Cystic endometrial hyperplasia. **(a)** Ultrasonographic appearance. Note the cystic endometrial glands and the hypoechoic content within the uterine lumen. (Courtesy of Alfort Veterinary College) (continues) ▶

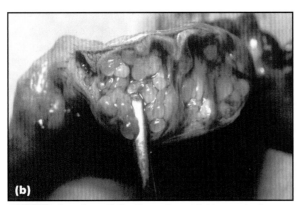

18.18 (continued) Cystic endometrial hyperplasia. **(b)** Following surgical removal and opening of a part of the uterine horn. (Courtesy of Alfort Veterinary College)

Treatment

Surgery: Ovariohysterectomy is the treatment of choice in non-breeding bitches. In cases of severe systemic disease stabilization and postoperative care should include appropriate intravenous therapy and broad-spectrum antibiotics, which may have to be continued for some time after surgery. Administration of anti-progestogens (e.g. aglepristone) before surgery may be valuable in cases of closed-cervix pyometra, because these drugs induce opening of the cervix and permit partial uterine emptying. This can lead to a marked clinical improvement before surgery and facilitates surgery by reducing the volume of the uterus to be removed.

Medical treatment: In young reproductively active bitches pharmacological treatment is also an option. Other indications are elderly or critically ill bitches where surgery is considered high risk, and where the owners do not give consent for surgery. The decision to proceed with non-surgical rather than surgical treatment should be weighed carefully and discussed with the owners. Recovery may take longer than with surgery, and if the treatment fails ovariohysterectomy will still be necessary.

Prior to the beginning of treatment, a careful clinical and ultrasound examination, blood count and biochemistry assays should always be carried out. Peritonitis following uterine rupture is an absolute contraindication for medical treatment. It is important to correct any fluid and electrolyte imbalances in the bitch. Where possible ultrasonography should be used to check for the presence of ovarian cysts, because the hormonal secretions of these cysts may increase the risk of relapse after treatment. Bitches with ovarian cysts should be ovariohysterectomized because they have a higher risk of further fertility problems.

In cases of closed-cervix pyometra, a purulent vulvar discharge is observed from 36–48 hours after the start of treatment, usually accompanied by a significant improvement in the general condition of the bitch. The most objective criteria of treatment efficacy are:

- Improvement in the general wellbeing of the bitch within 48 hours
- Reduction in the diameter of the uterine lumen. A decrease in the lumen of at least 50% on day 8 is seen as a positive result
- A significant decrease in the leucocyte count during the first week of treatment (Gobello *et al.*, 2003).

The different medical treatment protocols for pyometra are described in Figure 18.19.

Aglepristone: This is one of the most important drugs in the treatment of pyometra because of its marked affinity for progesterone receptors in the uterus. This treatment is particularly useful when serum progesterone concentrations are >1–2 ng/ml. Bitches that are in anoestrus at the time of treatment (with basal progesterone concentrations) may be more difficult to treat. Aglepristone also aids the opening of the cervix, uterine contractions and emptying of the uterine content, thus reducing the size of the uterus and the toxic effects. In the USA, where this product is not licensed, luteolysis is achieved using prostaglandin (PG) F2α or dopamine agonists (cabergoline).

Drug	Dosage	Reference
Aglepristone Note: in cases of closed-cervix pyometra, a purulent vulvar discharge is observed 36–48 hours after the start of treatment, usually accompanied by a significant improvement in the general condition of the bitch	10 mg/kg s.c. every 7 days: day 1, 2, 8, 15, 22, 29 26 bitches Success rate: • closed-cervix pyometra = 61.5% • open-cervix pyometra = 83.3%	Fieni (2006)
	10 mg/kg s.c. on day 1, 2, 7, 15, 23, 29 52 bitches Success rate: 92.3% The diameter of uterine lumen decreased by 43% in 48 hours, 56% after a week, 87% after 15 days, and it was invisible after 29 days	Trasch *et al.* (2003)

18.19 Published protocols for medical treatment of pyometra in the bitch. (continues) ▶

Drug	Dosage	Reference
Aglepristone + prostaglandins (PG) Note: due to their uterotonic effect, prostaglandins provide faster emptying of the uterus whilst their specific luteolytic activity strengthens the effect of aglepristone	Aglepristone 10 mg/kg s.c. on day 1, 2, 8, 15, and eventually 29 The dose of aglepristone is the same for the various treatment protocols with prostaglandin F2α that have proved satisfactory	
	Natural PGF2α (dinoprost) 25 µg/kg s.c. q8h, day 3 to 7, or 25 µg/kg s.c. on days 3, 6 and 9	
	Synthetic PGF2α (cloprostenol) 1 µg/kg s.c. q24h from day 3 to 7, or 1 µg/kg s.c. on days 3 and 8 The addition of cloprostenol to the aglepristone therapy, compared with aglepristone alone, significantly improved the overall success rate in treating pyometra in bitches: 84.4% *versus* 63% with aglepristone alone	Fieni (2006)
	PGE1 (misoprostol) 10 µg/kg orally q12h, day 3 to 12 Success rate: 75% at day 29 Significant clinical improvement in 75% of cases, without the side effects of PGF2α	Romagnoli *et al.* (2006)
Cabergoline + prostaglandins	Cabergoline 5 µg/kg orally q24h + cloprostenol 1 µg/kg s.c. or 7–14 days 29 bitches Success rate: 82.8% by day 14	Corrada *et al.* (2006)
	Cabergoline 5 µg/kg orally q24h + cloprostenol 5 µg/kg s.c. every third day + sulphonamides 21 bitches Success rate: 90.5% within 10 days Rapid clinical improvement Rapid improvement of haematological and biochemical profiles Next oestrus: 7/11 pregnancies; 4 recurrences of pyometra	England *et al.* (2007)

Notes
Broad-spectrum antibiotics should be used until the uterine lumen is no longer visible on ultrasonography and the leucocyte count has returned to normal.
An Elizabethan collar should be used to prevent the bitch licking the purulent vulvar discharge that occurs during the course of treatment.

18.19 Published protocols for medical treatment of pyometra in the bitch.

Prostaglandins: Different protocols for treatment with PGF2α are possible (Figure 18.19). Prostaglandins cause strong uterine contractions and emptying of the uterus, whilst their specific luteolytic activity strengthens the effect of aglepristone. The addition of cloprostenol to the aglepristone therapy significantly improves the overall success rate in treating pyometra in bitches. Prostaglandins should be avoided in cases of renal failure, pneumopathy and congestive heart failure.

Antibiotics: Septicaemia is a potential complication of pyometra. Broad-spectrum antibiotics should *always* be given during the whole period of treatment. The antimicrobial resistance of *E. coli* strains isolated from cases of pyometra seems to be low. *In vitro* sensitivity studies and clinical evidence suggest that amoxicillin, amoxicillin with clavulanic acid, cephalosporins and potentiated sulphonamides are good first choices (Verstegen *et al.*, 2008). It is recommended that antibiotic therapy be continued for at least 10 days after complete resolution of the pyometra, as assessed by ultrasonography and blood tests.

Prognosis
Even after successful medical treatment the recurrence rate varies between 10 and 25%. Recurrence is most likely to occur during dioestrus of the next season; therefore, it is recommended that the bitch be mated at the next oestrus, as pregnancy may be the best prevention against recurrence. The use of an androgen receptor agonist such as mibolerone may help the endometrium to heal by prolonging the anoestrous stage (Verstegen *et al.*, 2008). Pregnancy rates after non-surgical treatment of pyometra are between 50 and 70%, and reports confirm that several litters may be produced from successfully treated bitches (Fontaine *et al.*, 2009b).

Mucometra
Mucometra (hydrometra) is the accumulation of mucoid fluid in the uterus. It is rarely diagnosed in the bitch. However, 13% of patients in one study were found to have varying degrees of mucometra (Pretzer, 2008). The pathogenesis is unclear. Some authors have associated the disease with CEH and/or progesterone stimulation of the endometrium,

which causes a high degree of mucus hydration (Kennedy and Miller, 1992). As the cervix is closed the fluid accumulates inside the uterus (sometimes up to 500 ml of fluid). Repeated administration of progestogens may sometimes play a role in the development of this disease. Mucometra is often asymptomatic. Diagnosis is made by ultrasonography (Figure 18.20). The treatment of choice is ovariohysterectomy. The use of PGF2α alone has sometimes been successful as a non-surgical treatment (Verstegen *et al.*, 2008).

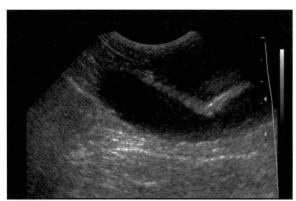

18.20 Mucometra. Note the anechoic appearance of the mucoid uterine contents. Mucus cannot be distinguished from the pus seen in cases of pyometra. (© A. Fontbonne)

Uterine serosal inclusion cysts
Uterine serosal inclusion cysts are fluid-filled structures within the uterine serosa that may lead to grape-like aggregated cysts, usually <1 cm in diameter. Sometimes only the cervix is affected. Serosal cysts form as a result of rapid involution of the myometrium with mesothelial infolding during the postpartum period, especially in older pluriparous bitches. Vulvar discharge is often the only clinical sign that leads to a veterinary investigation.

Uterine tumours
Growth of endometrial tissue into the lumen of the uterus can occur locally and, when associated with fibrosis, may develop into benign endometrial polyps. However, the most common uterine tumours in the bitch are leiomyomas. These can take three forms: intramural, intraluminal or expanding from the serosal surface into the peritoneum (Schlafer, 2003) (Figure 18.21). Mixed tumours that resemble angiolipoleiomyomas have also been found. Leiomyosarcomas and endometrial carcinomas are very rare. Haemangiosarcomas may also be found in the uterus, leading to ascites and a haemorrhagic vulvar discharge.

Miscellaneous conditions

Developmental abnormalities
Developmental abnormalities of the uterus include shortened uterine horns and segmental aplasia, which results in the distension of the proximal part of the horn and the accumulation of fluid. In West Highland White Terriers and related Scottish breeds (Cairn Terrier, Scottish Terrier), uterine segmentations of the right horn (most common) may occur,

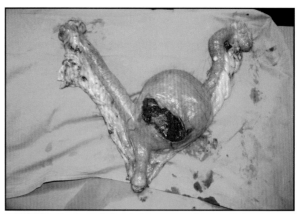

18.21 Benign uterine leiomyoma after ovariohysterectomy and incision of the horn. The affected bitch was just bleeding slightly from the vulva and was in good general health. (Courtesy of Alfort Veterinary College)

resembling 'white heifer disease' in cattle, in which an autosomal recessive gene *MDH01* has been found (Bamas, unpublished data).

Adenomyosis
Endometrial glandular tissue can penetrate into the adjacent myometrium, resulting in adenomyosis. When the glandular tissue proliferates it can weaken the uterine wall, leading to uterine rupture during pregnancy.

Segmental endometrial hyperplasia
Focal, placenta-like proliferations of the endometrium occur rarely and lead to a segmental distension of the uterine wall. This disease is referred to as segmental or focal endometrial hyperplasia.

Uterine torsion
Uterine torsion is rare in the bitch (only 15 cases have been described in the literature) compared with other species. Most cases occur during pregnancy (60%), but pyometra may also be a predisposing cause.

Pathology of the ovaries

There are few data in the literature regarding the incidence of ovarian disease in the bitch. Improved methods of diagnostic imaging will bring more results in the future and may show that these diseases are not uncommon.

Ovarian tumours
Ovarian neoplasia is relatively uncommon in the bitch. Most ovarian tumours are not hormonally active and clinical signs are not always obvious. Several different types of ovarian tumour are found in the bitch:

- Epithelial cell tumours:
  o Most are malignant tumours such as adenocarcinomas (including papillary adenocarcinomas, cystadenocarcinomas)
  o Non-malignant tumours such as adenomas, cystadenomas and fibromas.

- Stromal tumours such as granulosa cell tumours (GCTs), which are malignant in 50% of cases and, very rarely, Scrtoli cell-like tumours have been described
- Germ cell tumours such as dysgerminomas (sometimes called ovarian seminomas because it has been suggested that they may originate from male germ cells), teratomas and mixed tumours. Teratomas often contain differentiated tissues from two or three cell lines, such as ectoderm (hair, sweat and sebaceous glands, nervous tissue), mesoderm (cartilage, bone, teeth, muscle, even crystalline lens) and/or endoderm (respiratory and intestinal epithelium) (Figure 18.22). Dermoid ovarian cysts are a form of teratoma, containing cysts lined with an epiderm. All these germ cell tumours are malignant
- Non-specific soft tissue tumours such as haemangiomas, haemangiosarcomas, leiomyomas and leiomyosarcomas
- Metastatic lymphosarcomas or other carcinomas (e.g. mammary gland tumours) may occur in the ovary.

18.22 Teratoma from a Boxer bitch after surgical removal and incision. Note the hair inside the ovary. (© A. Fontbonne)

Epidemiology

The reported incidence of ovarian neoplasia is <6% of cancers that affect the bitch. According to Johnston *et al.* (2001), they represent around 20% of ovarian disease in the bitch. The incidence of different types of tumours is: epithelial tumours 45%, stromal tumours 34% (mostly GCTs) and germ cell tumours around 20%. Dysgerminomas and teratomas are rare and occur at equal frequency.

The average age of diagnosis is around 6–8 years, slightly younger in the case of teratomas (mean age of 4.3 years). Adenocarcinomas and adenomas often occur bilaterally (>30% of cases; Ettinger, 1983). Even if not neoplastic, the contralateral ovary is often inactive in cases of cystadenoma. In contrast, GCTs and teratomas are usually unilateral. Purebred bitches may be predisposed, especially Boxers, Boston Terriers and German Shepherd Dogs. English Bulldogs may be predisposed to GCTs and English Pointers to epithelial cell tumours.

Pathogenesis

A genetic predisposition has been suggested, as well as environmental risk factors. Some hormonal treatments may increase the risk of development of ovarian tumours (e.g. mibolerone; Ettinger, 1983).

Clinical signs

The clinical signs depend on the histology of the tumour.

- Epithelial tumours may be asymptomatic. If clinical disease develops, these tumours often lead to ascites (due to obstruction of the lymphatic vessels by the tumour or secondary obstructions caused by metastases) and abdominal distension. Neoplastic cells may be present in the abdominal exudate. Accumulation of pleural fluid may be found, due to lymphatic obstruction or to the presence of pulmonary metastases. Lameness may be seen in cases with bone metastases.
- GCTs may also induce ascites and abdominal distension. Non-specific signs include vomiting, diarrhoea and weight loss. GCTs may produce oestrogen, and sometimes also progesterone, thus inducing oestrogen-related clinical signs such as persistent oestrus and prolonged bloody or purulent vulvar discharge. In the most severe cases anaemia and/or leucopenia may be seen, on rare occasions with thrombocytopenia (due to bone marrow suppression). Oestrogen-related alopecia may develop, affecting the perineum and the caudal aspect of the hindlegs (Figure 18.23). Mammary gland development or even lactation may occur. In cases in which the tumour secretes progesterone long anoestrous phases occur, which when left undiagnosed develop into uterine CEH/pyometra in 95% of cases.

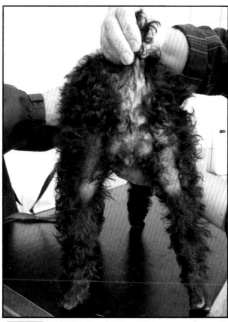

18.23 Alopecia of the perineum and caudal aspect of the hindlegs due to excessive oestrogenic stimulation. Such localization should initiate a search for an ovarian problem. (© A. Fontbonne)

- Germ cell tumours are often asymptomatic or accompanied by abdominal distension. They do not normally alter cyclicity. Some cases have been associated with oestrogen-induced clinical signs (alopecia, oestrous vaginal smear, anaemia) for unclear reasons. Teratomas may induce autoimmune haemolytic anaemia, attributable to an antibody reaction between red blood cells and some of the elements of the mucus produced by these tumours.

Metastasis

Epithelial cell tumours have a rate of metastasis of up to 50%. They disseminate mostly into the peritoneum and to the serosal surfaces of the abdominal organs. The degree of malignancy of epithelial cell tumours depends on size, mitotic index and degree of invasion into the surrounding tissues. The rate of metastasis of GCTs is <20%, mostly into the abdomen, but the lungs may also be affected. The metastases may remain hormonally active even after ovariectomy. Dysgerminomas may metastasize in 10–20% of cases, most commonly in the abdomen. The rate of metastasis in teratomas is higher, at 30–50%. Most are intra-abdominal, but they may also metastasize to the bones and lungs.

Diagnosis

Large ovarian tumours may be palpable abdominally. Radiography reveals a soft tissue mass that displaces the caudal pole of the kidney ventrally and the colon medially. Mineralization may be found in teratomas. Ultrasonography is the best method to diagnose ovarian tumours (Diez-Bru *et al.*, 1998); they appear as rounded or irregular structures with a heterogenous density. Some masses are hypoechoic and others hyperechoic, and either solid, solid with cystic components or cystic (Figure 18.24). Ovarian tumours are often large (5–10 cm, up to 30 cm in some germinomas) and contain cysts of variable sizes. Ultrasonography can also be used to look at uterine changes such as CEH at the same time.

In cases of GCT, vaginal cytology may help to detect the presence of oestrogen. It is useful also to assay plasma progesterone, because it is often increased above basal concentrations (range: 0.5–11 ng/ml). Hypercalcaemia and an elevated concentration of blood parathyroid hormone-related protein have been described in a case of papillary adenocarcinoma (Hori *et al.*, 2006).

Treatment and prevention

Ovariectomy is the treatment of choice. During surgery tumours should be handled with extreme care because shedding of cells from the tumour into the abdominal cavity must be avoided. Thorough lavage of the abdomen at the end of surgery is recommended. In breeding bitches where only one ovary is affected, unilateral ovariectomy may be considered. Complications may result when long-term damage to the uterus has already occurred as a result of persistantly elevated hormone concentrations.

In cases of unilateral GCT the contralateral ovary is inactive in 31% of cases (Johnston *et al.*, 2001). In cases of teratoma both ovaries and the uterus should be removed owing to the high risk of metastasis. With epithelial cell tumours, adjuvant chemotherapy is recommended in cases of metastasis (Figure 18.25). In cases of GCT, immunotherapeutic protocols have also been described (Hayes and Harvey, 1979).

Routine ultrasonography of the ovaries at least once a year is recommended to enable early detection of ovarian tumours in elderly bitches.

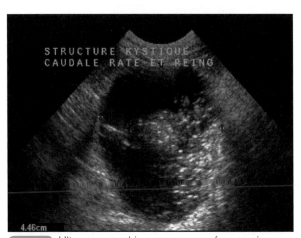

18.24 Ultrasonographic appearance of an ovarian teratoma. Note the solid and liquid parts, which is mucus present within the ovary. (© A. Fontbonne)

Chemotherapeutic agent	Dosage	Comments
Doxorubicin	Six injections of 30 mg/m², one slow intravenous administration every 3 weeks	First administration 10 days after ovariectomy
Cyclophosphamide + Chlorambucil	Cyclophosphamide 50 mg/m² three times weekly plus chlorambucil 8 mg/m² twice weekly	
Cisplatin	Six injections of 60–70 mg/m², one slow intravenous administration every 3 weeks	**Do not use in bitches with renal dysfunction** Intracavitary administration is one of the potential options for the management of diffuse abdominal metastases
Carboplatin	300 mg/m² every 3 weeks In a perfusion of isotonic glucose solution: 100 mg carboplatin in 100 ml of solution	Risk of transient bone marrow aplasia Vomiting/renal failure (low risk)

18.25 Suggested chemotherapy protocols for malignant ovarian tumours. (McEntee, 2002; Lévy, 2005; Tierny *et al.*, 2008).

Ovarian cysts

Ovarian cysts are fluid-filled structures that are distinct from and generally larger than normal follicles. Follicular cysts are well known to veterinary surgeons, but in fact there are many different kinds of cyst.

Follicular cysts

Follicular cysts are structures >8 mm in diameter that are present during pro-oestrus or prior to ovulation in oestrus, or follicles of any size that are present during late oestrus (postovulation), dioestrus or anoestrus (Johnston *et al.*, 2001).

Epidemiology: The reported incidence of follicular cysts varies greatly (3–62%, depending on the study). Follicular cysts may be single or multiple, but they seem to be unilateral in the majority of cases. They can measure up to 20 cm or more. Follicular cysts are often discovered after pyometra has occurred and the average age at diagnosis is 8 years. According to the author's clinical data, cysts can also be found in very young bitches. The clinical signs seem to appear earlier in cases with single larger cysts than in those with multiple cysts. There may be a predisposition in large breeds. The heritability of the condition in the bitch is unknown.

Pathogenesis: Most cysts are found without any clinical signs or obvious cause. Some drugs, such as oestrogens, progestogens and anti-oestrogens, may induce the development of follicular cysts. In these cases cysts are follicles that have not undergone atresia. As such, they are lined primarily with granulosa cells. The follicular fluid contains variable but often high concentrations of oestrogens, progesterone and testosterone.

Clinical signs: Follicular cysts can contain high quantities of oestradiol (up to 143 pg/ml), and therefore release oestrogens into the blood. The clinical signs may include those of CEH and/or pyometra (57% of cases), mammary gland, ovarian or uterine neoplasia, symmetrical alopecia (see Figure 18.23) and hyperkeratosis. The oestrous cycle is usually irregular, with shortened interoestrous intervals and prolonged oestrus. Mammary gland enlargement and lactation may occur, even during oestrus. Rarely, in cases with large cysts, abdominal distension may be present.

Diagnosis: Given the clinical signs, the main differential diagnoses are ovarian tumours, especially GCTs. Elevated concentrations of serum oestrogens are difficult to detect owing to the pulsatile nature of their release. Vaginal cytology is a quick and easy way of detecting the presence of oestrogens. Ultrasonography will usually confirm the diagnosis, with the cysts appearing as round hypoechoic or anechoic structures with a thin wall (Figure 18.26).

Treatment: Non-breeding bitches should undergo ovariectomy or ovariohysterectomy. In young bitches that are still intended for breeding, induction of luteinization of the cysts by hormones has been described in the literature. In the author's experience

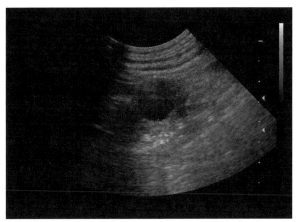

18.26 Ultrasonographic appearance of a follicular cyst in a Golden Retriever bitch suffering from shortened interoestrous intervals. (© A. Fontbonne)

this can promote the development of pyometra, because the treatment increases plasma progesterone concentrations, which then act on the uterus after a prolonged oestrogen phase.

In the author's clinical department, some success has been achieved using ultrasound-guided aspiration of the cysts. In women, repeated aspirations are often necessary. Alternatively, 95% ethanol or tetracyclines may be injected to induce fibrosis of the internal cell layer of the cyst. However, the best outcome seems to result from surgical removal of the cysts (Lévy *et al.*, 2007; Stratmann and Wehrend, 2007) (Figure 18.27).

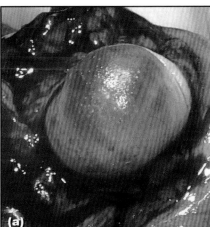

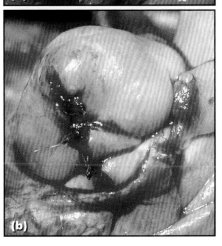

18.27 A follicular cyst **(a)** just before and **(b)** just after surgical removal (cystectomy). (Courtesy of X I évy)

Luteal cysts

Luteal cysts are luteinized anovulatory follicles. They represent <10% of ovarian cysts in the bitch. On ultrasound examination they look quite similar to follicular cysts, but with a thicker wall. They secrete progesterone, inducing long periods of anoestrus. In some cases they can also secrete oestrogens, resulting in clinical signs that include bilateral alopecia.

Cystic corpora lutea

Cystic corpora lutea occur rarely in the bitch and are asymptomatic.

Cystic subepithelial structures

Cystic subepithelial structures (SES) are microscopic folds of the external ovarian epithelium (modified peritoneal cells) that protrude into the ovarian stroma. The cells lining these structures commonly undergo hyperplasia and cystic distension. They can also undergo neoplastic changes and develop into cystadenomas or cystadenocarcinomas.

Cystic rete ovarii

Cystic rete ovarii are quite common. They are cystic enlarged tubules, derived from mesonephric tubules, around the ovary. They may sometimes replace the surrounding normal ovarian tissue, without resulting in any overt disease.

Paraovarian cysts

Paraovarian cysts develop from remnants of the mesonephric and paramesonephric structures surrounding the ovaries. They do not impair ovarian function. It is not always easy to differentiate them from ovarian follicular cysts using ultrasonography.

Ovarian remnant syndrome

Ovarian remnant syndrome is defined as the persistence of ovarian activity in a neutered bitch, which usually becomes obvious during pro-oestrus and oestrus.

Aetiology

The possible origins of ovarian remnant syndrome are:

- A fragment or an entire ovary left *in situ* at the time of neutering (this occurs more frequently when the ovarian bursa is opened before removing the ovary)
- Ectopic tissue – referred to as 'extra-ovarian tissue'. This is very uncommon, but may account for some cases (Schlafer, 2003). Ectopic adrenal tissue is sometimes located within the mesovarium near the ovary
- Oestrogenic drugs administered for urinary incontinence in an excessive dosage
- The owner's oestrogenic medication (for example patches or gels to treat the symptoms of menopause), if consumed by the bitch.

Clinical signs

Ovarian remnant syndrome may appear 3 months to 7 years after ovariectomy (average 1.5 years). The clinical signs are:

- Oestrus with oedema of the vulva, attraction of males, vulvar discharge and varying degrees of oestrous behaviour
- Perineal alopecia (see Figure 18.23)
- Lactation in a neutered bitch.

Diagnosis

Vaginal cytology shows typical pro-oestrus or oestrus cells. A progesterone concentration of >2 ng/ml confirms luteal activity and the existence of ovarian tissue. Basal progesterone concentrations do not exclude the possibility of an ovarian remnant, because the luteal phase may not be present at the time of blood sampling. It is very difficult to locate the ovarian remnant using ultrasonography, owing to its small size, but it can sometimes be found caudal to one of the kidneys as an anechoic or hypoechoic round cyst-like structure (Figure 18.28).

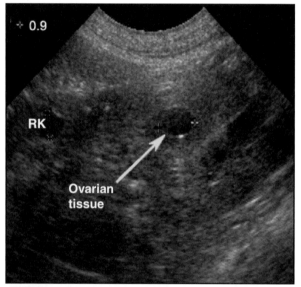

18.28 Ultrasonogram of the area just caudal to the kidney (RK) in a bitch with ovarian remnant syndrome. Note the round hypoechoic structure (arrowed). (Courtesy of X. Lévy)

Treatment

The owner should be carefully questioned to ensure that there is no iatrogenic origin. Exploratory laparotomy is the treatment of choice because lifelong medical suppression of oestrus is not recommended. The remnant is located more easily in the follicular or luteal phase because the follicles or corpora lutea enlarge the size of the tissue. Bilateral ovarian remnants are present in 35% of cases and both ovarian pedicles should be examined carefully. The excised tissue should be submitted for histopathology (Figure 18.29).

Miscellaneous conditions

- *Ovarian agenesis* is rare and often associated with other abnormalities of the genital tract (e.g. uterus unicornus).
- *Ovarian hypoplasia* may occur in bitches with chromosomal abnormalities (e.g. X monosomy: 77X0; X trisomy: 79XXX). In these cases the

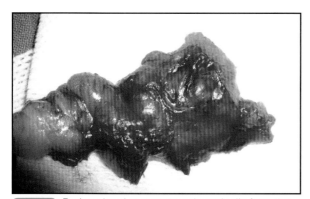

18.29 Periovarian tissue removed surgically from the bitch in Figure 18.28. Note the presence of a small remnant cyst. (Courtesy of X. Lévy)

external genitalia are often normal. One case of a clinically prolonged pro-oestrus period in a 77X0 Miniature American Esquimo has been described (Lofstedt *et al.*, 1992).
- Rare cases of parasitic *hydatid cysts* of the ovary have been described (Usha *et al.*, 1994).

Endocrine abnormalities

Pseudopregnancy
Pseudopregnancy, also called false pregnancy, phantom pregnancy, pseudocyesis or galactorrhoea, is a frequent clinical finding among intact bitches. Of intact bitches, 87% will exhibit signs of false pregnancy at some time during their lives; 64% of intact bitches develop signs of pseudopregnancy on a regular basis.

In the bitch, the long luteal phase, which mimics the endocrinological changes of a real pregnancy, is sometimes called 'covert pseudopregnancy'. Clinical pseudopregnancy (also called 'overt pseudopregnancy') is characterized by mammary gland development combined with lactation and behavioural changes (nesting, mothering of inanimate objects, aggression to other dogs).

Pseudopregnancy usually develops 1.5–3 months after oestrus (at the end of normal dioestrus), when the decrease in progesterone causes a rise in prolactin and oestrogen. Other physiological or pathological conditions may also induce pseudopregnancy. It can sometimes even be seen in male dogs, which show mammary gland development and lactation.

For information on clinical signs, diagnosis and treatment of pseudopregnancy, see Chapter 17.

Effect on mammary gland tumours
Whether repeated pseudopregnancies play a role in the development of mammary gland tumours is controversial. Only 30% of malignant mammary gland tumours in dogs have prolactin receptors, but in women more than 40% of breast cancers that show metastasis are associated with hyperprolactinaemia. In addition, prolactin induces mitosis in 80% of cancer cells, either *in vitro* or *in vivo*.

Some authors (Donnay *et al.*, 1994; Verstegen and Onclin, 2003) believe that there is an increased risk of mammary gland tumours in animals that have

shown previous pseudopregnancies. According to these authors, such tumours are detected earlier and are more often malignant. Verstegen and Onclin (2003) report a significant risk after a bitch has undergone three clinical pseudopregnancies, which is associated with continuous distension of the mammary gland acini and local accumulation of carcinogenic products caused by milk retention. Furthermore, prolactin may play a role in promoting pre-neoplastic lesions. These observations indicate that the prevention and treatment of pseudopregnancy may have added health benefits.

Other endocrine abnormalities
Any endocrine disease may have an effect on the genital tract. In particular, hypothyroidism has been reported to cause variable interoestrous intervals, prolonged anoestrus, infertility, abortion and stillbirth.

Acknowledgement

The author wishes to thank Mrs Felicity Leith-Ross for her great help in improving the English language of this Chapter.

References and further reading

Bilgliardi E, Parmigiani E and Cavirani S *et al.* (2004) Ultrasonography and cystic hyperplasia–pyometra complex in the bitch. *Reproduction in Domestic Animals* **39**, 136–140

Campbell BG (2004) Omentalization of a nonresectable uterine stump abscess in a dog. *Journal of the American Veterinary Medical Assocation* **224**(11), 1799–1803

Corrada Y, Arias D, Rodriguez R *et al.* (2006) Combination dopamine agonist and prostaglandin agonist treatment of cystic endometrial hyperplasia complex in the bitch. *Theriogenology* **66**(6–7), 1557–1559

De Bosschere H, Ducatelle R, Vermeirsch H, Simoens P and Coryn M (2002) Estrogen-α and progesterone receptor expression in CEH and pyometra in the bitch. *Animal Reproduction Science* **70**, 251–259

De Bosschere H, Ducatelle R, Vermeirsch H, van den Broeck W and Coryn M (2001) CEH–pyometra complex in the bitch: should the two entities be disconnected? *Theriogenology* **55**, 1509–1519

Diez-Bru N, Garcia-Real I, Martinez EM *et al.* (1998) Ultrasonographic appearance of ovarian tumors in 10 dogs. *Veterinary Radiology and Ultrasound* **39**(3), 226–233

Donnay I, Rauis J and Verstegen J (1994) [Influence des antécédents hormonaux sur l'apparition clinique des tumeurs mammaires chez la chienne. Etude épidémiologique.] *Annales de Medicine Veterinaire* **138**, 109–117

Downs M, Miller-Liebl D, Fayrer-Hoskin R and Caudle A (1994) Obtaining a useful uterine biopsy specimen in dogs. *Veterinary Medicine* **89**, 1055–1059

England GC, Freeman SL and Russo M (2007) Treatment of spontaneous pyometra in 22 bitches with a combination of cabergoline and cloprostenol. *Veterinary Record* **160**(9), 293–296

Ettinger SJ (1983) The reproductive system. In: *Textbook of Veterinary Internal Medicine: Diseases of the Dog and Cat, 2nd edn*, ed. SJ Ettinger, pp.1818–1819. WB Saunders, Philadelphia

Fieni F (2006) Clinical evaluation of the use of aglepristone, with or without cloprostenol, to treat cystic endometrial hyperplasia–pyometra complex in bitches. *Theriogenology* **66**, 5–10

Fontaine E, Bassu G, Levy X *et al.* (2009b) Fertility after medical treatment of uterine diseases in the bitch: a retrospective study on 24 cases. *Proceedings of the 6th EVSSAR Meeting, Wroclaw, Poland*, p.66

Fontaine E, Lévy X, Grellet A *et al.* (2009a) Diagnosis of endometritis in the bitch: a new approach. *Reproduction in Domestic Animals* **44** (Suppl. 2), 196–199

Gobello C, Castex G, Broglia G and Corrada Y (2003) Coat colour changes associated with cabergoline administration in bitches. *Journal of Small Animal Practice* **44**, 352–354

Gobello C, Concannon PW and Verstegen J (2001) Canine pseudopregnancy: a review. In: *Recent Advances in Small Animal*

Reproduction, ed. PW Concannon *et al.* IVIS (www.ivis.org)

Gobello C, de la Sota L, Castex G, Baschar H and Goya R (2001) Diestrous ovariectomy: a model to study the role of progesterone in the onset of canine pseudopregnancy. *Journal of Reproduction and Fertility* **Suppl.57**, 55–60

Groppetti D, Pecile A, Arrighi S and Cremonesi F (2007) Endometrial cytology in the bitch: nuclei measurement in different phases of the estrous cycle and in uterine disorders. *Theriogenology* **68**, 499 (abstract)

Groppetti D, Pecile A, Arrighi S, Di Giancamillo A and Cremonesi F (2010) Endometrial cytology and computerized morphometric analysis of epithelial nuclei: a useful tool for reproductive diagnosis in the bitch. *Theriogenology*, **73(7)**, 927–941

Günzel-Apel AR Wilke M, Aupperie H and Schoon HA (2001) Development of a technique for transcervical collection of uterine tissue in bitches. *Journal of Reproduction and Fertility* **Suppl. 57**, 61–65

Hayes A and Harvey HJ (1979) Treatment of metastatic granulosa cell tumor in a dog. *Journal of the American Veterinary Medical Association* **174**(12), 1304–1306

Hori Y, Uechi M, Kanakubo K, Sano T and Oyamada T (2006) Canine ovarian serous papillary adenocarcinoma with neoplastic hypercalcemia. *Journal of Veterinary Medical Science* **68(9)**, 979–982

Johnston SD, Root-Kustritz M and Olson P (2001) *Canine and Feline Theriogenology.* WB Saunders, Philadelphia

Kennedy PC and Miller RB (1992) The female genital system. In: *Pathology of Domestic Animals, 4th edn*, ed. K Jubb *et al.*, pp. 349–470. Academic Press, San Diego

Lawler DF, Johnston SD, Keltner DG, Ballam JM, Kealy RD *et al.* (1999) Influence of restricted food intake on estrous cycles and pseudopregnancies in dogs. *American Journal of Veterinary Research* **60**, 820–825

Lévy X (2005) [Prise en charge des tumeurs ovariennes.] In: *Cancérologie du Chien et du Chat*, pp. 112–115. Le Point Vétérinaire, Paris

Lévy X, Fontainne E and Gellet A **(2007)** Surgical cysts removal: a new technique for the treatment of ovarian cysts in the bitch. *Proceedings of the 16th APMVEAC Congress, 5th Biannual EVSSAR Symposium.* Estoril, Portugal, p.120

Lévy X, Marseloo N, Cot S *et al.* (2004) Study of the aerobic bacterial flora in the anterior vagina of breeding bitches with reproductive disorders and its sensibility to marbofloxacine. *Proceedings of the European Veterinary Society for Small Animal Reproduction Congress.* Barcelona

Lofstedt RM, Buoen LC, Weber AF, Johnson SD, Huntington A and Concannon PW (1992) Prolonged proestrus in a bitch with X chromosal monosomy (77, X0). *Journal of the American Veterinary Medical Association* **200**(8), 1104–1106

McEntee M (2002) Reproductive oncology. *Clinical Techniques in Small Animal Practice* **17**, 133–149

Okkens AC, Dieleman SJ, Kooistra HS and Bevers MM (1997) Plasma concentrations of prolactin in overtly pseudopregnant Afghan hounds and the effect of metergoline. *Journal of Reproduction and Fertility Supplement* **51**, 295–301

Pretzer SD (2008) Clinical presentation of canine pyometra and mucometra: a review. *Theriogenology* **70**, 359–363

Romagnoli S, Fieni F, Prats A *et al.* (2006) Treatment of canine open-cervix and closed-cervix pyometra with combined administration of aglepristone and misoprostol. *Proceedings of the 5th Biannual EVSSAR Congress*, Budapest, Hungary, p. 287

Root-Kustritz MV (2006) Collection of tissue and culture samples from the canine reproductive tract. *Theriogenology* **66**, 567–574

Schlafer DH (2003) Pathology of the canine ovary and uterus: what's important, what's not. *Proceedings of the SFT Annual Conference*, Colombus, Ohio, pp.212–220

Siliart B (1992) Endocrionologie Clinique. In: *Reproduction du chien et du chat*, ed. PMCAC, Paris, France

Stratmann N and Wehrend A (2007) Unilateral ovariectomy and cystectomy due to multiple ovarian cysts with subsequent pregnancy in a German shepherd dog. *Veterinary Record* **160**, 740–741

Tierny D *et al.* (2008) [Chimiothérapie et radiothérapie des tumeurs de l'appareil génital.] *Le Nouveau Praticien Vétérinaire* **38**, 38–46

Trasch K, Wehrend A and Bostedt H (2003) Follow-up examinations of bitches after conservative treatment of pyometra with the antigestagen aglepristone. *Journal of Veterinary Medicine A Physiology Pathology Clinical Medicine* **50**(7), 375–379

Troxel MT, Cornetta AM, Pastor KF, Hartzband LE and Besancon MF (2002) Severe hematometra in a dog with CEH/pyometra complex. *Journal of the American Animal Hospital Association* **38**, 85–89

Usha, Tiwari PV, Khanna A and Garg M (1994) Hydatid cyst of the ovary: a case report. *Indian Journal of Pathology and Microbiology* **37**(3), 349–351

Verstegen J, Dhaliwal G and Verstegen-Onclin K (2008) Mucometra, cystic endometrial hyperplasia and pyometra in the bitch: advances in treatment and assessment of future reproductive success. *Theriogenology* **70**, 364–374

Verstegen J and Onclin K (2003) Etiopathogenesis, classification and prognosis of mammary tumours in the canine and feline species. In: *Proceedings of the SFT Annual Congress*, Colombus, Ohio

Verstegen J, Onclin K and Romagnoli S (2000) [Prolattina: funzioni fisiologiche dell'ormone ed applicazioni cliniche dell'impiego degli antiprolattinici in cagna e gatta.] *Riv. Zoot. Vet.* **28**(1), 3–18

Watts JR and Wright PJ (1995) Investigating uterine disease in the bitch: uterine cannulation for cytology, microbiology and hysteroscopy. *Journal of Small Animal Practice* **36**, 201–206

19

Clinical approach to conditions of the non-pregnant and neutered queen

Eva Axnér

Introduction

Ovariohysterectomy of queens is one of the most common veterinary procedures and has a protective effect against most diseases of the female reproductive organs. Ovariohysterectomy is also curative for many of the conditions that can occur in the female genitalia. However, treatment of reproductive problems is not limited to ovariohysterectomy. Problems related to the reproductive tract can occasionally occur in spayed females, and some owners may want to conserve the breeding value of the queen, which is why other treatment options may be required. Knowledge of normal reproductive physiology and techniques for evaluation of reproductive organs are essential to make a correct diagnosis and to differentiate between normal and pathological conditions.

Vulvar, vestibular and vaginal abnormalities

In contrast to the bitch, very little as been described about vulvar, vestibular and vaginal abnormalities in the queen. The anatomy and size of the queen make these abnormalities more difficult to diagnose than in the bitch.

Developmental abnormalities

Abnormalities of chromosomal, genetic or gonadal sex may lead to malformation of the reproductive tract, often referred to as 'intersex'. Intersex animals often have ambiguous external genitalia. Male pseudohermaphrodites with male gonads and feminization of the external genitalia have been described. Diagnosis is made by careful visual inspection and palpation of the gonads, establishment of the karyotype and determination of gonadal sex.

Whilst detection of changes in the external genitalia is relatively simple, developmental abnormalities can also occur within the internal genitalia. Failure in the development of the Müllerian duct or in the fusion between the Müllerian duct and the urogenital sinus during embryogenesis can cause atresia or the formation of septa in the female tubular genital organs. Vaginal atresia causes fluid accumulation in the uterus as the primary clinical sign and this may be so significant as to ultimately cause abdominal distension.

Masculinization

Female external genitalia that are initially normal may become masculinized subsequently, and the queen may develop an enlarged barbed clitoris or vaginal hyperplasia. These signs can be caused by a testosterone-producing ovarian thecoma, or overproduction of sex steroids by the adrenal glands (Boag *et al.*, 2004).

Vaginitis

Primary bacterial or viral vaginitis is rare in the queen but may be associated with *Chlamydophila felis* infection, which causes a mucopurulent vulvar discharge. When investigating cases of purulent vulvar discharge, the presence of fetal remnants within the vagina should be excluded because this is a not uncommon cause of vaginal inflammation (Figure 19.1). However, in most cases when there is a vulvar discharge it is related to uterine disease (commonly inflammation) rather than to vaginitis.

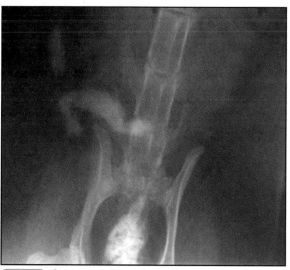

19.1 Contrast radiograph showing the fetal remnant in the vagina of a queen with purulent vaginal discharge. Note also the enlarged uterine horn.

Vaginal tumours

There are very few reports of vaginal tumours in the cat and such tumours appear to be extremely rare. Fibromas and leiomyomas of the vagina may cause constipation as the primary clinical sign as a result of dorsal pressure on the descending colon and rectum.

Uterine abnormalities

Developmental abnormalities

Failure of normal development of the Müllerian duct can result in uterus unicornis or segmental aplasia of a uterine horn. Usually both ovaries and oviducts are present, whilst renal agenesis is common on the affected side in cats with uterus unicornis. Segmental aplasia of a uterine horn can cause accumulation of fluid cranial to the aplastic part. Uterine aplasia is usually diagnosed at surgery. Fluid accumulation as a result of segmental aplasia can cause clinical signs related to the distension of the uterus, whilst uterus unicornis is usually an incidental finding, although it can result in smaller litter sizes.

Cystic endometrial hyperplasia

Cystic endometrial hyperplasia (CEH) is common in the cat. It is characterized by hyperplasia of the endometrial epithelium and cystic dilatation of the endometrial glands (Figure 19.2). The uterus may be lined with translucent cysts. Cats with CEH usually do not have any clinical signs, unless the condition has been complicated by infection; although, rarely vaginal haemorrhage may occur as a result of torsion of large cysts that are polypoid in nature.

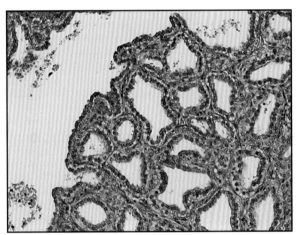

19.2   Endometrium sample from a queen with cystic endometrial hyperplasia. Note the multiple enlarged and dilated glands. (H&E stain.)

The incidence of CEH increases with age, probably because of the cumulative nature of the condition. Oestradiol and progesterone induce cyclical endometrial growth that normally regresses during inter-oestrus or anoestrus (Chatdarong et al., 2005). It is believed that repeated hormonal stimulation predisposes to abnormal endometrial proliferation. Uncomplicated CEH is not associated with any specific stage of the oestrous cycle. Repeated oestrous cycles, spontaneous ovulation and treatment with progestogens are likely to be predisposing factors because of the stimulatory effect of oestrogens and progesterone on the endometrium. Ultrasonography may reveal an enlarged uterus with uneven echogenicity of the uterine wall. The diameter of healthy uterine horns (which contain no fluid) is usually 0.4–0.7 cm depending on the stage of the cycle, whilst the diameter is usually increased in a uterus with CEH.

Metritis, endometritis and pyometra

Inflammation of the uterus is referred to as metritis, endometritis or pyometra depending on the nature of the lesions. In a queen with pyometra the syndrome has resulted in accumulation of pus within the uterine lumen. Signs reported by the owners include vulval discharge, anorexia, abdominal distension, lethargy, vomiting and weight loss. Polydipsia and polyuria are seen in approximately 6–9% of affected cats. The most common clinical sign found on physical examination is vulval discharge, which is seen in 68–100% of cases. Other common (30–100%) clinical signs are abdominal distension, a palpably enlarged uterus and the presence of dehydration. Pyrexia is seen in approximately 20% of cases. Leucocytosis and neutrophilia with a left shift are common. However, some cats with pyometra have leucopenia or a normal white blood count. Anaemia, hyperproteinaemia and hyperglobinaemia may also be present. Radiography can be used to diagnose uterine enlargement. Abdominal ultrasonography reveals enlarged fluid-filled uterine horns containing anechoic to hypoechoic fluid.

Aetiology and predisposing factors

The bacteria isolated from the uteri of queens with pyometra are of the same species as those found in the normal vaginal flora. The uterus is usually a sterile environment but during oestrus the cervix is open and bacteria may enter from the vagina. The most common bacterium isolated is *Escherichia coli*, but other bacteria such as *Streptococcus* spp., *Staphylococcus* spp., *Pasteurella* spp., *Klebsiella* spp. and *Moraxella* spp. may also be isolated. Normally these bacteria are probably eliminated by natural local defence mechanisms, but if the bacteria persist then metrititis, endometritis or pyometra may develop.

There is an association between pyometra and cyclicity; pyometra is often diagnosed within 2 months of the last oestrus and is less common during the winter months when cats often are in anoestrus. A high proportion of cats with pyometra have ovaries in the luteal phase (42–77% of cases), but the disease can sometimes also be found together with ovaries in the cystic or follicular phase, as well as in cats with inactive ovaries. CEH is considered to be a predisposing factor for the development of pyometra, and they are often seen together. Considering the association between luteal phase ovaries and cyclicity, it is likely that frequent oestrous cycles as well as spontaneous ovulation or treatment with exogenous progesterone are predisposing factors for the condition.

Treatment

The first choice treatment for pyometra is ovariohysterectomy together with supportive treatment. However, it is often possible to treat the condition medically, although there are no drugs authorized for this use in the cat. The best candidates for medical treatment are queens in good body condition without any cardiac or pulmonary disease. The aims of treatment are to terminate the luteal phase or to block the effect of progesterone, and to cause evacuation of the uterus. In addition, supportive measures may be necessary and queens may need to be hospitalized during treatment. The effect of treatment should be monitored by repeated ultrasound examinations. If the uterine diameter does not decrease in size, the queen should undergo ovariohysterectomy.

Administration of prostaglandins to achieve uterine evacuation and to promote luteolysis has been demonstrated to be an effective treatment. However, high doses can cause uterine rupture in queens with closed-cervix pyometra. Treatment with 0.1 mg/kg natural prostaglandin F2α (PGF2α) s.c. q12–24h for 3–5 days has been recommended for queens with open-cervix pyometra. Recurrence of the disease may occur but queens can often be treated successfully again. Fertility rates after treatment are close to 90%. Side effects after injection of PGF2α include vocalization, panting, restlessness, grooming, tenesmus, salivation, diarrhoea, mydriasis, emesis, urination and lordosis. Some cats do not show any physical reactions at all and the total gamut of side effects is not observed in all cats. The side effects usually resolve within 1 hour. The severity of the side effects diminishes after each subsequent injection, which is why it is recommended to give a lower initial dose. It has been suggested that lower doses (50 μg/kg natural PGF2α) should be used and administered more frequently (3–5 times/day for 5–7 days) to reduce the side effects further (Verstegen, 2000).

The anti-progestogen aglepristone (RU46534) binds to progesterone receptors and thereby inhibits the effects of progesterone. Administration of 10 mg/kg bodyweight aglepristone subcutaneously on days 1, 2, 7 and 14 was shown to be an effective short-term and long-term (2 years) treatment for pyometra in the cat; the overall recovery rate was 90% (Nak *et al.*, 2009). It is also possible to use aglepristone and PGF2α together to have the combined effect of blocking the effect of progesterone (aglepristone) and promoting efficient uterine evacuation via stimulation of myometrial contractions (PGF2α) (Figure 19.3). It is recommended that medical treatment to evacuate the uterus is combined with the use of antimicrobial agents. Suitable first choice antibiotics are enrofloxacin or trimethoprim/sulfadiazine.

Drug	Dose	Reference
Prostaglandin (PGF2α)	0.1 mg/kg s.c. q12–24h for 3–5 days	Davidson *et al.*, 1992
	50 μg/kg s.c. 3–5 times/day for 5–7 days	Verstegen, 2000
Aglepristone	10 mg/kg s.c. on days 1, 2, 7 and 14	Nak *et al.*, 2009

19.3 Medical treatment of pyometra. Treatment with PGF2α and aglepristone can be combined.

Mucometra

Mucometra (hydrometra) is characterized by a uterus that is filled with a clear and watery to mucinous non-infectious fluid. The main clinical sign is abdominal distension caused by the enlarged uterus. Usually the queen remains in good health except

for the uterine enlargement. Mucometra can be the result of increased secretory activity of the endometrial glands (e.g. in queens with CEH) or caused by a congenital or acquired obstruction of the female tubular organs. The uterus is often thin walled as a result of the distension. On ultrasonographic evaluation a distended uterus filled with anechoic fluid is seen. The recommended treatment is ovariohysterectomy. However, spontaneous recovery has been reported. Treatment with aglepristone at a rate of 10 mg/kg s.c. twice 24 hours apart may also cause remission. If the mucometra is caused by an obstruction and/or there is fibrosis of the uterine wall, recovery of fertility is not likely.

Uterine adenomyosis

Uterine adenomyosis is characterized by the presence of endometrial tissue in the myometrium. The cause has not been completely identified. The condition causes uterine enlargement but may be linked to exogenous hormonal treatment. Treatment in most cases involves ovariohysterectomy.

Uterine tumours

Tumours of the female genital tract are very rare, but occur most commonly in the uterus, accounting for 0.29% of all female neoplasms. The most common uterine tumour is adenocarcinoma. Other tumours that have been reported include adenosarcoma, leiomyosarcoma, fibrosarcoma, carcinosarcoma, lymphosarcoma, fibroma, lipoma and leiomyoma. Although uterine tumours are most common in intact queens, previously spayed females may also develop adenocarcinomas in the uterine stump. Clinical signs vary but include abdominal enlargement, a palpable abdominal mass, weight loss, lethargy, haemorrhagic vaginal discharge, stranguria, constipation and infertility. A presumptive diagnosis is made by abdominal palpation and ultrasonography, but often the tumour is diagnosed during surgery or post-mortem examination. In the absence of metastases the treatment of choice is ovariohysterectomy. The prognosis depends on the type of tumour.

Ovarian abnormalities

Abnormal oestrous cycle

The causes of absence of oestrus and prolonged oestrus are described in Chapter 7 (see also Persistent cyclicity, below).

Ovarian remnant syndrome

Occasionally, spayed females continue to show oestrous behaviour. The cause is almost always that there are remnants of the ovaries left *in situ*. Alternative causes have also been suggested, and it is plausible that small pieces of ovarian tissue that are dropped in the abdomen during surgery can become revascularized and functional. Supernumerary ovaries have also been described, but they are rare. Sometimes nodular paraovarian adrenocortical tissue can be found in the queen. This tissue is not of any clinical significance but can be misdiagnosed as ovarian tissue unless histological examination is performed. Irrespective of the reason for the presence of ovarian tissue in a neutered queen, the recommended treatment is surgical removal of the ovarian remnants. Localization of the small pieces of ovarian tissue can be facilitated if surgery is performed during oestrus or dioestrus when the ovarian follicles or corpora lutea increase the size of the remnant.

Diagnosis

In queens with ovarian remnant syndrome, the interval between ovariohysterectomy and the first occurrence of oestrous behaviour may vary between 17 days and 9 years. Sometimes there may be uncertainty as to whether or not the queen is truly displaying oestrous behaviour. For these reasons it is imperative to make an accurate diagnosis before surgery is attempted. A summary of the tests available for diagnosing the presence of ovaries is given in Figure 19.4.

Vaginal cytology: This can be helpful when performed during the peak of oestrous behaviour; a cornified smear is diagnostic of oestradiol secretion and indicates the presence of ovarian tissue (for details of the collection and interpretation of vaginal smears see Chapter 5). However, vaginal cytology may be inconclusive in some queens because oestrous behaviour often starts before the epithelium is fully cornified.

Hormone concentrations: The basal concentration of oestradiol in intact queens overlaps with values found in neutered females, except during oestrus. However, oestrous behaviour may continue for some days after oestradiol has returned to basal concentrations, and this is why evaluation of oestradiol may not be conclusive even when samples are collected when the queen is displaying oestrous behaviour. Induction of ovulation during oestrus with 25 µg of gonadotrophin-releasing hormone (GnRH) or 500 IU human chorionic gonadotrophin (hCG) and measurement of serum progesterone 1–3 weeks later is the most precise method to diagnose ovarian

remnants. A progesterone concentration above basal confirms the presence of ovarian tissue. Surgery can then be performed during the luteal phase to facilitate localization of the ovarian remnants.

The diagnosis of ovarian remnants by vaginal cytology, basal oestradiol concentrations or induction of ovulation and progesterone analysis requires that the queen is in oestrus. An alternative in a non-oestrous queen could be to induce oestradiol release with the GnRH analogue buserelin (0.4 µg/kg i.m.) and measure plasma oestradiol 2 hours later. This test has 100% specificity and sensitivity to differentiate previously ovariohysterectomized cats from intact queens. Intact females had oestradiol concentrations ≥12 pmol/l and ovariohysterectomized females had concentrations ≤9 pmol/l. However, this test has not been evaluated for diagnosis of ovarian remnant syndrome but only to differentiate between intact and ovariohysterectomized females (Axnér *et al.*, 2008b).

Ovarian tumours

Ovarian tumours are uncommon in the cat, probably at least partly because a large proportion of females are ovariohysterectomized at a young age. Tumours are most common in females >5 years of age, but have also been reported in kittens. The most common ovarian neoplasm in the cat is the granulosa cell tumour. Queens with granulosa cell tumours often demonstrate signs of continuous oestrus and hyperoestrogenism such as CEH and alopecia. Granulosa cell tumours are often malignant in the cat with reports of 44–60% having metastasized at the time of diagnosis.

Other ovarian tumours that have been described in the cat include dysgerminoma, cystadenoma, luteoma and teratoma. Dysgerminomas are usually hormonally inactive but may be associated with both masculine behaviour and continuous oestrus. Luteomas may also be associated with persistent oestrus as well as masculinization, even at the same time. Clinical signs related to ovarian tumours include abnormal oestrous or masculine behaviour, a palpable abdominal mass, abdominal distension, anorexia and depression. A presumptive diagnosis is

Test	Stage of oestrous cycle at which to perform the test	Interpretation
Observation of oestrous behaviour	Oestrus	Pronounced oestrous behaviour usually indicates presence of ovaries or ovarian remnants
Vaginal cytology		Cornified smear indicates presence of ovaries or ovarian remnants
Serum oestradiol concentration		Elevated concentrations indicate presence of ovaries or ovarian remnants
Induction of ovulation with 25 µg GnRH or 500 IU hCG and measurement of serum progesterone concentration 1–3 weeks later		Elevated progesterone concentration indicates presence of ovaries or ovarian remnants
Injection of 0.4 µg/kg buserelin and measurement of serum oestradiol concentration 2 hours later	All	Serum oestradiol concentration >12 pmol/l indicates presence of ovaries. It has not been tested for ovarian remnants but the endocrinological principle is sound

19.4   Diagnostic tests for the presence of ovaries.

made by abdominal palpation and ultrasound examination of the ovaries. For a definitive diagnosis of the tumour type, histological evaluation after ovariohysterectomy is required. The recommended treatment is ovariohysterectomy unless the tumour has metastasized.

Ovarian cysts

Normal ovarian follicles are approximately 3 mm in diameter when mature and are usually present only during oestrus. However, smaller, growing follicles can also be seen during other stages of the cycle. Other cystic structures (i.e. not follicular cysts) can also be identified; cystic distension of the rete ovarii tubules is more common in cats and dogs than in other species, and cystic dilatation of remnants from the cranial mesonephric tubule may be confused with intraovarian cysts. These cystic structures are usually asymptomatic and can be considered as incidental findings unless they have become large and give rise to clinical signs related to their size. However, follicular cysts may produce oestradiol, leading to prolonged oestrus and infertility. The recommended treatment for ovarian cysts that cause clinical signs is ovariohysterectomy. For information on treatment in queens intended for breeding, see Chapter 7.

Endocrine abnormalities

There is surprisingly scant information about reproductive problems related to other endocrine abnormalities in the cat. The adrenal glands produce reproductive hormones as well as corticosteroids. Overproduction of androgens by the adrenal glands has been reported to result in masculinization.

Clinical approach

Vulvar discharge

Vulvar discharge is not normal in the queen except for a small amount of clear discharge during oestrus, the discharge noted during parturition, and the postpartum lochia which persists for approximately 3 weeks (usually a shorter period of time). Primary vaginitis is rare in the cat, although vaginal discharge can be related occasionally to vaginal inflammation. Most often vaginal discharge in the queen is caused by uterine inflammation. Causes of vaginal discharge in the non-pregnant queen include:

- Vaginal irritation caused by trauma (mating)
- Vaginal inflammation caused by chlamydophilosis
- Vaginal inflammation caused by retained fetuses
- Metritis, endometritis and pyometra
- Mucometra
- Uterine neoplasia.

The diagnostic approach should include a thorough clinical evaluation, complete blood count and ultrasonography of the uterus. Radiography of the vagina can be used to detect retained fetuses. A uterine biopsy may be required to make a diagnosis, but

it is a very invasive procedure and rarely performed. If the queen is not going to be used for breeding, the recommended treatment for most cases is ovariohysterectomy. It may be possible to treat mild endometritis with antibiotics following bacterial culture from the vagina. Medical treatment of pyometra usually also requires treatment with PGF2α and/or aglepristone (see above). Vaginal discharge in the postpartum period is described in Chapter 14.

Persistent cyclicity

Continuous oestrous behaviour for >19 days is considered to be abnormal. The differential diagnoses for persistent oestrus include:

- Normal ovarian function with oestrous behaviour in interoestrus
- Normal ovarian function with short interoestrus (i.e. rapid return to oestrus)
- Follicular cysts
- Ovarian neoplasia.

Causes of persistent oestrus are described in more detail in Chapter 7. If the queen is not going to be used for breeding, ovariohysterectomy is curative for all conditions (except ovarian tumour metastases), and a diagnosis can be made after surgery by inspection and histology of the ovaries. The occurrence of metastases should, if possible, be ruled out before surgery.

If the owner wants to retain the breeding potential of the queen, the diagnostic approach includes repeated vaginal cytology to differentiate between a true prolonged oestrus and oestrous behaviour in interoestrus. Abdominal palpation and ultrasonography may reveal enlarged ovaries and ovarian cysts. It is not recommended to try to save the breeding potential of a queen if the condition is caused by ovarian neoplasia (for information on treatment of ovarian cysts and frequent oestrus in queens intended for breeding, see Chapter 7).

Abdominal distension

A number of different conditions affecting the female reproductive organs may result in abdominal distension as the main or only clinical sign. In addition, abdominal distension can be seen together with other clinical signs in some conditions. Differential diagnoses not related to the reproductive tract are not described in this chapter but should also be considered. Pregnancy should be ruled out. Abdominal distension related to pathology of the female reproductive organs can be attributed to either accumulation of fluid in the uterus or to a uterine or ovarian tumour mass. Abdominal palpation and ultrasonography may be used to diagnose a fluid-filled uterus or uterine/ovarian neoplasia.

Developmental abnormalities, such as vaginal atresia or segmental aplasia of the uterus, which cause obstruction of the lumen of the tubular genitalia may result in fluid accumulation in the uterus. The uterine enlargement may eventually result in abdominal distension. Mucometra is often not associated with clinical signs other than uterine enlargement, and is frequently not diagnosed until the uterine enlargement has caused abdominal distension. Abdominal distension is a common sign of pyometra noted by owners. However, pyometra is often associated with other clinical signs of disease together with the abdominal distension.

Lethargy and systemic illness

Lethargy and other signs of systemic illness such as anorexia, alopecia, vomiting, fever and weight loss can be caused by various conditions, including:

- Metritis, endometritis and pyometra
- Functional ovarian follicular cysts
- Ovarian neoplasia
- Uterine neoplasia.

A thorough clinical examination, complete blood count and ultrasound examination of the ovaries and uterus should be performed in queens with signs of illness that raise suspicion of reproductive tract pathology.

Masculinization

Male behaviour has been described in queens, including signs such as weight loss, aggression and a 'tom cat' smell of urine, prominent jowls and an enlarged barbed clitoris. These can be associated with testosterone-producing ovarian tumours such as ovarian dysgerminomas, luteomas and thecomas. Rarely these tumours are found in remnant ovarian tissue left *in situ* following ovariectomy, and may be associated with clinical signs in queens that have been surgically neutered many years previously. It is uncommon for there to be overproduction of sex steroids by the adrenal glands, but this may occur and result in clinical signs of male-type behaviour, aggression, vulvar hyperplasia and thickened skin.

The recommended approach to a queen with signs of masculinization involves abdominal palpation and ultrasound examination of the ovaries. Evaluation of adrenal gland function and serum androgen concentrations may also be warranted. Karyotyping can be used to diagnose male pseudo-hermaphroditism. Exploratory laparotomy may be necessary for a definitive diagnosis.

References and further reading

Altera KP and Miller LN (1986) Recognition of feline paraovarian nodules as ectopic adrenocortical tissue. *Journal of the American Veterinary Medical Association* **189**, 71–72

Axnér E, Ågren E, Båverud V *et al.* (2008a) Infertility in the cycling queen: seven cases. *Journal of Feline Medicine and Surgery* **10**, 566–576

Axnér E, Gustavsson T and Ström Holst B (2008b) Estradiol measurement after GnRH stimulation as a method to diagnose the presence of ovaries in the domestic cat. *Theriogenology* **70**, 186–191

Boag AK, Neiger R and Church DB (2004) Trilostane treatment of bilateral adrenal enlargement and excessive sex steroid hormone production in a cat. *Journal of Small Animal Practice* **45**, 263–266

Chatdarong K, Rungsipipat A, Axnér E *et al.* (2005) Hysterographic appearance and uterine histology at different stages of the reproductive cycle and after progestagen treatment in the domestic cat. *Theriogenology* **64**, 12–29

Davidson AP, Feldman EC and Nelson RW (1992) Treatment of pyometra in cats using prostaglandin F2 alpha: 21 cases (1982–1990). *Journal of the American Veterinary Medical Association* **200 (6)**, 825–828

Meyers-Wallen VN, Wilson JD, Fisher S *et al.* (1989) Testicular feminization in a cat. *Journal of the American Veterinary Medical Association* **195**, 631–634

Nak D, Nak Y and Tuna B (2009) Follow-up examinations after medical treatment of pyometra in cats with the progesterone-antagonist aglepristone. *Journal of Feline Medicine and Surgery* **11**, 499–502

Verstegen J (2000) Feline reproduction. In: *Textbook of Veterinary Internal Medicine, 5th edn*, ed. SJ Ettinger and EC Feldman, pp.1585–1598. WB Saunders, Philadelphia

20

Clinical approach to conditions of the male

Cheryl Lopate

Introduction

There are numerous conditions of the dog and tom cat that may present without infertility. These animals may have changes in behaviour, external appearance, comfort, or in normal urination and defecation patterns. A summary of the differential diagnoses for various clinical signs related to male reproductive anatomy and physiology is given in Figure 20.1.

Diseases and disorders of the scrotum

Infectious orchitis and epididymitis

Aetiology

Infectious orchitis/epididymitis is the invasion of the testes or epididymides with bacteria, viruses or fungi. Infections may ascend from the prepuce, urethra, bladder or prostate gland and occur more commonly in young dogs (average 4 years of age). The disease may be acute or chronic. In the dog, one or several of the following bacteria are usually present: *Escherichia coli*, *Klebsiella* spp., *Pseudomonas* spp., *Proteus vulgaris*, *Bacillus* spp., *Staphylococcus* spp., *Streptococcus* spp., *Mycoplasma* spp. and *Ureaplasma* spp. *Blastomyces dermatitidis* and other fungi, or canine distemper virus, may also disseminate to the testes or epididymides. Infection typically results in neutrophilic inflammation. *Brucella canis*, as well as *B. abortus*, *B. melitensis* and *B. suis*, may also cause orchitis/epididymitis and are the only true venereally transmitted organisms that affect the dog. Non-infectious lymphocytic–plasmacytic orchitis may also occur as a result of breakdown of the blood–testes barrier. There may be an association between autoimmune orchitis and epididymitis and autoimmune lymphocytic thyroiditis.

Infection may be ascending, haematogenous or introduced via trauma or puncture to the scrotum or spermatic cord. Initially, small microabscesses form, which may coalesce into larger areas of inflammation or become walled off by fibrosis. Chronic infection results in atrophy and fibrosis. Dogs that suffer blunt abdominal trauma with a full bladder may have

Presenting sign	Cause
Increase in scrotal size	Neoplasia, testicular abscess, hydrocele, haematocele, trauma, scrotal oedema, epididymitis, orchitis, varicocele, testicular torsion
Decrease in scrotal size	Age-related testicular degeneration, immune-mediated testicular degeneration, exogenous anabolic steroid administration, sterol ingestion in food/supplement, vascular infarct
Preputial discharge	Balanitis, posthitis, balanoposthitis, urethritis, cystitis, prostatitis, benign prostatic hyperplasia
Urethral discharge	Balanitis, posthitis, balanoposthitis, urethritis, cystitis, prostatitis, benign prostatic hyperplasia
Stranguria	Prostatic enlargement (any cause), urolithiasis, urethritis, cystitis, perineal hernia
Haematuria	Benign prostatic hyperplasia, prostatitis, cystitis, urolithiasis, urethritis, perineal hernia
Incontinence	Urethral sphincter incompetence (ageing or neurological), prostatic neoplasia
Constipation	Prostatic disease, perineal hernia
Fever and/or abdominal pain	Prostatic abscessation/rupture, acute prostatitis, orchitis, epididymitis, schirrous cord, vascular infarct, testicular torsion, testicular neoplasia
Back/hindlimb pain	Prostatic enlargement (any cause), acute prostatitis, metastatic cancer, lumbosacral spondylosis, perineal hernia, scrotal hernia, orchitis, epididymitis
Persistent male behaviour following neutering	Behavioural (learned behaviour), cryptorchidism, polyorchidism, exposure to testosterone patches or creams, adrenal gland tumours or hyperadrenocorticism
Attraction of other males	Sertoli cell tumours, administration of exogenous steroid hormones, exposure to oestrogen patch/creams, adrenal gland tumours, hyperadrenocorticism

20.1 Differential diagnoses for clinical signs related to male reproductive anatomy and physiology.

urine contamination of the scrotal contents due to backflow of urine through the vas deferens at the time of impact.

Diagnosis

Physical examination: This typically reveals enlargement, erythema and pain in one or both epididymides or testes during the acute stage of infection (Figure 20.2). Swelling may extend up the spermatic cord and involve the vaginal tunic. Scrotal size may exceed five times normal. If the disease is acute there may be fever, lethargy, hindquarter lameness, scrotal pain, erythema and swelling, scrotal oedema and/or a purulent preputial discharge. If the disease is chronic there may be swelling or contraction of all or part of the scrotal contents, which is typically non-painful and firm. If only one testis is affected, the contralateral testicle may be smaller than the normal testis.

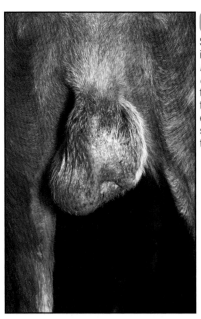

20.2

Scrotum of a dog infected with *Blastomyces dermatitidis*. Note the enlarged testicle, epididymis and spermatic cord on the right side.

Semen collection: This will often reveal pyospermia. In acute cases, animals may be in too much pain to ejaculate or to mate. If semen can be collected, it should be submitted for aerobic, *Mycoplasma* spp. and *Ureaplasma* spp. culture. Cultures for fungi may be indicated in certain cases. Care must be taken when interpreting the results of culture because they do not localize infection to the testes or epididymides. Bacteria may originate from anywhere in the genitourinary tract. If the infection appears to be ascending, the site of initial infection should be cultured instead. In addition, it is important to remember that low numbers of bacteria can be cultured from healthy animals due to the presence of the normal preputial flora (see below).

Imaging: Ultrasonography typically reveals heterogenous tissue with inflamed areas being hypoechoic compared with the normal parenchyma. There may be areas of abscessation (Figure 20.3) within the parenchyma or epididymides, which appear as anechoic or hyperechoic fluid-filled areas. In chronic cases,

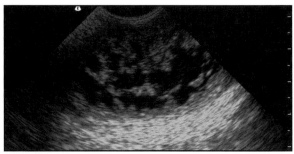

20.3 Testicular abscess. Note the multiple hypoechoic areas of fluid accumulation within the testicular parenchyma.

hyperechoic pockets or regions of the parenchyma may be apparent due to fibrosis or mineralization in the testicular interstitium. Doppler ultrasonography may reveal increased blood flow to the affected side.

Cytology: Cytological examination of aspirates or biopsy samples from the affected areas reveals neutrophilic inflammation in cases with infectious bacterial or fungal causes, and lymphoplasmacytic inflammation in association with autoimmune causes. Aspiration of fluid pockets is possible and this fluid should be submitted for cytology and culture. Sperm granulomas may be associated with the inflammation, either primarily, as in the case of non-infectious diseases, or secondarily, as a result of fibrosis and local inflammation in cases with infectious causes.

Brucellosis: In countries in which *Brucella canis* is present (rare in the UK) all dogs and cats with scrotal enlargement should be tested for brucellosis. In the dog, the rapid card agglutination test (RCAT) and tube agglutination test (TAT) are appropriate for screening because they have high sensitivity (they are accurate if negative) and lower specificity (they are accurate if positive). If the results of either test are positive, further testing is required. The RCAT can be made more specific by the addition of 2-mercaptoethanol (2-ME) to the test protocol. Alternatively, there is an enzyme-linked immunosorbent assay (ELISA) that may be used for screening which has a higher sensitivity than the RCAT or TAT. If the RCAT, TAT or ELISA is positive, the agar gel immunodiffusion (AGID) test should be performed. The AGID is the gold standard of brucellosis testing. It is important to remember that all these tests are serological studies and thus require time after infection for antibodies to be produced. None of them will be positive for at least 30 days after infection. The ELISA requires 45–60 days to become positive, whilst the AGID requires 8–12 weeks. In the first 30 days of infection, blood culture for the organisms is the most accurate method of diagnosis.

Treatment and prognosis

Cytology with or without Gram staining should be performed on any collected fluid or aspirates to help determine the initial choice of antibiotics. If no samples are available for cytology, either enrofloxacin or ciprofloxacin is a good first choice pending culture results, because they are both effective against the majority of reproductive pathogens.

If the disease process is unilateral, hemicastration is effective and may salvage the remaining testis. Removal of the affected side may prevent local extension of the infection and ascension up the tract to other genitourinary structures. The affected testis causes thermoregulatory disturbances in the adjacent testis, resulting in abnormal spermatogenesis. If the animal is affected bilaterally and future breeding is desired, then appropriate antibiotics should be administered long term (minimum 4 weeks). If response to therapy is not immediate, castration should be considered early on. Non-steroidal anti-inflammatory drugs (NSAIDs) and cold packing of the scrotum are recommended to reduce inflammation and swelling. Sexual rest should be provided until resolution of the clinical signs and return to spermatogenic function has occurred. A minimum of 4 weeks of antibiotic therapy is usually necessary, and in some cases treatment may be needed for up to 12 weeks or more.

Return to normal semen quality can be expected in no less than 62 days, and may take several months depending on the degree of injury to the remaining tissues. After unilateral castration, moderate compensatory hypertrophy of the remaining testis may occur, bringing total sperm numbers up to two-thirds of the prior total.

The prognosis for return to normal function of affected testes is poor. Sperm production and morphology are often affected permanently. This is particularly true if there is already palpable evidence of degeneration (softening, atrophy or fibrosis) in the remaining testis. Relapse or ascension to the prostate gland or bladder are common sequelae because bacterial foci frequently become walled off in the epididymides or testes and provide a source of future infection.

Tom cats

Orchitis and epididymitis are rare in the cat. In the tom cat, tuberculosis, feline coronavirus, *Brucella* (rarely), and any of the above-mentioned aerobic bacteria may cause infectious orchitis and/or epididymitis. Cats are generally believed to be resistant to *Brucella* infection. Infection of the testes and epididymides with feline coronavirus is associated with fluid accumulation in the vaginal space from peritoneal overflow into the vaginal cavity via the vaginal process. Lymphocytic–plasmacytic inflammation has also been described in the cat. Clinical signs of orchitis and epididymitis include enlargement of the testes, heat, pain and erythema. Excessive licking may result in self-mutilation. Broad-spectrum antibiotics should be administered for 2–4 weeks. If this is not successful, castration is indicated.

Scrotal trauma and dermatitis

Aetiology

Trauma or dermatitis may result from:

- Physical injury (road accident, fight, puncture wound)
- Environmental trauma (lying on rough surfaces or irritating bedding)
- Chemical trauma (contact with caustic substances)
- External parasites (fleas, lice, mites)
- Allergic reactions
- Infectious agents (bacterial or fungal dermatitis, *B. canis*, Rocky Mountain spotted fever)
- Immune-mediated disorders (pemphigus, purpura)
- Sperm granuloma.

Contact dermatitis may be an allergic hypersensitivity reaction (type IV) or caused by direct contact with an irritant substance. Common allergic-type reactions may occur in response to plant pollens, shampoos, soaps or cleaners, or insect bites (particularly flea or spider bites). *B. canis* is commonly associated with epididymitis and scrotal dermatitis (due to self-mutilation as a result of pain). Scrotal dermatitis may occur as a drug eruption caused by an allergic reaction to an ingested medication. Scrotal trauma or dermatitis typically results in thickening of the scrotal skin, which in turn may disrupt scrotal thermoregulation and result in derangement of spermatogenesis.

Diagnosis

The most common presenting sign is an enlarged scrotum. In addition there may be erythema and pain. It is important to investigate whether the swelling is attributable to testicular enlargement or fluid in the vaginal tunic, or is within the scrotal skin itself. This may be accomplished via a combination of digital palpation and ultrasonography. If there is fluid present in the vaginal cavity or the testes are enlarged, aspiration of a sample may be performed using a 22-gauge needle. Cytological assessment is necessary to differentiate inflammation from infection and neoplasia. With *B. canis* infection, there is a transient lymphadenopathy, variable fever, epididymitis and dermatitis.

Treatment and prognosis

Applying cold water several times a day helps reduce the swelling, and NSAIDs may be used for their anti-inflammatory effects. Topical antibiotic ointments may be applied if the scrotal skin is damaged. Care should be taken if the area must be clipped to prevent razor burn or heat trauma. Prevention of self-mutilation requires use of an Elizabethan or neck collar. If a bacterial dermatitis is present, oral antibiotics should be administered. If there is a contact allergen or surface irritant, the patient should be bathed in oatmeal or other soothing shampoo to remove the offending substance. If there is pruritus associated with an allergic reaction, topical or systemic antihistamines or glucocorticoids may be beneficial. In countries where *B. canis* is present, it is a notifiable disease, and the treatment, eradication and prevention will follow government guidelines. Euthanasia of *Brucella*-positive animals may be the safest way to prevent zoonotic transmission.

A return to normal spermatogenesis typically occurs 60–90 days after resolution of the oedema or thickening. If the thickening persists, there may be permanent impairment of spermatogenesis.

Tom cats

Scrotal trauma is uncommon in the tom cat because the scrotum is held in close apposition to the body wall, but it may occur as a result of blunt trauma (e.g. a road accident) or in fights when the animal is retreating. Cleaning and debriding the wound, plus systemic and/or topical antibiotic therapy, cold packing the scrotum, NSAIDs for oedema or inflammation, and drainage of any surrounding abscessed tissue are indicated.

Testicular tumours

Testicular neoplasia is the second most common tumour type in entire dogs, following skin cancer, and is usually found in older animals. Neoplasia of the testis occurs in approximately 1% of entire male dogs and comprises >90% of all tumours of the genital system. The mean age at diagnosis is 9–11 years. The Boxer is at increased risk, while the Beagle, Dachshund, Labrador Retriever and cross breeds are at decreased risk.

The normal testicular parenchyma may be partially or completely obliterated by neoplastic tissue. Thermoregulation is often altered in the affected testis owing to the aberrant cell growth and in the non-affected testicle due to the proximity to the altered temperature of the diseased testis. Sertoli cell tumours (SCTs), interstitial cell tumours (ICTs), seminomas (SEMs), undifferentiated carcinomas, epidermoid cysts, fibrosarcomas, lymphomas, haemangiomas, gonadoblastomas, embryonal carcinomas, sarcomas, granulosa cell tumours and teratomas have all been described. In addition, sarcoma of the spermatic cord and leiomyoma of the tunica vaginalis have been described. Multiple types of tumours may be present in the same testis. Retained testes or those that were retained at some point are at 9–11 times greater risk of neoplasia compared with testicles that descended normally.

Sertoli cell tumours

SCTs comprise 44% of testicular tumours. These can be endocrinologically active neoplasms, involving the sustentacular or nurse cells that line the seminiferous tubules. Boxers and Weimaraners are at increased risk. It is the most common tumour of retained testes. The tumour size varies from 1–12 cm. They are typically solitary, white to pale yellow and firm on cut section. The unaffected testis is often atrophied. Sertoli cell tumours rarely metastasize (2–6%).

An SCT may be associated with oestrogen secretion and paraneoplastic syndrome. Signs of hyperoestrogenism in the male include bilaterally symmetrical alopecia, hyperpigmentation, an easily epilated hair coat, gynaecomastia, pendulous prepuce, squamous metaplasia of the prostate gland, attraction of male dogs and keratinization of the preputial epithelium. Haematological abnormalities may include non-regenerative anaemia, leucopenia, thrombocytopenia and/or pancytopenia. Signs associated with these haematological abnormalities may include pale mucous membranes, lethargy, petechiae, epistaxis, haematemesis, melaena and haematuria. Serum oestrogen concentrations may be normal to elevated, and gonadotrophin concentrations may be decreased. Retained testes with SCTs are more likely to have hyperoestrogenism than scrotal testes with SCTs. Resolution of feminization occurs within 3–4 weeks after tumour removal.

Seminomas

SEMs comprise 31% of testicular tumours. These are typically benign tumours of the germ cells of seminiferous tubules. German Shepherd Dogs are predisposed. Approximately 25% will be found in retained testes and 75% in scrotal testes. They may be single or multiple and small to large (1–10 cm) in size. They are white to cream or tan to pink–grey, and homogenous to lobulated in texture (Figure 20.4). The patient may present with an enlarged testis, a palpable testicular mass or infertility (abnormal sperm morphology), or the tumour may be found incidentally at the time of castration. Seminomas metastasize locally in about 15% of cases, and distantly in 6–10% of cases.

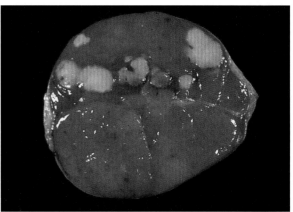

20.4 Cross-section of a testis showing testicular seminonas. Note the multiple small, white–tan smooth or slightly lobulated masses within the parenchyma of the testis.

Interstitial cell tumours

ICTs (Leydig cell tumours) comprise 25% of testicular tumours. They are mostly benign tumours of the interstitial or Leydig cells of the testis. The Leydig cells are the endocrine cells of the testis. The tumours are more common in scrotal testes than in retained testes. These tumours tend to be small (<1 cm) and may be single, but are often multiple. They are usually soft, round and tan, yellow or orange. Atrophy of the contralateral testis may be noted. They are most commonly found incidentally at necropsy; ICTs rarely metastasize.

A paraneoplastic syndrome similar to that seen with SCTs may occur with ICTs. There may be elevated concentrations of oestrogen or testosterone. With elevated oestrogen concentrations there may be bilaterally symmetrical alopecia of the trunk and flanks, pale mucous membranes, fever, petechiae, prolonged bleeding times, bone marrow hypoplasia and aplastic anaemia. With elevated testosterone concentrations, signs may include prostatic disease, perianal adenoma, perianal gland hyperplasia and perineal hernia.

Diagnosis

Physical examination may reveal an enlarged testis or a palpable mass. The surrounding testicular tissue tends to be softer than the tumour. Semen evaluation commonly reveals abnormalities in morphology with primary defects predominating. Ultrasonography may be used to better delineate the tumour(s). Sometimes, a tumour may not be palpable due to its small size, but is easily visualized via ultrasonography (Figure 20.5a). Usually, the neoplastic tissue is hypoechoic compared with normal parenchyma and sometimes, if the tumour is very large, no normal tissue can be visualized (Figure 20.5b). The mediastinum testis may be obliterated or displaced by neoplastic tissue. Testicular aspiration or biopsy may be used to obtain a diagnosis, and histopathology will confirm the findings.

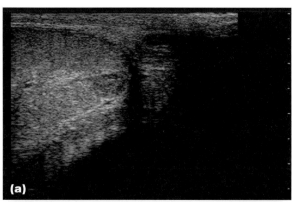

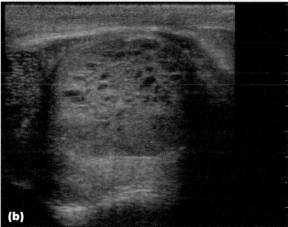

20.5 **(a)** Ultrasonogram showing a well delineated testicular mass in the caudal portion of the testis. **(b)** Ultrasonogram showing a very large testicular mass, encompassing most of the testicular area and pushing the normal parenchyma to the cranial margin. The mass is of irregular echogenicity, probably as a result of its large size and poor blood flow.

Treatment and prognosis

In cases in which both testes are affected, or the individual is not a breeding male, complete castration is recommended. In a breeding animal, hemicastration may be the preferred method of treatment if the lesion is unilateral. Following hemicastration, if there are no complications, a return to normal spermatogenesis should be expected within 60–90 days. Compensatory hypertrophy may occur following hemicastration, with a return to two-thirds of original

spermatogenic function possible in some cases within 6 months. It is important to warn owners of the possibility of small tumours in the apparently normal testis, which may grow over the ensuing months and result in failure to return to normal spermatogenesis. Bilateral castration is recommended for patients with SCTs, regardless of breeding status. In valuable breeding animals, unilateral castration may be chosen. Chemotherapy with cisplatin, vinblastine, cyclosphosphamide and methotrexate is used for metastatic tumours.

Cases of metastatic disease hold a poor prognosis, while the prognosis is good in cases of benign tumours, even when they occur as multiple masses.

Tom cats

Testicular neoplasia is rare in tom cats but is more likely to occur in older individuals. Undifferentiated carcinoma, SCT, ICT, SEM and teratoma have all been described in the tom cat. They may be single or multiple and small to large in size. Lymphosarcoma has been shown to metastasize to the testes. Diagnosis is via palpation of an enlarged testis. Sometimes a distinct mass may be palpable, but at other times the tumour may not be distinguishable from the remaining parenchyma. The definitive diagnosis is based on histopathology. Neither hormonally active tumours nor paraneoplastic syndromes have been described with ICT or SCT in the cat. Orchiectomy is the treatment of choice. In breeding males, hemicastration may be performed, with an expected return to normal spermatogenesis in the remaining testis within 2–3 months. Compensatory hypertrophy of the remaining testis may occur within 6 months.

Scrotal tumours

The most common types of neoplasia of the scrotal skin are squamous cell carcinoma (SCC), melanoma and mast cell tumour (MCT). An SCC may present as a proliferative lesion or as an ulcerated raised nodule. They may be solitary or multiple, are locally invasive and metastasize slowly. Melanomas are usually either pigmented raised nodules that ulcerate readily, or pigmented macules. They are usually solitary masses that are locally invasive. Similarly, MCTs are usually raised nodules that are erythematous and commonly ulcerated. They are locally invasive and may metastasize locally or distantly.

Diagnosis

Aspiration cytology may be performed and is often diagnostic for melanoma and MCT, but may not be for SCC. Biopsy with histopathology is required for definitive diagnosis and to determine whether the entire tumour has been removed.

Treatment

An SCC should receive wide surgical excision. Cryotherapy or radiation may be performed adjunctively. Melanoma requires wide surgical excision and postoperative chemotherapy or autogenous vaccination protocols. Similarly, MCTs require wide surgical excision. If they are markedly ulcerated, pre-treatment with an antihistamine may minimize degranulation during surgery. Adjunctive cryosurgery, radiation or chemotherapy may be needed depending on the surgical margins.

Vascular insult and infarct

Vascular insult or infarct may occur as a result of a breakdown of the blood–testis barrier, trauma, inflammation or immune-mediated disease. It may occur unilaterally or bilaterally. Testicular degeneration is a common sequel to infarct and may occur in both testes owing to impairment of thermoregulation or in just the affected testicle. Infarcts may be unilateral or bilateral and may occur in dogs of any age.

Diagnosis

There will be abnormal sperm morphology and/or total sperm numbers in the ejaculate. Physical examination and diagnostic tests may lead to the diagnosis of immune disease, with the testicular lesions being found incidentally. Ultrasonography may be useful; a typical triangular hypoechoic lesion is seen, with the point of the triangle located nearer the mediastinum testis and the base of the triangle at the surface of the testis (Figure 20.6). Alternatively, the diagnosis may be made at the time of castration, when the typical radiating lesion is seen to be darker than the normal parenchyma.

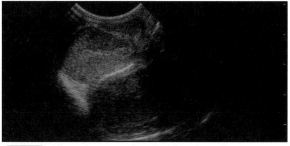

20.6 Ultrasonogram showing the triangular hypoechoic lesion in the dorsum of the testis, radiating from the mediastinum testis towards the parietal vaginal tunic.

Treatment and prognosis

Hemicastration may be attempted in animals with unilateral lesions, but the risk of a similar infarct occurring in the remaining testis is high. Castration is the alternative treatment. If hemicastration is performed and the other testis remains unaffected, the patient should return to normal spermatogenesis within 60–90 days.

Spermatic cord torsion

Spermatic cord (testicular) torsion is an uncommon condition that most frequently affects middle-aged to older dogs (6–8 years; range 5 months to 10 years). It is more common in retained testes than in scrotal testes. It is believed to occur as a result of the increased mobility of abdominal testes. Neoplastic testes are heavier and therefore may be more mobile. Testicles that are twisted may be enlarged prior to the torsion and this will worsen after the torsion. When enlargement occurs after the torsion, it is caused by venous occlusion, oedema and inflammation associated with vascular compromise.

Diagnosis

Dogs present typically with signs of acute abdominal pain, abdominal splinting, depression, lethargy, anorexia, vomiting, fever, haematuria, stranguria and stiff hindlimb gait. Rarely, torsion may be found incidentally with no obvious clinical signs. However, these clinical signs plus cryptorchidism make testicular torsion highly suspect. The enlarged testis may be palpable abdominally. Ultrasonography may reveal a testis with decreased echogenicity compared with normal. Colour Doppler ultrasonography reveals diminished or complete lack of blood flow to the testis. Exploratory surgery is necessary for a definitive diagnosis.

Treatment

Surgical removal is curative. Bilateral castration is recommended because the patients typically have cryptorchidism. If a paraneoplastic syndrome exists due to SCTs or ICTs of the torsed testicle, complications may result from surgery (see above).

Tom cats

Testicular torsion is rare in the tom cat and typically occurs unilaterally. It is seen with equal frequency in retained and descended testes. Acute abdominal pain is the main clinical sign and castration is curative.

Inguinal or scrotal hernia

Defects in closure of the internal and external inguinal ring may result in inguinal or scrotal hernia following evagination of abdominal contents or fat through the rings and into the vaginal process. The hernia may result from congenital defects or from blunt trauma with increased abdominal pressure. There may be some connection between scrotal hernia and testicular neoplasia. Shar Peis and chondrodystrophic breeds may be predisposed. Fat is commonly herniated along with intestinal contents. If the herniated structure becomes incarcerated this can become a life-threatening condition. If the hernia is large enough or if only a small amount of tissue is protruding through the rings, the herniated tissue may be reduced into the abdominal cavity manually. Dogs may present with an enlarged scrotum and a filling defect that runs up into the inguinal region.

Diagnosis

Physical examination is required, along with the use of ultrasonography to confirm the type of tissue that is herniated. Patients with herniated intestinal contents may present with signs of abdominal pain, vomiting, diarrhoea, dehydration, depression, anorexia and/or shock. Patients with herniation of the bladder may present with stranguria, dysuria, haematuria, anuria, dehydration, vomiting, anorexia or lethargy. Confirmation is via exploratory surgery. These hernias may be confused with enlarged testes (due to neoplasia or inflammation), hydrocele, varicocele and/or scrotal oedema or inflammation.

Treatment

Scrotal and inguinal hernias should be treated by surgical correction. If the tissue is not incarcerated at the time of diagnosis, surgery should be performed as soon as possible because incarceration may occur at any time. Even if only fat is herniated, surgical correction is necessary to prevent fat necrosis. If other tissues are incarcerated, emergency surgery is required. Bilateral castration should be performed at the time of hernia repair because the condition is typically bilateral. It is also hereditary, so these individuals should not be used for breeding.

Fluid within the vaginal tunic

Fluid within the vaginal tunic may be a transudate, modified transudate or exudate depending on the cause. It may originate from the abdominal cavity or from the scrotum itself. It may be caused by trauma, inflammation, infection, neoplasia or foreign body migration. The presence of bloody fluid is termed a haematocele and is typically the result of trauma or vascular disease with capillary rupture. Hydrocele (serous fluid) may be caused by accumulation of abdominal fluid (peritonitis, ascites) with leakage through the vaginal rings into the vaginal tunic, poor lymphatic blood flow often associated with scrotal or inguinal herniation, inflammation, parasite migration or neoplasia.

Diagnosis

The patient may present with malaise, abdominal enlargement, fever and anorexia or simply with an enlarged scrotum. The diagnosis is made on the basis of physical examination, blood chemistry, complete blood count, abdominal and/or thoracic radiography, and ultrasonography of the abdomen and scrotum. The type of fluid within the vaginal tunic can be determined via needle aspiration and cytology. If the fluid is inflammatory, culture (aerobic, anaerobic, *Mycoplasma* spp. and *Ureaplasma* spp.) should be performed. If caused by trauma or infection within the scrotum itself, fluid accumulations may be unilateral, but if the cause is abdominal (ascites, parasite migration, abdominal neoplasia) typically fluid accumulates bilaterally.

Treatment

Treatment of the primary disease is first and foremost. A return to normal spermatogenesis may be expected if the fluid accumulation is not longstanding, but up to 90–120 days may be required. Purulent exudate or transudates in the vaginal tunic may result in adhesions between the vaginal tunic of the testicle and the parietal vaginal tunic. These adhesions may affect the ability of the testicle to thermoregulate properly. If only one side is affected, hemicastration may be indicated to salvage spermatogenesis in the contralateral testicle. Broad-spectrum antibiotics chosen on the basis of the results of culture and sensitivity testing should be administered when indicated.

Diseases and disorders of the penis and prepuce

Balanitis and posthitis

Inflammation of the penis (balanitis), the prepuce (posthitis) or a combination of the two (balanoposthitis) may be caused by concurrent cystitis, urethritis

or prostatitis, trauma, bacterial sepsis or lack of mucosal immunity from immunosuppressive conditions. Most affected dogs are >4 years of age and may be intact or neutered. Dogs living in unhygienic conditions may develop posthitis from environmental contact. In intact dogs, the presence of a non-pathological posthitis (smegma) is commonly noted and is not of clinical significance. The normal preputial flora includes *E. coli*, *Staphylococcus* spp., *Streptococcus* spp., *Pasteurella* spp., *Proteus mirabilis*, *Klebsiella pneumoniae*, *Haemophilus* spp., *Corynebacterium* spp., *Moraxella* spp., *Mycoplasma* spp. and *Ureaplasma* spp. *Mycoplasma canis* is a common normal inhabitant, whilst *Ureaplasma* spp. tend to be pathogenic. Viral causes may include canine herpesvirus and calicivirus. Herpesvirus may be transmitted by dog to dog contact either via respiratory aerosol or venereally. Herpesvirus may recrudesce in individuals when they are stressed or become immunocompromised. Balanoposthitis may be a component of atopic dermatitis or may be a result of self-mutilation due to anxiety disorders or pruritus.

Diagnosis

Physical examination may reveal a green, yellow or tan purulent, fetid discharge emanating from the preputial opening. Bloody discharge may be noted with some infections. Pathological inflammation also presents with erythematous and hyperaemic mucous membranes involving the inner preputial fold and the penile mucosa (Figure 20.7). There may be erosions or plaques on the mucosal surfaces or there may be lymphoid hyperplasia. There may be pain associated with manipulation of the penis either within the preputial sheath or on eversion. Culture of the inner preputial fold using a guarded swab for aerobic bacteria, *Mycoplasma* spp. and *Ureaplasma* spp. should be performed. Cytological evaluation of the mucosal surface or the discharge may reveal heavy overgrowth of a particular type of bacteria (rods or cocci) and neutrophils with intracellular bacteria. This differs from normal smegma, which has the typical appearance of high numbers of neutrophils, low numbers of rods and cocci, and no intracellular bacteria.

20.7 Balanoposthitis. Note the raised granular lesions, hyperaemia and purulent exudate over the surface of the penis. (Reproduced from *Johnson et al.*, 2001 with permission from the publisher)

Treatment

Excessive licking of the prepuce should be discouraged because this may result in abrasion of the preputial orifice followed by a clinical posthitis. Use of an Elizabethan or neck collar may be necessary. Gentle douching of the preputial orifice with warm saline may be helpful. Care should be taken not to destroy the normal genital flora because this may result in overgrowth of bacteria and worsening of the condition. Oral antibiotics, with the choice based on culture and sensitivity, are recommended. In addition, topical antibiotics may be helpful. Use of systemic probiotics may help to re-establish the normal flora. If the preputial opening is too small, enlarging this opening may be necessary for resolution. If atopy is a causative factor, treatment with antihistamines or glucocorticoids is necessary. Self-mutilation should be treated with behaviour modification or anxiolytic medication.

Tom cats

The normal preputial flora of the cat includes *E. coli*, *Pseudomonas aeruginosa*, *Proteus mirabilis*, *Klebsiella oxytoca*, *Streptococcus* spp., *Enterococcus* spp., *Bacillus* spp. and *Staphylococcus* spp. Balanoposthitis is rare in the cat but may follow trauma at mating, fights or blunt trauma (e.g. road accidents). Complications include the formation of preputial–penile adhesions, which may result in urinary tract obstruction or prevent mating. Feline herpesvirus may cause vesicular lesions on the penile shaft following experimental infection (Verstegen, 1998), but this has not been noted following natural infection.

Paraphimosis

Paraphimosis is the inability to retract either the erect or the non-erect penis into the prepuce (Figure 20.8). It accounts for 7% of the penile and preputial abnormalities reported in the dog. Paraphimosis may occur as a result of rolling inwards of the preputial orifice following masturbation or semen collection, entrapment of hair around the penis, oedema of the penis as a result of chronic balanitis or balanoposthitis, oedema of the penis following purposeful entrapment with string or a rubber band, a small preputial

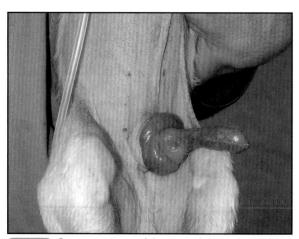

20.8 Severe oedema of the preputial membrane in a dog with prolapse of the prepuce and paraphimosis.

orifice, a short prepuce, preputial trauma, or a neurological deficit caused by encephalitis or intervertebral disc disease (IVDD) in the lumbar or sacral segments. It is commonly seen in toy breeds and peripubertal animals that masturbate frequently. Once the penis is entrapped in the prepuce it becomes ischaemic, dry and excoriated. The longer the condition continues the more severe the damage, creating a vicious cycle of injury. Self-mutilation may occur owing to the pain associated with the condition.

Diagnosis

The diagnostic procedure is based primarily on physical examination, including neurological examination, but may include radiography, magnetic resonance imaging (MRI) or a computer-assisted tomographic (CAT) scan of the pelvic region if IVDD is suspected, and a spinal tap if encephalitis is involved.

Treatment

Lubrication of the penis and bulbus glandis, with elevation of the penis towards the body wall to relieve the pressure of the preputial orifice on the proximal penis, should be attempted first. Sedation or anaesthesia may result in detumescence. Wrapping the penis with a cool, flat compress (applying pressure starting from the distal end and wrapping proximally) that has been soaked in hyperosmolar dextrose solution may result in a reduction in oedema. Application of sugar granules directly on to the penis should be avoided because they may puncture the oedematous mucosa and cause more damage. If the preputial orifice is too tight it may be opened surgically and the penis replaced. Preventing masturbation after replacement is critical to prevent relapse. Keeping intact males away from oestrous females is recommended. If the condition recurs after behaviour modification has been attempted, myorrhaphy of the preputial muscles may be performed. In cases that are unsuccessfully treated in any other manner, penile shortening via amputation may be the only remaining option.

Tom cats

This condition occurs uncommonly in the tom cat. Diagnosis is based on physical examination. Failure to treat the condition in a timely manner results in penile excoriation and ischaemia, and ultimately gangrenous necrosis. Treatment involves replacing the penis, following lubrication or surgical widening of the preputial orifice if the penis cannot be otherwise replaced. If gangrenous necrosis occurs, penile amputation and perineal urethrostomy are necessary.

Priapism

Priapism is a persistent erection without sexual arousal, with the subsequent inability to retract the penis into the prepuce. It is very rare in the dog. Priapism may be caused by neurological disorders that result in prolonged parasympathetic stimulation or by decreased venous outflow from the penis as a result of a coagulation disorder or mass effect. It may occur secondarily to diabetes mellitus, neoplasia, trauma or coagulopathy, or as an idiosyncratic reaction to the phenothiazine tranquillizers, to hypertensive medications or to general anaesthesia. In some cases it may be idiopathic. In dogs it has been reported to be a sequel to distemper encephalitis and thromboembolic disease. Dysuria or stranguria are commonly noted clinical signs.

Diagnosis

A complete history should be obtained, including medication administration and accidental ingestion of human or veterinary drugs. The diagnosis is made via physical examination, and a complete diagnostic work-up is required to reveal underlying metabolic disease or neoplasia. A complete blood count, chemistry, coagulation panel and urinalysis may be necessary. Radiography or ultrasonography of the abdomen or thorax may be needed to rule out neoplasia.

Treatment

Treatment of any underlying disease is of primary concern. The condition of the penis should be addressed promptly because priapism results in ischaemia of the glans penis, and this may cause permanent damage requiring amputation if resolution is not obtained early on. Flushing the corpus cavernosum penis (CCP) with heparinized 0.9% saline solution has been described in humans, bulls and stallions. The CCP may be infused with a sympathomimetic drug such as phenylephrine after saline flushing. The CCP may also be incised longitudinally to allow for venous drainage via manual pressure.

Benzotropine mesylate, an anticholinergic antihistamine, has been used in stallions within the first 6–8 hours following the onset of the erection at a dose of 0.015 mg/kg i.v. In most cases in the dog, the penis has become excoriated and ischaemic by the time of presentation, such that amputation and perineal urethrostomy are the only treatments available. These animals should be castrated if they are not already neutered.

Tom cats

Priapism is uncommon in the tom cat, but Siamese cats may be overrepresented, indicating a hereditary predisposition. The condition may occur in both neutered and intact animals. The clinical signs include a persistent erection and dysuria. Priapism has been reported in cats following castration and in those with penile thromboembolism. Thrombosis of the CCP and erosion or ulceration of the penile mucosa is found in many cases. Treatment includes penile amputation and perineal urethrostomy.

Phimosis

Phimosis is the inability to protrude the penis out of the preputial orifice. It is an uncommon occurrence in the dog (0.5% of all penile problems). It may be caused by a persistent penile frenulum (particularly in pre- or peripubertal animals), a congenitally short penis, as a part of an intersex condition, by a small preputial orifice or preputial stricture resulting from trauma or wound healing, as a result of neoplasia of the prepuce or penis, or by apprehension and/or pain associated with erection. The cause may be anatomical or a result of chronic inflammation from trauma or balanoposthitis.

A persistent penile frenulum is a thin band of connective tissue that joins the ventrum of the glans penis to the penile shaft or prepuce (Figure 20.9). It is a remnant of the balanopreputial fold. Retention of the frenulum results from inadequate testosterone in the pre- or peripubertal animal or from an abnormal response to testosterone production at puberty. It may also be due to a hormonal imbalance, testicular hypoplasia, testicular agenesis or chromosomal abnormalities (i.e. intersex conditions) where testosterone concentrations are not normal. Other cases are caused by an abnormally thick frenulum, which simply does not break down at puberty as expected.

20.9 Persistent penile frenulum in a dog.

Diagnosis

Physical examination reveals that the penis cannot be extended fully out of the preputial sheath; attempting to do so may cause pain or anxiety. Sedation or general anaesthesia may be needed to determine whether the condition is real or induced by apprehension. Semen collection in the presence of a bitch in oestrus may help determine whether the condition is behavioural or physical.

In cases of persistent penile frenulum the clinical signs include excessive licking of the penis and prepuce, dermatitis from urine scalding on the hindlegs or inguinal area, phallocampsis (ventral flexion of the penis), pain on breeding or masturbation, and an inability to penetrate the bitch or achieve a copulatory tie. Physical examination of the extruded penis provides a diagnosis. Most dogs are young to middle-aged at diagnosis but some who are never used for breeding may not be diagnosed until they are older, or not at all.

Treatment

A persistent penile frenulum can be transected using a combination of local anaesthesia and light to heavy sedation, or general anaesthesia. The frenulum may be vascular and require ligation or cauterization, although some may simply be cut with scissors without complications. Dogs that show no clinical signs do not require treatment. This condition is not considered to be hereditary.

There is no treatment for a congenitally short penis and these animals should not be used for breeding owing to the hereditary nature of this condition. Adhesions of the penis and prepuce may be broken down surgically and topical antibiotic plus steroid creams applied to prevent re-adhesion. If inflammation is present, cultures should be obtained and the cause of the inflammation treated or relapse is highly likely. In animals with fear issues, behaviour modification training with regard to the collection process, and the use of teaser animals in oestrus (and of similar breed) will be beneficial in getting them over their fears. There is no treatment for animals with chromosomal abnormalities and they should be gonadectomized.

Tom cats
Phimosis is reported predominantly in young tom cats with a congenitally small preputial opening. It may result in dysuria or an outflow obstruction with haematuria and a distended urinary bladder. It must be differentiated from persistent penile frenulum following gonadectomy at an early age, hair rings, and adhesions to the prepuce as a result of penile trauma. In addition, this condition must be differentiated from penile hypoplasia, which may also occur as a result of gonadectomy at an early age. The treatment involves enlarging the preputial opening surgically by removing a V-shaped wedge from the prepuce.

Urethral prolapse and urethritis
Urethral prolapse typically occurs in younger dogs (9 months to 5 years) and the English Bulldog is predisposed. It is considered to be hereditary in some lines. In other breeds it may occur secondarily to urethritis or sexual arousal or masturbation. Usually the entire distal end of the urethra is prolapsed and becomes oedematous, owing to exposure and licking stimulated by irritation.

Diagnosis
A report of intermittent drops of blood from the end of the penis is common. Physical examination reveals a doughnut- or pea-shaped structure of pink tissue protruding from the end of the penis (Figure 20.10). Pollakiuria may be noted. In some cases prolapse may be noted only when the dog is masturbating or has an erection.

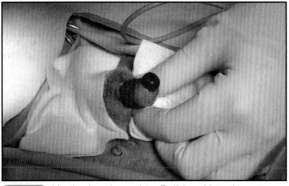

20.10 Urethral prolapse in a Bulldog. Note the doughnut-shaped hyperaemic urethral mucosa which has prolapsed out of the urethral orifice at the end of the glans penis.

Treatment
Sexual rest and tranquillization may be utilized until surgery can be performed. An Elizabethan or neck collar is recommended to prevent self-trauma and worsening of the oedema prior to surgery. Two surgical approaches are commonly used. The first involves resection of the prolapsed tissue followed by suturing of the normal urethral mucosa to the penile mucosa. This approach involves resection of the prolapsed tissue from one-quarter of the circumference of the urethra, followed by suturing with 0.7 or 1 metric (6/0 or 5/0 USP) absorbable suture in a simple interrupted pattern. This is followed by removal and suturing of the next quarter of the prolapsed tissue, and so on, until all the prolapsed tissue has been removed. Significant haemorrhage can occur with this surgical technique, and bleeding from the excision site commonly occurs for a few days postoperatively. Care needs to be taken to prevent self-trauma in the postoperative period.

The second approach is simpler and utilizes a special urethral guide to push the prolapsed tissue back into the urethra. The urethra is then sutured to the shaft of the penis, using the trough in the guide to avoid passing full thickness sutures through both sides of the urethra. Only two tack sutures are required with this approach and there is minimal bleeding afterwards. If the tack sutures do not result in the formation of proper adhesions, relapse is possible, but this is not common. If surgery is successful and no relapse is noted within the following 6–12 months, these individuals may be used for breeding. Bulldogs with urethral prolapse should never be used for breeding because the condition is considered to be hereditary in this breed.

Fracture of the os penis
Fracture of the os penis is uncommon (2% of all penile conditions). There may be a history of prior trauma. The os penis is susceptible to various degrees of damage, from a simple fracture to compound and/or comminuted fractures. Acute fractures present with haematuria, dysuria, stranguria, pain during manipulation of the penis, gross deformation of the penis and abdominal pain. Chronic fractures with abnormal apposition of the fragments may present with dysuria, inability to empty the bladder, post-renal azotaemia and abnormal deviation of the penis. This injury may be suspected on the basis of physical examination, but radiography is needed to confirm the diagnosis.

Treatment
Catheterization of the urinary bladder should be performed with caution in cases with comminuted or compound fractures so that the urethra is not lacerated. If the fracture is recent, reduction may be attempted under anaesthesia. After reduction is achieved, the largest urethral catheter possible should be passed and the catheter left *in situ* to act as a splint along the fracture. A closed catheter system is needed to minimize the likelihood of ascending infection of the bladder. The catheter

should be left in place for 5 days and broad-spectrum antibiotics administered as prophylaxis against cystitis and urethritis.

In cases where closed reduction cannot be achieved, open reduction may be necessary. A bone plate is placed with the screws dorsal to the os penis to avoid damage to the urethra. A urinary catheter is typically placed following open reduction to facilitate urination until the immediate postoperative swelling subsides. In cases of malunion, wedge osteotomy can be performed. Alternatively, penile amputation with scrotal or perineal urethrostomy may be necessary or desirable.

Penile trauma

Trauma may occur from a road accident or other blunt injuries, dog fights, during mating, or as a result of a hair ring or placement of string or rubber bands on the penis. Bleeding from the penis, abdominal pain, hindlimb pain, pain during breeding or pain on manipulation of the penis may be noted.

Diagnosis

Physical examination may reveal swelling of the prepuce, intermittent or continuous haemorrhage from the preputial orifice, pain on manipulation of the prepuce or penis, hindquarter pain or lameness, stranguria, dysuria or haematuria. Direct examination of the penis provides the diagnosis.

Treatment

Lacerations of the glans penis or penile shaft may be sutured. Abrasions should be gently cleansed with warm saline solution. Topical antibiotics may be infused into the preputial opening and NSAIDs may be administered for a few days until the swelling and pain subside. If there is penile oedema or urethral swelling or injury, an indwelling urinary catheter with a closed collection system should be placed. Prevention of sexual arousal is very important until healing is complete. In cases of severe penile trauma, amputation with perineal or scrotal urethrostomy may be necessary.

Tom cats

Causes of penile trauma in the tom cat include hair rings at the base of the penis, trauma from fights, blunt trauma (road accident), haematoma following attempts to clear a blocked urethra, or damage to the urethral mucosa by urethral stones or urinary crystals. Hair rings may be suspected in tom cats that mount repeatedly with excessive pelvic thrusting. These rings are generally very easily slipped off the base of the penis, thus relieving the anxiety and pain associated with them. Care should be taken when passing a catheter to treat urethral obstruction so as to prevent haematoma formation. If a haematoma does occur, it should be treated with topical and/or systemic antibiotics and anti-inflammatory drugs.

Neoplasia

Tumours of the penis and prepuce may occur anywhere on the mucosal surfaces. Squamous cell carcinoma (SCC) usually occurs on the surface of the glans penis, outer prepuce or preputial orifice. These tumours may be small to large, single or multiple, pigmented or non-pigmented, papillary, sessile or dermal. They may metastasize or be themselves metastases. Papillomas are most common in young and old animals. In young animals they may be transient in nature, result from exposure to papilloma virus and regress within 1–3 months as immune competence develops. They tend to be small, but can grow to be larger tumours. Transitional cell carcinoma (TCC) tends to occur at the urethral orifice, within the urethra or bladder, or in the prostate gland. It may metastasize or be metastatic. Mast cell tumours (MCTs) may occur on the penis and are commonly malignant. They are often raised and ulcerated.

In some countries, transmissible venereal tumours (TVTs) are the most common type of tumour associated with the penis and prepuce. They may be pedunculated or lobulated. They are locally invasive and may metastasize, and tend to occur in young, sexually mature, intact animals (average age 2–4 years). They may be single or multiple, small to large, irregular and erosive tumours, and may occur on the inner or outer preputial membranes or on the surface of the prepuce or penis. The incidence of TVTs is highest in animals that come from hot, humid tropical environments, and the tumours are transmitted venereally. These tumours may be transferred to oral or nasal mucous membranes when the dog licks the mass.

Diagnosis

Bloody discharge from the prepuce, excessive licking, blood in the urine, pain on ejaculation or breeding, or the physical presence of a mass may all be presenting signs of any of the tumours. The diagnosis is based on the location and physical appearance of the tumour. Impression smears or aspiration cytology may be used to make a cytological diagnosis, but histopathology is generally needed for confirmation. TVTs tend to exfoliate readily and are easily diagnosed from an impression smear or aspiration cytology.

Treatment

Wide surgical resection is recommended for SCCs, TCCs and TVTs. Following surgical removal, topical and systemic antibiotics and daily extrusion of the penis to prevent adhesions are recommended.

The rate of recurrence of TVTs is high, so chemotherapy is recommended. Vincristine, cyclophosphamide and methotrexate are the chemotherapeutic agents most commonly used for TVTs. In addition, radiation therapy and cryotherapy may be utilized following surgical excision. Castration does not decrease the recurrence rate. No breeding should be allowed because the neoplasm is transmitted venereally. In about 20% of cases, spontaneous remission will occur. Systemic chemotherapy may be necessary for TCCs and SCCs, especially if they are metastatic.

Tom cats

Tumours of the penis are rare in the tom cat. One case each of carcinoma and sarcoma have been reported.

Diseases and disorders of the prostate gland

Benign prostatic hyperplasia and hypertrophy

Benign prostatic hyperplasia/hypertrophy (BPH) occurs as a result of long-term exposure to dihydrotestosterone (DHT), the active metabolite of testosterone. DHT is the androgen active at the cellular level of the prostate gland. As cell numbers and size increase, the prostate gland enlarges. The enlargement tends to be uniform throughout both lobes of the prostate gland. The hyperplastic gland has increased vascularity, which results in vascular leakage or haemorrhage into the gland. This blood is carried by the ducts and lymphatic system of the gland, and is excreted via glandular secretion through the prostatic portal system. Enlargement will continue as long as there is systemic exposure to testosterone, and the gland may increase significantly in size. The swelling may result in pressure on the ducts in the gland, leading to cyst formation. These cysts may be small (<2 mm) or large (up to 3 cm), single or multiple. The capsule surrounding the gland is fibrous and so increased pressure within the gland can lead to compression of the urethra. More than 50% of intact dogs will have signs of BPH by 5 years of age, and senile involution of the prostate gland begins around 11 years of age. It is by far the most commonly diagnosed prostatic disorder.

Diagnosis

Clinical signs: A history of sanguineous or serosanguineous discharge from the penis is the most common presenting complaint. This may occur only when the dog is exposed to bitches in oestrus, or may be intermittent or continuous. The blood may be fresh or digested. Pollakiuria, stranguria and haematuria are also common signs, whilst incontinence and inability to urinate are less common. Constipation and production of a ribbon-like stool may occur once the prostate gland has become markedly enlarged. Infertility may be another complaint.

Physical examination: Digital palpation reveals an enlarged but smooth and symmetrical prostate gland (Figure 20.11). In some cases, enlargement may result in the gland being transposed cranially into the abdominal cavity, such that it cannot be reached by

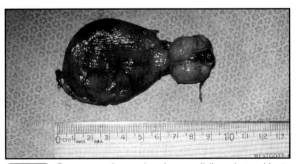

20.11 Gross specimen showing a mildly enlarged but smooth and symmetrical prostate gland immediately caudal to the urinary bladder.

digital palpation. The complete blood count and serum chemistry are normal, unless acute infection accompanies the disorder.

Diagnostic imaging: Radiography may reveal an enlarged soft tissue opacity in the caudal abdomen or cranial pelvic canal, just caudal to the bladder. In some cases, the bladder may be markedly enlarged owing to inadequate emptying. Retrograde cystourethrography may be necessary to illustrate prostatomegaly and to visualize the urethral architecture. There is often a diffuse uptake of contrast medium from the urethra into the glandular tissue. The distance in mm of this uptake into the gland can be used subjectively to assess the severity of disease.

Ultrasonography reveals a uniform hypoechoic to hyperechoic parenchyma with mild heterogeneity. The gland may be normal to increased in size but is usually symmetrical and may have smooth or irregular margins. Intraprostatic cysts may be present and can be single or multiple and small to large with irregular borders. If there are intraprostatic cysts, the symmetry between the lobes may be distorted. The fluid within the cysts is typically anechoic.

Culture and cytology: Cytological examination of prostatic secretions assists with the diagnosis and may be obtained via prostatic massage or semen collection. Collection of the third fraction of the ejaculate usually reveals no increase in white blood cells, but the fraction may contain many red blood cells or may be tinged dark brown as a result of lysis of red blood cells and digestion of the haem pigment.

If semen cannot be collected, a prostatic massage or wash can be performed. The dog should be allowed to empty his bladder prior to the procedure. The dog should be placed in lateral recumbency, usually without sedation unless the prostate gland is painful on palpation. A polyethylene catheter of the largest size possible for the dog is passed (using sterile technique) into the urethra as far as the bladder. The bladder is emptied, flushed with sterile saline, and emptied again. A gloved finger is then placed in the rectum and the catheter withdrawn until the tip can be felt just distal to the prostate gland. Using the other hand, the prostate gland is pushed caudally into the pelvic canal and massaged gently for 1–2 minutes. A 12 ml syringe filled with 5–10 ml sterile saline is attached to the end of the catheter. The fluid is injected and the catheter advanced while aspirating. At least 2–3 ml of fluid is usually retrieved. This fluid may be cultured (aerobic bacteria, *Mycoplasma* spp., *Ureaplasma* spp.) and then centrifuged for cytological examination. If infection of the prostate gland occurs concurrently with BPH, the cultures may be positive, but if BPH occurs alone cultures will be negative. Alternatively, cytology samples may be obtained using a flexible cytobrush system. Diagnosis may also be made following ultrasound-guided aspiration cytology or biopsy.

Treatment

The mode of treatment depends on whether the dog is used for breeding. In older, non-breeding patients

indicates prostatomegaly or prostatic displacement. Retrograde cystourethrography is helpful in defining prostatomegaly. Mineralization of prostatic cysts is common and may be noted on radiographs. Ultrasonography provides a definitive diagnosis, allowing determination of the size, location and nature of the cystic structure(s). The fluid within the cysts is usually hypoechoic or anechoic (hyperechoic fluid should alert the operator to possible prostatic abscessation, see below). Large pelvic cysts must be differentiated from the urinary bladder. Passage of a catheter into the bladder, with or without instillation of saline solution to make the urine obviously heterogenous in echotexture, may be necessary.

Treatment
Small cysts associated with BPH alone may decrease in size with medical treatment, but typically do not resolve completely. Needle aspiration of fluid from retention cysts may be performed, but the effect tends to be temporary because the cystic cavity remains and refills over time. Solitary extraprostatic cysts can be resected surgically. Surgery may be complicated if adhesions to adjoining structures are present. For intraprostatic cysts, removal is often difficult, so drainage should be established and the central cavity packed with omentum to prevent refilling. Large cysts may be marsupialized to the abdominal wall for drainage if they cannot be removed completely. Relapse is common once the stoma closes. Complications include urinary incontinence, urethral fistulas, oedema around the surgical or stoma site, hypoproteinaemia, hypoglycaemia, hypokalaemia and anaemia.

Squamous metaplasia of the prostate gland
Squamous metaplasia occurs secondary to endogenous or exogenous oestrogen exposure. Endogenous oestrogens may derive from the testes, adrenal glands or a hormonally active SCT or ICT, whilst exogenous oestrogens may be administered for behaviour modification or for treatment of prostatic enlargement or other conditions. Histopathology reveals normal epithelial cells of low cuboidal to tall columnar morphology, changing into flattened cells that form concentric circles around the prostatic acini. Proteinaceous debris and leucocytes are commonly present in the fluid within the acini. Squamous metaplasia is typically asymptomatic because the cells are not metabolically active, but it predisposes to infection owing to glandular stasis, and thus the presenting signs may be those of prostatitis.

Diagnosis
Enlargement of the prostate gland is typically not marked but may be unilateral or bilateral, and the gland has a texture that is slightly firmer than normal. If the prostate gland is markedly enlarged (usually due to concurrent prostatic disease) it may be drawn cranially into the abdominal cavity and not palpable *per rectum*. Radiography may reveal mild to moderate prostatomegaly. Ultrasonography reveals a more heterogenous architecture in affected areas. Definitive diagnosis requires prostatic biopsy or exfoliative cytology.

Treatment
Castration is the recommended treatment for squamous metaplasia. Hormone therapy should not be administered because it may exacerbate the disease process. Given that metaplasia is commonly diagnosed in conjunction with prostatitis, care should be taken when obtaining a biopsy sample from a possibly infected gland and appropriate antibiotic therapy administered at the time of sampling.

Tom cats
Squamous metaplasia in tom cats can be induced by exogenous administration of oestrogens.

Prostatic adenocarcinoma
The most common neoplastic condition of the prostate gland is adenocarcinoma. Adenocarcinoma is more common in neutered than in intact males. Castrated males are at 2.4% greater risk for the development of prostatic neoplasia than intact males. Dogs of medium to large breeds may be predisposed. Metastasis is common by the time a diagnosis is made, owing to the aggressive nature of the tumour. Lung, lymph nodes, liver, urethra, spleen, rectum or colon, bladder, bone, heart, kidney and adrenal glands are all possible metastatic sites. Other neoplastic conditions of the prostate gland include TCC, lymphosarcoma, haemangiosarcoma and SCC. All can affect the prostate gland either by direct extension or metastasis.

Diagnosis
Signs of prostatic neoplasia include tenesmus, constipation, obstipation, ribbon-like faeces, stranguria, haematuria, weight loss, hindlimb ataxia and occasionally neck pain. The prostate gland may be enlarged and painful on rectal palpation, and there is typically asymmetry of the gland with the affected lobe being firmer than normal. Leucocytosis and neutrophilia may be present. Elevations in alkaline phosphatase (ALP) are common, as are pyuria and haematuria.

Radiography (plain films or contrast retrograde cystourethrography) may reveal prostatomegaly. Prostatic mineralization may occur. If the urethra is involved or compromised by invasion of tumour cells, contrast studies may reveal urethral irregularity. Ultrasonography reveals a heterogenous architecture with the neoplastic tissue being hyperechoic with respect to the normal prostatic tissue. Mineralization is common with neoplastic lesions.

The diagnosis is confirmed via cytology or biopsy. A prostatic wash or massage may be performed or brush cytology preparations may be obtained. Alternatively, fine-needle aspiration may be performed with ultrasound guidance. A lack of neoplastic cells within cytological samples does not rule out neoplasia because the tumour may be deep in the gland or poorly exfoliative. In cases in which neoplasia is suspected and cytology samples are negative, prostatic biopsy should be performed.

Treatment
Castration will not affect the course of the disease, so is not necessary in intact males diagnosed with

adenocarcinoma. Orthovoltage radiation and chemotherapy may be administered, but increases in survival times are small because metastasis is common by the time the disease is diagnosed. If no metastasis is evident, total prostatectomy may be performed, although serious complications including urinary or faecal incontinence, stranguria, haematuria and hindlimb oedema are not uncommon. For invasive or metastatic TCC, piroxicam may be administered at a rate of 0.3 mg/kg orally q24h.

Tom cats

Prostatic adenocarcinoma has been described in the cat. As in dogs, it may be more common in neutered individuals *versus* intact males. The signs include haematuria, dysuria, stranguria, pollakiuria and outflow obstruction of the urethra. Radiography with retrograde cystourethrography and ultrasonography may be helpful in the diagnosis.

Prostatitis

Infection of the prostate gland is very common and is seen in all age groups from young dogs to older animals. It is seen more commonly in intact than in neutered animals. The presence of BPH, prostatic cysts, prostatic neoplasia or squamous metaplasia may predispose to prostatic infection. Chronic prostatitis is more common than acute prostatitis and is more insidious in onset and clinical signs. It may result from ascending infection from the urinary tract, preputial flora, epididymides or testes, haematogenous spread, or venereal transmission of infection (e.g. brucellosis). The most common organism to infect the prostate gland is *E. coli*. Other bacteria isolated include *Klebsiella* spp., *Pseudomonas* spp., *Pasteurella* spp., *Streptococcus* spp., *Staphylococcus* spp., *Proteus mirabilis*, *Enterobacter* spp., *Haemophilus* spp., *Brucella* spp., *Mycoplasma* spp., *Ureaplasma* spp. and *Blastomyces dermatitidis*. It is rare for anaerobes to be involved. Infection is usually caused by a single species and typically ascends up the urethra from the prepuce, or from the bladder secondary to cystitis. Failure of the local genitourinary mucosal immune system may predispose to infection.

Diagnosis

Acute prostatitis:

Clinical signs: The type and extent of clinical signs present depend on whether the prostatitis is acute or chronic. With acute prostatitis, if inflammation is profound and/or toxins are released, signs may be severe. Swelling and oedema of the prostate gland occurs, along with various combinations of fever (often >40°C), malaise, abdominal pain, vomiting, dysuria, stranguria, haematuria, tenesmus, constipation, obstipation and ribbon-like faeces, along with an unwillingness to breed or pain on semen collection. The prostate gland is typically painful, but may be soft or firm, and is usually swollen and asymmetrical. In some cases, prostatomegaly may be minimal, but if abscessation occurs, alterations in prostatic size and symmetry become much more obvious. The size of the prostate gland depends somewhat on whether there is concurrent prostatic disease or prostatitis alone. Given that the blood–prostate gland barrier is breached in acute prostatitis, haematogenous spread of the bacteria may occur resulting in sepsis, disseminated intravascular coagulation (DIC) or multi-organ failure. In some cases abscessation of the prostate gland may occur. If undiagnosed and untreated, these abscesses may rupture intra-abdominally resulting in peritonitis, sepsis and rapid death.

Physical examination: This may reveal dehydration, fever, shock, sepsis and multi-organ disease. These animals may have mild to severe dehydration with pre- or post-renal azotaemia, elevated hepatic enzymes (particularly ALP), hypoproteinaemia or electrolyte imbalance. Complete blood count reveals a leucocytosis and neutrophilia, often with a left shift. Urinalysis may reveal pyuria and haematuria.

Imaging: Abdominal radiography may demonstrate mild to moderate prostatic enlargement, particularly if concurrent prostatic disease is present. Retrograde cystourethrography may show diffuse uptake of contrast medium into the gland, which is often more severe than that seen with BPH. Ultrasonography reveals a focal to diffuse heterogenous echotexture with or without cavitating lesions filled with hyperechoic fluid. Infected areas tend to be hypoechoic with respect to the normal parenchyma, although if mineralization is present they may be hyperechoic.

Chronic prostatitis:

Clinical signs: With chronic prostatitis, there may be stranguria, haematuria, tenesmus, constipation, obstipation, ribbon-like faeces and a normal to enlarged, symmetrical to asymmetrical, non-painful prostate gland. Typically, these animals are not systemically ill. Prostatitis may be found incidentally during investigation of infertility, during routine physical examination, or during breeding soundness examination.

Physical examination: Patients with chronic prostatitis are typically healthy on physical examination but may present for urinary signs or signs of rectal or colonic obstruction. They may be diagnosed incidentally based on pyospermia or pyuria found during routine examination. Complete blood count and chemistry are typically normal. Urinalysis may reveal pyuria or haematuria. When trying to differentiate cystitis from prostatitis, administration of ampicillin for 24 hours prior to collection of prostatic fluid for culture may help, because many organisms that cause cystitis are sensitive to ampicillin and there is good penetration of the bladder mucosa and urine. In contrast, ampicillin does not cross the intact blood–prostate gland barrier at all, so will be ineffective at reducing bacterial load if the source is the prostate gland.

Imaging: Radiographic findings are similar to those in acute prostatitis, except that there is typically minimal uptake of contrast medium on retrograde cystourethrography. Ultrasonography will usually reveal an asymmetrical gland with heterogenous architecture (Figure 20.14a). Abscessation may occur with chronic prostatitis (Figure 20.14b). It can be difficult to differentiate a prostatic abscess from a prostatic cyst via ultrasonography alone, so cytology of the fluid may be necessary.

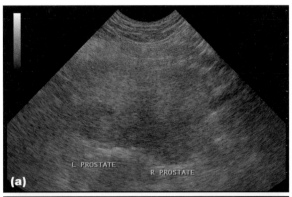

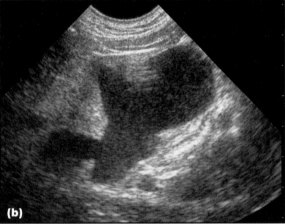

20.14 **(a)** Ultrasonogram showing asymmetrical prostate gland lobes with mild heterogeneity of the prostatic parenchyma. **(b)** Ultrasonogram showing a prostatic abscess. Note the large fluid-filled structure within the prostate gland. The fluid within the structure is moderately hyperechoic with flocculent material visible.

Cytology: Prostatic wash or massage fluid may be cultured for aerobes, anaerobes (acute prostatitis only), *Mycoplasma* spp. and *Ureaplasma* spp. Cytology of this fluid usually reveals a marked inflammatory response, with intra- and extracellular bacteria often present. Gram staining may be performed to provide direction for antibiotic therapy. Semen collection is difficult in dogs with acute prostatitis because of their physical condition and pain on ejaculation, but is an excellent source of prostatic fluid in cases of chronic prostatitis. Cytology brushes may also be used to obtain samples. Caution should be shown when considering prostatic aspiration if prostatitis is suspected because bacteria may be drawn along the needle tract, resulting in dissemination of infection through the soft tissues.

Treatment

Acute prostatitis requires aggressive therapy. Intravenous fluids should be administered based on the degree of dehydration and electrolyte imbalance.

Balanced electrolyte solutions may be used pending laboratory results. If there is fever and renal function is normal, NSAIDs may be administered. If renal function is compromised, rehydration should be accomplished prior to NSAID administration.

The choice of antibiotics should be based on the degree of cytology and Gram stain until culture results are obtained. With acute prostatitis, the blood–prostate gland barrier is breached, allowing rapid penetration of most antibiotics into the gland regardless of prostatic fluid pH. With chronic prostatitis, the choice of antibiotics is far more critical. Antibiotics with high lipid solubility and high pK_a (acid dissociation constant) and those that are weakly alkaline will diffuse most readily across the prostatic membrane. The antibiotics that best fit these characteristics include trimethoprim, clindamycin, chloramphenicol (high end of dose range required) and erythromycin. Alternatively, zwitterion antibiotics, the fluoroquinolones enrofloxacin and ciprofloxacin have multiple pK_as, so they diffuse into the prostate gland regardless of the surrounding tissue and fluid pH. As enrofloxacin (5 mg/kg orally q24h) and ciprofloxacin (10 mg/kg orally q12h) are effective against most prostatic pathogens, they are frequently the first choice of antibiotics pending culture results for both acute and chronic prostatitis. With acute prostatitis, addition of amoxicillin (15 mg/kg orally q12h), ampicillin (20 mg/kg orally q8h) or amoxicillin plus clavulanate (13.75 mg/kg orally q12h) is often used to ensure full coverage of Gram-positive organisms. Fungal prostatitis may be treated with ketoconazole or itraconazole plus amphotericin B.

Antibiotics should be continued for a minimum of 4 weeks, and 8–12 weeks of treatment is often necessary for chronic cases or in cases with abscessation. Long-term antibiotic therapy may cause resistance if an antibiotic of too broad a spectrum is continued without amendment when the culture results are obtained. Long-term administration of trimethoprim may cause keratoconjunctivitis sicca, hypothyroidism, urolith formation, liver dysfunction, anaemia and immune-mediated arthropathy, so care should be taken when using this medication for more than 30 days.

Animals with concurrent BPH should be castrated or treated pharmacologically. Oestrogens are not recommended owing to the possibility of exacerbation of any component of squamous metaplasia in the prostatic disease. Progestogens may be used but are not recommended for long-term therapy of BPH. If prostatic cysts are present they should be managed as outlined above. Castration may help to resolve prostatitis up to 4–5 weeks more quickly than antibiotic therapy alone. Animals with acute prostatitis should be stabilized metabolically prior to surgery.

Prostatic abscesses may need to be drained surgically because they are typically well encapsulated and run the risk of rupturing and causing a fatal peritonitis. Surgical management of abscesses may include marsupialization, drainage with omentalization of the remaining cavity, or partial or complete prostatectomy. Partial prostatectomy can be performed using an ultrasonic aspirator, and shows good results with a low complication rate. Complications include recurrence, abdominal contamination, urinary incontinence, hypoproteinaemia and hindquarter oedema. There is a high mortality rate associated with prostatic abscessation with or without surgical treatment. Drainage and omentalization may be the safest surgical technique. Drainage via needle aspiration is rarely curative and has a high relapse rate because the abscess capsule is left behind and is too thick for antibiotics to penetrate, thereby leaving bacteria *in situ* to re-grow in a short period of time. In addition, bacteria may be seeded in the soft tissue on the way out from the prostate gland, leaving an infected needle tract.

Disorders of the bulbourethral glands

The tom cat has a paired set of bulbourethral glands. Abnormalities of these glands are uncommon and have little clinical consequence. Inflammation of the bulbourethral glands may occur as a result of ascending infection from the urinary tract. Cystic changes may be noted in older intact males but have been found only incidentally.

Perineal hernia

Perineal hernias are most common in aged intact male dogs and are caused by weakening of the muscular walls of the pelvic diaphragm of the anal and perineal region. They may occur in females, associated with abdominal or pelvic trauma. The average age at occurrence in dogs is 9–10 years with increasing prevalence to 14 years of age. Boston Terriers, Boxers, Welsh Corgis, Pekingese, collies, poodles and cross breeds are predisposed. Hernias may be unilateral or bilateral.

With increased pressure from abdominal contents, including an enlarged prostate gland, the muscular and connective tissue in the wall of the perineum may rupture, resulting in herniation of the rectum, perineal body, prostate gland or bladder into the hernial sac. The presence of steroidogenic hormones predisposes to this type of muscular weakness, as do age-related weakening and straining.

Diagnosis

The diagnosis is often made by physical examination alone, with detection of a protruding soft tissue bulge lateral to or lateral and slightly ventral to the anal opening. Animals may be noted to strain when defecating. If the bladder or bowel is incarcerated, the patient may present with an acute abdomen, abdominal pain, vomiting, stranguria, obstipation, dyschezia or rectal prolapse. Digital examination of the rectum will typically reveal evidence of the hernial ring. The opposing wall should also be examined because it too may be weakened. Ultrasonography can be used to assess the contents of the hernia. Radiography, with or without barium contrast medium, can be used to visualize the caudal abdominal contents and determine the position of the colon and rectum. Contrast cystography may be needed to assess bladder position if the bladder is thought to be involved in the hernia.

Treatment

Stabilization of the patient is the primary aim and treatment of prostatic disease should be instituted, if it is present. Neutering may be recommended to decrease the size of the prostate gland associated with BPH. Extraprostatic cysts may be aspirated to reduce their size. Softening of the stool is beneficial to minimize straining during defecation. Surgery is recommended to repair the hernia (and to remove cysts) and may require placement of a mesh to fill the hernial defect. It is not uncommon to have the contralateral side herniate shortly after surgical repair of the affected side if the predisposing cause of the hernia is not addressed.

Tom cats

Perineal hernias are rare in cats. They are more common in queens than tom cats, and when they occur in males they are typically **neutered**. Hernias tend to be bilateral. The average age of occurrence is 10 years.

Genetic and chromosomal disorders

Intersex conditions

Chromosomal defects may result from alterations in the sex chromosomes (additions, deletions or translocations). These animals are called intersexes, chimeras and mosaics. The most common chromosomal abnormality is 79XXY trisomy. Other described abnormalities include 78XX/78XY and 78XX/79XXY. In dogs, the XXY chromosomal complement typically results in male pseudohermaphroditism. Male animals with pseudohermaphroditism have ambiguous or male phenotype, with either remnants of the uterine tubes internally or vasa deferentia that lead to ovaries. True hermaphrodites have ambiguous or either male, female or some combination of external genitalia, an internal tubular tract reminiscent of that of either the dog or bitch, and either one ovary and one testis or ovotestes. Testicular feminization syndrome is associated with a normal male karyotype and functional testes, but non-functional androgen receptors elsewhere in the body. These dogs appear phenotypically female but lack the normal tubular tracts of either male or female animals.

Sexual differentiation occurs during fetal development and depends on a normal chromosomal complement followed by normal development of the gonads and external genitalia. The Y chromosome is the sex determining chromosome and carries the *SRY* gene. If *SRY* is present and normal, the fetus will develop into a male, whilst if it is abnormal or absent the fetus will develop into a female or something between male and female. Normal male dogs have the 78XY karyotype. The gonads develop from the genital ridge and in the male migrate caudally during development. Once the gonads have developed, the phenotypic sex of the fetus develops as a result of the hormones being secreted by the gonads. If the fetal gonad is a testis or ovotestis and secretes testosterone, then the external genitalia will be male or tend towards the male; whereas if the gonads are ovaries or ovotestes that secrete oestrogens, then the external genitalia will be female or tend towards female.

In the normal male, the paramesonephric ducts degenerate and the mesonephric tubules develop. The Sertoli cells in the fetal testis produce Müllerian-inhibiting hormone (MIH) and this is the hormone that causes regression of the paramesonephric ducts. In males with ovaries (male pseudohermaphrodites) there is inadequate MIH, and so some or all of the paramesonephric ducts remain. The rest of the female tract will not develop in the presence of dihydrotestosterone (DHT), which is produced by the fetal testicular Leydig cells. Therefore, in the absence of normal fetal testes, some or all of the caudal vagina, vestibule, clitoris and vulva may develop.

Diagnosis
Physical examination is often sufficient to raise the suspicion of a chromosomal abnormality. The prepuce may be located in a more posterior position than normal (closer to where the vulvar lips normally reside) and the penis may be smaller than expected for an animal of similar age. These animals are often unilateral or bilateral cryptorchids. The diagnosis may be made on exploratory laparotomy with direct visualization of the reproductive tract. A definitive diagnosis requires karyotyping and histopathology of the reproductive tract. Inevitably, most of these animals are sterile.

Treatment
Neutering or ovariohysterectomy is recommended because these animals will not be used for breeding. Animals with an os penis for a clitoris or hypospadias may require corrective surgery for urinary incontinence or for cosmetic reasons.

Tom cats
Common abnormalities of sexual differentiation include 39XXY trisomy and 38XX/38XY (which are chimeras or mosaics), true hermaphrodites, and male and female pseudohermaphrodites. Male tortoiseshell cats with these abnormal chromosomal complements are typically sterile, although a small percentage may be fertile. It is possible for some individuals to be fertile because they have two sets of seminiferous tubules, one being functional and the other not. Cystic rete testes are reported in cats and appear clinically as a large cystic area in the centre of the testis. Testicular hypoplasia has been reported following infection with panleucopenia virus or in cryptorchid testes.

Hypospadias
Hypospadias occurs when there is improper termination of the urethral orifice along the ventrum of the penile shaft. It results from abnormal closure of the median raphe anywhere along the ventrum, from the scrotum to the preputial opening. Boston Terriers appear to be predisposed. Cryptorchidism commonly occurs concurrently.

Diagnosis
The most common complaint is urinary incontinence, and affected individuals may have a juvenile appearance to the penis (shortened) and prepuce or an abnormal preputial opening. The glans penis may be underdeveloped or deviated ventrally.

Treatment
If there are no clinical signs, no treatment is necessary. If incontinence or penile exposure with desiccation of the mucosal surface is a problem, surgical correction is required. Surgical closure of the defect in the urethra and serosal oversewing of the penis, followed by closure of the defect in the preputial skin, may be necessary. Castration is also recommended. When hypospadias occurs close to the scrotum, penile amputation and scrotal or perineal urethrostomy may be required to correct the defect.

Diphallus
Diphallus involves duplication of the penis. Other genitourinary defects, duplications or hypoplasia may occur concurrently. Duplication of the cloacal membrane typically results in diphallus alone, whilst duplication, aplasia or hypoplasia of the hindgut may be involved in more complicated cases.

Diagnosis
The clinical signs typically include haematuria, inappropriate urination, pollakiuria or recurrent urinary tract infection.

Treatment
Determination of urethral patency to the urinary bladder is performed first, to ascertain whether one or both penises are fully functional. Amputation of the shaft of one of the penises with closure of any defect in the remaining urethra may correct the problem. If the urethral sphincter is compromised, incontinence may persist.

Cryptorchidism
Unilateral or bilateral cryptorchidism is a common finding in the dog. The fetal testis is derived from the gonadal ridge and is moved caudally during development by the outgrowth of the gubernaculum, which attaches to the external inguinal ring. The gubernaculum shortens as the fetus grows, forcibly pulling the testicle towards the vaginal ring. Once the testis enters the vaginal space (internal inguinal ring) it continues towards the scrotum. After leaving the external ring, the testis is drawn further down into the scrotum via shortening of the gubernaculum. The descent of the testes is regulated by testosterone and other factors released by the fetal and neonatal testes. Testosterone is particularly important in the descent of the testes from the external inguinal ring to the scrotum.

Neither lack nor excess of hormones (testosterone, luteinizing hormone (LH), follicle-stimulating hormone (FSH) or oestrogens) is likely to contribute to cryptorchidism. This is corroborated by the fact that cryptorchid animals do not have different concentrations of these hormones from normal males, and that treatment with gonadotrophic medications does not resolve the condition successfully, regardless of the age at which they are administered. However, cryptorchid males do have a less marked response to gonadotrophin-releasing hormone (GnRH) stimulation testing than their intact counterparts.

Cryptorchidism is considered to be hereditary and an autosomal recessive trait. This means it is carried by both maternal and paternal lines. If the testes begin to enlarge prior to entry into the internal ring or if their descent is slowed they may be retained abdominally. If they begin to enlarge after entering the internal ring but before leaving the external ring, or if testicular descent is altered or stopped, the animal may be an inguinal cryptorchid. Normally both testes are in the scrotum by 2 weeks postpartum. The right testis is retained more commonly than the left owing to its more cranial starting position in the abdomen.

There are a number of breeds at risk for cryptorchidism, including the Boxer, English Bulldog, Cairn Terrier, Chihuahua, Miniature Dachshund, Maltese, Pekingese, Pomeranian, Toy, Miniature and Standard Poodle, Miniature Schnauzer, Old English Sheepdog, Shetland Sheepdog, Siberian Husky and Yorkshire Terrier. Other inherited conditions associated with cryptorchidism include inguinal and umbilical hernia, hip dysplasia, patellar luxation, and penile and/or preputial defects.

Retained testes are smaller than scrotal testes, and the location of retention is positively correlated with size. Abdominally retained testes (Figure 20.15) are smaller than inguinally retained testes. There is a marked reduction in spermatogenesis in retained testes that worsens with the degree of retention. Whilst spermatogenesis does not progress normally in retained testes, steroidogenesis does. Thus, although cryptorchid dogs may show interest in and mate with bitches, unless they have one descended testis they will be sterile. In cases of inguinal retention, dogs may have lower than normal numbers of spermatozoa per ejaculate and increased abnormal forms. Retained testes are predisposed to neoplasia (the risk is 9–14 times higher than normal) and testicular torsion.

20.15 Small abdominally retained testis seen on post-mortem examination. The gubernaculum is visible attached to the caudal pole of the testis and the coiled spermatic cord and vas deferens are also clearly visible. The caudal pole of the right kidney is on the far right.

Lack of two testes in the scrotum by 8 weeks of age is considered to be suspicious for cryptorchidism. Classically, the diagnosis is accepted by the time an animal reaches 6 months of age: if it does not have two scrotal testes, it is considered a cryptorchid. However, realistically, with the vast differences in age at puberty between breeds this is probably not a reasonable criterion. Based on the average age at puberty for any given breed, it could be suggested that dogs which have both testes in the scrotum within 2 months of puberty are considered normal. Descent of a testis after 1 year of age is very rare.

Diagnosis

The inguinal region and the area just lateral to the penis and prepuce should be palpated carefully when attempting to locate a retained testis. Lymph nodes and fat are commonly mistaken for testes, as is the gubernacular outgrowth from the scrotum. Testes should be freely movable and have an attached epididymis. Ultrasonography may be used to locate the cryptorchid testis and is more successful in locating inguinal than abdominal testes. This is due to the smaller size of abdominal testes and the fact that they may be located anywhere in the abdominal cavity (although they tend to lie along a line drawn between the caudal pole of the kidney and the external inguinal ring). Abdominal testes appear to have the same echotexture as scrotal testes, with an obvious epididymis and mediastinum testis. However, if the abdominal testis is neoplastic, it may be larger and show abnormal echogenicity.

In cases where there is no history of neutering and bilateral cryptorchidism is believed to exist, a human chorionic gonadotrophin (hCG) or GnRH stimulation test can be performed. A blood sample for baseline testosterone is taken, followed by intramuscular injection of 500–1000 IU hCG or 2 µg/kg GnRH. A second blood sample is obtained 1 hour after GnRH or 4 hours after hCG administration to measure testosterone concentration post-stimulation. The baseline concentration should more than double if testicular tissue is present. Measurement of serum LH utilizing the commercial assay available is not accurate for identification of castrated males owing to the overlap of resting ranges in cryptorchid and neutered males.

Treatment

Typically, no treatment is recommended for cryptorchidism beyond castration. These individuals should be neutered and not used in any breeding programme owing to the hereditary nature of the disease. There are a number of published and anecdotal protocols that involve multiple injections of either hCG or GnRH to stimulate precocious puberty and encourage testicular growth and descent, but there is no strong scientific evidence that any of them is consistently effective. Two published protocols are: 50–750 µg GnRH i.m. at weekly or biweekly intervals for up to 6 weeks; and 100–1000 IU hCG i.m. once or twice weekly for 4 weeks. Various degrees of success are associated with these treatments, proving the multifactorial and polygenetic nature of the disease.

Medical treatment may increase the fertility of an affected dog, thereby perpetuating the trait in his off-spring. In addition, the increased risk of neoplasia remains whether the testes descend or not, so in a successfully treated animal, failure to neuter at a young age increases the risk for neoplasia later in life. Anecdotal reports of massaging inguinally retained testes towards the scrotum to hasten their descent are unlikely to be true except in shy individuals who retract the testes towards the body when touched. Daily manipulation of the scrotum and perineal area may make these individuals more relaxed when the scrotum is palpated, thus allowing the testes to remain in their scrotal position.

Exclusion of all affected males, their offspring and their parents from the breeding programme is recommended. It is difficult to determine whether a bitch is a carrier of the cryptorchid gene owing to the large number of male puppies all required to have two scrotal testes by 6 months of age (at least 40 normal male puppies) to exclude carrier status.

Tom cats

Both testes are normally in the scrotum at birth in tom cats. The testes in some individuals may slide up and down from the scrotum to the inguinal canal until puberty at 7–8 months of age. Puberty in pure-bred cats occurs later than in Domestic Shorthaired or Longhaired cats, and thus a male should not be pronounced a cryptorchid until a minimum of 8 months after birth. The incidence of cryptorchidism in cats is reported to be between 0.37 and 1.7%. Cryptorchid testes have defective spermatogenesis but the Sertoli and Leydig cells are functional so that the male phenotype and behaviour are still present.

Cryptorchidism in the cat is more likely to be unilateral (90%) than bilateral (10%), with both testes being affected equally frequently when unilateral. Abdominal cryptorchids are more common than inguinal. Cryptorchidism is not considered to be hereditary in the tom cat; however, given its hereditary nature in many other species, it may have a hereditary predisposition in the cat but there are insufficient reported cases to document it. Some cats with cryptorchidism have been reported to have other congenital or developmental defects such as patellar luxation, tarsal deformities, kinked or shortened tails, microphthalmia, agenesis of the upper eyelids and tetralogy of Fallot. Retained testes may be at increased risk for neoplasia and therefore should be removed. Testicular torsion of a retained testis has not been reported in the cat.

Diagnosis: The diagnosis is based on a history of no prior castration and physical examination findings of either none or only one scrotal testis in a cat >7–8 months of age. In suspected bilateral cryptorchids that have attained a pubertal age, the presence of penile spines (barbs) is indicative of testosterone production and should raise the suspicion of cryptorchidism. The spines disappear within 6 weeks of castration. Serum testosterone testing can confirm the suspicion of cryptorchidism if the physical examination is not definitive. A baseline testosterone

concentration is obtained and then the cat is given either hCG (250–500 IU i.m.) or GnRH (25 µg i.m.). A sample for measurement of testosterone concentration is drawn 1 hour after GnRH or 4 hours after hCG administration. The testosterone concentration should increase to 3–10 times above baseline following hCG stimulation or 3–15 times following GnRH stimulation testing, to indicate a normal stimulation response. Ultrasonography can be used in an attempt to locate the retained testis but, owing to their small size, they are difficult to detect.

Treatment: Medical treatment using hCG or GnRH has not been reported to be successful in cats. Surgical exploration to find the retained testis is necessary. The retained testis can be located anywhere from the kidney to the scrotum, so a meticulous exploratory laparotomy is necessary. Following the vas deferens of an already descended testis is helpful. Sometimes both abdominal and inguinal approaches are necessary.

Behavioural disorders

Retention of male behaviour following neutering

Male behaviour may be learned or hormonally driven. Young males that have undergone neutering at an early age may still display some behaviour typical of intact dogs when they reach puberty (mounting and thrusting behaviours). Older males that have been neutered recently may continue to display male behaviour for months or years after neutering. Mounting behaviour between dogs is a display of dominance by the mounting animal over the mounted animal, and is frequently observed in animals of the same sex. Some behaviours (marking or pollakiuria) may be caused by infection of the genitourinary tract. Some animals may have a history of prior castration obtained from a previous owner when in fact the animal is a unilateral or bilateral cryptorchid, and so the behaviour emanates from continued testosterone production from a retained testis.

Diagnosis

In cases where males continue to display male behaviours (marking, roaming, breeding, aggression) following definite neutering, the first step of the diagnostic process is to determine whether the behaviours are physiological or behavioural in origin. The animal should be observed to determine which stimuli provoke the display of certain behaviours. It should be noted whether they are housed with intact females or males, by themselves, or only with humans. The frequency and amount of urination, as well as the urinary pattern and any signs of urge incontinence, should be monitored. Videotaping of the behaviour may be helpful in determining what is happening before, during and after the event.

Any infectious or inflammatory disease of the genitourinary tract should be ruled out via various diagnostic procedures (urinalysis, ultrasonography, radiography, prostatic wash/massage). If the neutering status of the animal cannot be confirmed, a testosterone stimulation test may be performed. Animals with elevated testosterone concentrations, either with or without stimulation testing, should be assessed for abdominal or inguinal testicles and exposure to exogenous steroid hormones. The owners should be questioned about foods, supplements and any exposure to human steroid hormone products or anabolic steroids administered to enhance performance. Finally, adrenal gland function should be evaluated and ultrasonography may be necessary to rule out steroid-secreting adrenal gland tumours. Pituitary gland or hypothalamic tumours are also possible sources but are uncommon.

Treatment

Cryptorchid animals should be neutered. Any exposure to exogenous hormones should be stopped. If a tumour is present, it should be surgically removed, if possible. In cases of behavioural conditioning, positive reinforcement and diversion training should be employed. Behaviour-modifying drugs may be beneficial during the retraining process.

Tom cats

In tom cats that have a history of prior castration, examination of the penis may show the distinct spikes (as seen on the penis of an intact male) and indicate that the animal is cryptorchid.

References and further reading

Barsanti JA (1997) Diseases of the prostate gland. In: *Proceedings of the Annual Meeting of the Society for Theriogenology*, Nashville, TN, pp. 72–80

Bassinger RR, Robinette CL, Hardie EM and Spaulding KA (2003) The prostate. In: *Textbook of Small Animal Surgery, Vol. 2*, ed. D Slatter, pp. 1542–1557. WB Saunders, Philadelphia

Bloom F (1962) Retained testes in cats and dogs. *Modern Veterinary Practice* **43**, 160

Doig PA, Ruhnke HL and Bosu WTK (1981) The genital *Mycoplasma* and *Ureaplasma* flora of healthy and diseased dogs. *Canadian Journal of Comparative Medicine* **45**, 233–238

Feeney DA, Johnston GR, Klausner JS *et al.* (1987) Canine prostatic disease – comparison of ultrasonographic appearance with morphologic and microbiologic findings: 30 cases (1981–1985). *Journal of the American Veterinary Medical Association* **190**, 1027–1034

Feldman EC and Nelson RW (2004) *Canine and Feline Endocrinology and Reproduction.* WB Saunders, Philadelphia

Fossum TW (2002) *Small Animal Surgery 3rd edition.* Mosby, St Louis

Freak MJ (1949) Discussion: clinical aspects of disease of the alimentary tract of the cat. *Veterinary Record* **61**, 679

Hayes HM and Pendergrass TW (1976) Canine testicular tumors: Epidemiologic features of 410 dogs. *International Journal of Cancer* **18**, 482–487

Herron MA (1988) Diseases of the external genitalia. In: *Handbook of Small Animal Practice*, ed. RV Morgan, pp. 673–678. Churchill Livingston, New York

Hornbuckle WE and Kleine LJ (1980) Medical management of prostatic disease. In: *Current Veterinary Therapy VII*, ed. RW Kirk, pp.1146–1150. WB Saunders, Philadelphia

Johnston SD, Root Kustritz MV and Olson PNS (2001) *Canine and Feline Theriogenology.* WB Saunders, Philadelphia

Lyle SK (2007) Disorders of sexual development in the dog and cat. *Theriogenology* **68**, 338–343

Madewell BR and Theilen GH (1987) Tumors of the urogenital tract. In: *Veterinary Cancer Medicine*, ed. GH Theilen and BR Madewell, pp. 567–600. Lea & Febiger, Philadelphia

McEntee K (1990) *Reproductive Pathology of Domestic Animals.* Academic Press, San Diego

Meyers-Wallen VN and Patterson DF (1989) Sexual differentiation and inherited disorders of sexual development in the dog. *Journal of Reproduction and Fertility Supplement* **39**, 57–64

Pugh CR and Konde LJ (1991) Sonographic evaluation of canine testicular and scrotal abnormalities: a review of 26 case histories. *Veterinary Radiology* **32**, 243–250

Ravaszova O, Mesaros P, Lukan M *et al.* (1995) Testicular descent in dogs and therapeutic measures. Problems of abnormal descending testes. *Folia Veterinaria* **39**, 45–47

Romagnoli SE (1991) Canine cryptorchidism. *Veterinary Clinics of North America: Small Animal Practice* **21**, 533–544

Root Kustritz MV, Johnston SD, Olson PN *et al.* (2005) Relationship between inflammatory cytology of canine seminal fluid and significant aerobic bacterial, anaerobic bacterial or mycoplasma cultures of canine seminal fluid: 95 cases. *Theriogenology* **64**, 1333–1339

Sojka NJ (1980) The male reproductive system. In: *Current Therapy in Theriogenology*, ed. DA Morrow, pp.821–844. WB Saunders, Philadelphia

Verstegen J (1998) Conditions of the male. In: *BSAVA Manual of Small Animal Reproduction and Neonatology*, ed. G England and M Harvey, pp.71–82. BSAVA, Gloucester

21

Clinical relevance of biotechnology advances

Gaia Cecilia Luvoni

Introduction

Biotechnology offers new strategies to improve the treatment of male and female infertility and to preserve biodiversity. In carnivores, the improvement of reproductive performance in breeding animals of high genetic merit and the preservation of rare wild species in danger of extinction are areas in which recent advances have been made.

In vitro systems of sperm–oocyte interaction have been developed in feline and canine reproductive biotechnology. Using in vitro fertilization (IVF) and intracytoplasmic sperm injection (ICSI) in conjunction with embryo transfer (ET), it is possible to produce offspring from animals that could not have conceived otherwise, as well as increasing reproductive rates from certain individuals. Recent successes in cloning make it possible to produce embryos without sperm–oocyte interaction by using the transfer of a somatic nucleus into a female germ cell. In vitro sperm–oocyte interaction can also be used to assess the fertility of male gametes. Zona binding assays and penetration tests can evaluate the functional integrity of spermatozoa and contribute to improved diagnosis in cases of male infertility.

Biotechnologies applied to reproduction are constantly evolving. Some are already available for clinical application, but others are still at an experimental level. The success of in vitro techniques depends on the ability to mimic in vivo conditions.

In vivo sperm–oocyte interaction: fertilization

The achievement of monospermic fertilization and embryo development depends on the fusion of competent oocytes and spermatozoa in the oviduct.

Oocyte competence

Oocytes acquire developmental competence progressively during the process of maturation, which is a highly coordinated process that involves morphological, ultrastructural and transcriptional changes of the nuclear and cytoplasmic compartments. The nucleus of the oocyte resumes meiosis after a long period of quiescence. From the immature stage (germinal vesicle; GV) it reaches metaphase II (MII), which is characterized by the extrusion of the first polar body and by the orientation of the chromosomes on the MII plate (Figure 21.1). The oocyte will remain at this

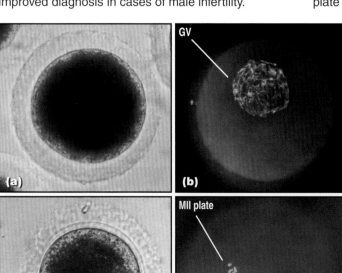

21.1 **(a)** Immature canine oocyte at the germinal vesicle (GV) stage of meiosis and **(b)** stained with fluorescent dye (Hoechst 33342) to visualize the GV nucleus. **(c)** Mature feline oocyte at the metaphase II (MII) stage of meiosis and **(d)** stained with fluorescent dye (Hoechst 33342) to visualize the MII plate and the first polar body.

stage until fertilization occurs. Cytoplasmic maturation consists of a long growth period, during which different molecules are accumulated that will not be used until later during early embryonic development. Communication between the oocyte and its surrounding cumulus cells is critical for supporting nuclear and cytoplasmic maturation and enhancing the development of a competent oocyte.

In most mammalian species, oocyte maturation occurs in the preovulatory follicle. Oestrogen dominates the preovulatory follicular environment and meiotic division is resumed shortly before ovulation as a consequence of the preovulatory surge in luteinizing hormone (LH). Ovulated oocytes are in the MII stage of meiotic division and are ready to be fertilized.

Canine female gametes undergo a different form of development from the oocytes of many other domestic mammals. In fact, in dogs, foxes and other canids, the ovarian follicles luteinize prior to ovulation, which exposes the oocytes to high concentrations of progesterone. After the LH surge, the oocytes are ovulated spontaneously as primary oocytes, at the beginning of the first meiotic division. Subsequent stages of meiotic maturation are resumed in the oviduct, and the oocytes take 2–3 days to achieve the MII stage and be ready for fertilization.

The reproductive physiology of the domestic cat is characterized by induced ovulation. Ovulation requires the release of LH as a consequence of copulation, although ovulation can also occur in group-housed queens in the absence of mating. As in other mammals, the ovulated feline oocyte is in the MII phase of meiosis and can be fertilized immediately after ovulation (see Chapter 1).

Spermatozoal competence

The competence of spermatozoa describes the fertilizing capacity acquired by the cell during its sequential maturational changes. The sperm acquires motility, the ability to undergo capacitation and the acrosome reaction, and the ability to bind to the zona pellucida and fuse with the oolemma. Maturation starts in the epididymis and continues with capacitation and the acrosome reaction in the female genital tract after ejaculation. In the female the dilution of seminal plasma with vaginal secretions neutralizes the activity of 'decapacitating' factors present in the ejaculate. Thus, during capacitation, surface components of the sperm are modified or removed, which allows the expression of binding sites, the activation of membrane receptors and the destabilization of the phospholipid bilayer of the cellular membranes.

This destabilization leads to the acrosome reaction, which is triggered by the influx of extracellular calcium. The fusion of the plasma membrane with the outer acrosomal membrane enables the release of hydrolytic enzymes (Figure 21.2). The enzymes facilitate penetration into the zona pellucida and fusion of the spermatozoa with the oolemma of the oocytes. Acquisition of full spermatozoa competence involves complex mechanisms that are not yet fully understood, but in dogs and cats these are similar to those known in other mammalian species.

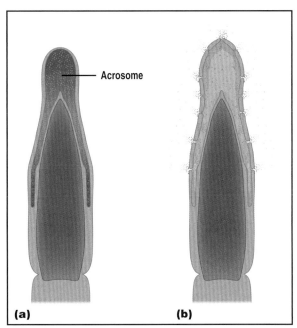

21.2 The acrosome reaction. **(a)** Spermatic head with intact acrosome. **(b)** Fusion and vesiculation of the plasma membrane and outer acrosomal membrane with release of hydrolytic enzymes.

Canine spermatozoa remain viable in the female genital tract for as long as 11 days following mating, compared with 1–2 days in cattle and 5–6 days in humans. The exact time at which spermatozoa penetrate canine oocytes is not certain, owing to the considerable asynchrony between ovulation, insemination and oocyte maturation. The duration of sperm survival in the genital tract of the queen is not known.

Embryo development

In most mammalian species, zygotes undergo early stages of cleavage in the oviduct before migration into the uterus. This takes place, in both cattle and women, about 4 days after ovulation. In dogs an extended period of time is needed. Embryos at the blastocyst stage enter the uterus 11–12 days after the LH peak and implantation occurs 4–7 days later. In the cat, the morulae migrate into the uterus 5 days after the LH peak and implantation occurs 7–9 days later.

In vitro sperm–oocyte interaction: *in vitro* embryo production

Extracorporal production of embryos can be achieved by IVF or ICSI. With IVF, several spermatozoa and oocytes are required, whilst ICSI uses a single spermatozoon which is injected into a single oocyte. In both cases, the zygotes that have been created are cultured *in vitro* to obtain embryos at optimal stages of development (late morulae or early blastocysts) for transfer or cryopreservation.

Collection and selection of oocytes

Oocytes can be collected from *in situ* or *ex situ* ovaries.

In situ ovaries

Growth: Collection from *in situ* ovaries requires stimulation of the ovaries through superovulation. This increases the number of follicles containing oocytes that can be harvested. Preovulatory oocytes are collected and, because they have matured *in vivo*, they can be fertilized without further culture for maturation. This procedure is not commonly used for recovering oocytes in the bitch because in this species the preovulatory oocytes are still meiotically immature.

Superovulation of queens for the recovery of oocytes from *in situ* ovaries involves treatment with 150 IU of equine chorionic gonadotrophin (eCG) and 100 IU of human chorionic gonadotrophin (hCG). It can also be achieved by using 1.5–6.0 mg of porcine follicle-stimulating hormone (p-FSH) for 4 days followed by 100–300 IU hCG. Following one of these protocols, 8–20 preovulatory oocytes can be collected.

The interval between the eCG and hCG treatments and the hCG dosage have been shown to affect oocyte quality, with optimal results obtained using 100 IU hCG given 80–84 hours after eCG. However, individual variability in ovarian response following gonadotrophin administration also affects the results. In the cat, repeated treatments with eCG and hCG at intervals of <4 months reduces follicular development as a result of an immune-mediated negative response to ovarian stimulation. One option is to alternate the use of different gonadotrophins (e.g. equine chorion *versus* pituitary gland derived).

Harvesting: The oocytes of preovulatory follicles can be recovered from donors by laparoscopy or laparotomy. Laparoscopy has been used effectively in numerous felid species, including the domestic cat, and is the least invasive procedure. During laparoscopy the ovaries are evaluated for the number and size of follicles, and the largest follicles are aspirated using a vacuum pump at low pressure. Collection tubes are emptied into separate culture dishes and examined under a stereomicroscope to retrieve the oocytes. Oocyte collection by laparotomy should only be considered in exceptional cases, owing to the increased risks of surgery and the stress that it may cause to the animal.

Transvaginal ultrasound-guided oocyte retrieval, commonly used in women, has not yet been used in carnivores. Owing to the specific anatomical features of the genital tract (orientation of the vagina, narrow, non-distensible cranial vagina, prominent dorsal medial fold, adipose and connective tissue of the ovarian bursa) it is difficult to visualize the ovaries in these species. Commercial probes for intravaginal use are not yet available.

Ex situ ovaries

Collection of follicular oocytes from *ex situ* ovaries, obtained *post mortem* or after ovariectomy, provides a plentiful source of potential embryos. The oocytes must be collected together with their surrounding cumulus cells, which play an important role in maturation. *Ex situ* oocytes are mostly at the GV stage of meiosis (immature oocytes) and need to be matured *in vitro* before fertilization.

Ovaries are usually processed within 2–6 hours of excision, but may be stored at 4°C for not more than 24 hours before processing. Mammalian oocytes are recovered by aspiration of the antral follicles, slicing of the ovarian cortex and follicle dissection. These techniques are performed under laminar flow hoods for sterility and under temperature control (35–38°C).

Aspiration is performed by using a syringe and needle and puncturing the antral follicles, allowing collection of the follicular fluid together with the oocytes. The follicular content is then examined in a dish under a stereomicroscope. The advantage of follicle aspiration in most species is the speed of oocyte recovery. This is not the case in dogs and cats for several reasons. The duration of anoestrus and dioestrus, when follicles are not visible on the ovarian cortex, is relatively long. The small size of canine and feline ovaries makes aspiration of individual follicles very difficult and blind aspiration of the ovarian surface is generally performed. This results in significantly lower recovery rates than may be obtained by slicing the ovaries.

Slicing of the ovarian cortex and follicle dissection are the most common procedures adopted in carnivores. The ovarian surface is cut repeatedly lengthwise and crosswise with a surgical blade and the ovaries are washed in a dish filled with culture medium. For follicle dissection, the ovarian cortex is divided into small pieces and the antral follicles are punctured with a needle to release the oocytes into the culture medium.

Selection

The recovered oocytes are a heterogeneous population of competent and non-competent oocytes derived from growing or atretic follicles. In order to select oocytes which might undergo successful *in vitro* maturation, they are assessed morphologically. Selection is based on oocyte morphology and diameter, as well as cumulus conformation.

Good quality (grade 1) canine and feline oocytes must have a spherical shape, a darkly pigmented and homogenous cytoplasm, an intact oolemma and zona pellucida, a diameter of >100 μm, and must be completely surrounded by compacted layers of cumulus cells (Figure 21.3). Oocytes that are misshapen, have lightly pigmented cytoplasm, are pale in colour or with aggregation of small lipid yolk droplets to form large vacuoles or fragmented cytoplasm, and those with incomplete layers or an absence of cumulus cells, should be discarded.

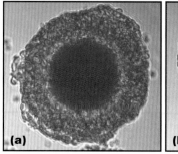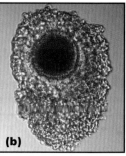

21.3 **(a)** Immature canine oocyte (grade 1). **(b)** Immature feline oocyte (grade 1).

Collection and selection of spermatozoa

Spermatozoa used for IVF can be ejaculated, epididymal or testicular in origin. Ejaculated spermatozoa are usually collected from dogs by digital manipulation or with use of an artificial vagina, and from cats by electroejaculation or urethral catheterization after administration of medetomidine (see Chapter 8).

Epididymal and testicular spermatozoa may be recovered surgically from *in situ* testes or collected from excised organs. The latter represents a valuable method to preserve genetic material from animals that have to be castrated for medical reasons, or even from dead animals.

The most commonly used techniques for collection of spermatozoa from *ex situ* testes in dogs and cats are flushing of the epididymal duct from the ductus deferens end, squeezing or dissection of the epididymides, and dissection of testicular tissue in a dish containing collection medium. The disadvantage of dissection is that contamination with epididymal or testicular tissue and blood cells usually occurs.

Selection

In order to assess semen quality and suitability for *in vitro* techniques, routine semen analysis, both macroscopic (volume, colour and pH) and microscopic (sperm motility, viability, concentration and morphology), must be performed. However, this does not always ensure that the tested sample is able to fertilize oocytes. Sperm function tests (see below) are more accurate at identifying the ability of the sperm to bind to the zona pellucida and penetrate the oocyte.

Although the presence of viable, motile and morphologically normal spermatozoa does not guarantee their competence to fertilize oocytes, selection of a population of spermatozoa showing the aforementioned characteristics is the first step towards successful IVF. Separation of motile sperm can be performed by sperm migration; a technique called 'swim-up'. Motile sperm migrate from a semen pellet placed at the bottom of a given volume of medium to the top during an incubation time of 30–60 minutes. The less motile, immotile or dead spermatozoa do not swim up.

Another method is separation by density gradients. This technique is based on the fact that normal sperm are denser than abnormal sperm. Layers (1 ml thick) of 90%, 70% and 50% colloidal silica solution are stratified in a vial and the sperm suspension is layered over the lower density gradient. After centrifugation the sperm pellet and the highest gradient are collected and washed with culture medium at the desired concentration.

In vitro fertilization

Procedure

Before the co-incubation of gametes can proceed it is important that oocyte maturation and sperm capacitation have been completed (Figure 21.4). Unless preovulatory oocytes collected after hormonal stimulation (i.e. matured *in vivo* and ready to be fertilized) are used, oocytes collected from *ex situ* ovaries need to be cultured to allow maturation before IVF.

In vitro oocyte maturation (IVM) is a complex process, which mimics the dynamic changes that occur in the preovulatory follicle and the oviduct, and achieves oocyte competence. A successful maturation system must support the cumulus–oocyte complex, including the cumulus cells, nuclear material and cytoplasm. Selected oocytes are incubated in microdroplets of culture medium under oil at a controlled atmosphere and temperature. Maturation medium consists of a bicarbonate-buffered solution,

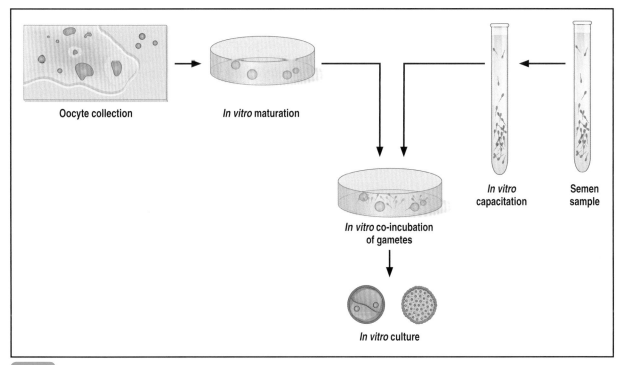

Oocyte collection   *In vitro* maturation

In vitro capacitation   Semen sample

In vitro co-incubation of gametes

In vitro culture

21.4 *In vitro* fertilization procedure.

or a commercially available medium supplemented with energy sources, protein and hormonal components. Maturation time varies between species (72–96 hours for canine oocytes and 24–48 hours for feline oocytes).

In vitro sperm capacitation involves the removal of seminal proteins and other substances that coat the sperm membrane (decapacitating factors). The sperm is centrifuged and the pellet re-suspended in a specific medium. It is then incubated in capacitating medium, which promotes the destabilization of the sperm membranes and the influx of extracellular calcium. Epididymal spermatozoa can be capacitated more easily than ejaculated spermatozoa because they have not been exposed to the decapacitating factors which are present in the seminal plasma.

After oocyte maturation and sperm capacitation, the gametes are co-incubated in fertilization medium and cultured for embryo development.

Results: dogs

In vitro maturation of canine oocytes is, at present, characterized by low and greatly variable success rates. Canine oocytes resume meiosis *in vitro* at a much lower rate than oocytes of other species. The MII stage is achieved in only about 20% of cultured oocytes, compared with blastocyst rates of up to 50% in cats. Consequently, the results of IVF and *in vitro* embryo development are very limited in dogs. Low maturation rates may be caused by the lower meiotic competence of the oocytes or suboptimal culture conditions. This has focused most of the reproductive laboratory research in the dog on oocyte IVM.

The source of oocytes, in terms of the reproductive status of the donor bitch and the diameter of the follicle from which the oocytes are retrieved, seems to affect the competence of the gametes, although the results are not conclusive. Several attempts have been made to improve the culture conditions for maturation, taking into account the unique features of the bitch (i.e. extrafollicular maturation that requires an extended period of time). To date, supplementation of media with hormones, proteins, energy substrates and antioxidants, and use of different culture systems and times have been extensively investigated. However, the optimal culture conditions for canine oocytes have not yet been defined.

The only reported successful canine pregnancy following the transfer of oocytes fertilized *in vitro* was a study (England *et al.*, 2001) in which preovulatory oocytes were collected from bitches in oestrus that had been induced pharmacologically. Thus, *in situ* preovulatory priming might be helpful for canine oocytes to improve developmental competence, and preovulatory oocytes may be more competent than those collected at any other time of the oestrous cycle.

Results: cats

In recent decades considerable progress has been made concerning *in vitro* embryo production in domestic cats, and these techniques have also been successfully applied to non-domestic feline species.

The developmental competence of feline oocytes and the cultural requirements for IVM/IVF procedures have been established, and kittens have been born following transfer of embryos produced *in vitro*. At present, under optimal culture conditions, an average of 40–70% of immature feline oocytes achieve nuclear maturation. The effect of the stage of the oestrous cycle at the time of collection on the developmental competence of oocytes is not conclusive. A significant reduction in cleavage and blastocyst rates has been reported in feline embryos produced from oocytes matured *in vitro* compared with those of embryos derived from oocytes matured *in vivo*. Nevertheless, cleavage rates (to the 2–4 cell stage) of up to 80% and embryo development of 40–60% into morula and blastocyst stages have been achieved from oocytes matured *in vitro* (Figure 21.5). Recently, IVF using sex-sorted feline spermatozoa resulted in the birth of kittens of predetermined sex (Pope *et al.*, 2009), creating new opportunities in cat breeding, especially in the preservation of rare felids.

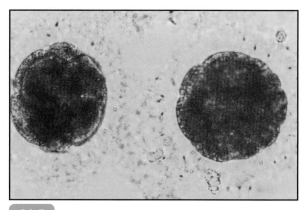

21.5 Feline embryos obtained by IVF.

Intracytoplasmic sperm injection

ICSI is the injection of a single spermatozoon through the zona pellucida directly into the ooplasm of the oocyte (Figure 21.6).

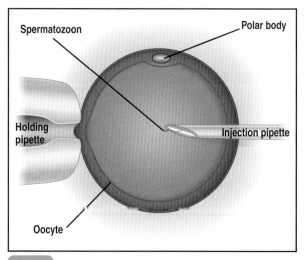

21.6 ICSI procedure.

Procedure

The procedure requires an oocyte that has been matured *in vivo* or *in vitro*, but capacitation and acrosome reaction of the spermatozoon is not required. The direct delivery into the ooplasm bypasses all the tasks, such as binding to the zona pellucida and penetration of the oocyte, that spermatozoa have to perform *in vivo*.

In order to perform ICSI, microtools (holding pipette and injection pipette) and a microscope equipped with micromanipulators and microinjectors are needed. Manipulators can be operated electrically or hydraulically, depending on the precision of movement that is required. To facilitate handling for ICSI, the oocyte is placed in a microdroplet of medium. After removal of the cumulus cells from the oocyte, the first polar body is visualized. The oocyte is held in place by applying gentle suction to the zona pellucida with the holding pipette, and rotated to locate the first polar body at either the 6 or 12 o'clock position.

Sperm are suspended under microdroplets of medium containing polyvinylpyrrolidone (PVP). Motile spermatozoa start to migrate into the medium, and are decelerated by the high viscosity. Their reduced speed allows careful observation and aspiration. Drawing the injection pipette across the midpiece, a single spermatozoon is immobilized.

The immobilized sperm is then aspirated (flagellum first) into the injection pipette. The injection pipette should be inserted into the cytoplasm of the oocyte at the 3 o'clock position in order to avoid damage to the meiotic spindle, which is located in the cytoplasm close to the polar body, but is not visible. Damage to the flagellum and gentle aspiration and ejection of small amounts of cytoplasm during injection induces greater permeability of the sperm membranes and activation of the oocytes, which improve oocyte fertilization. The injected oocyte is then washed in culture medium and cultured for embryo development.

Results: dogs

Canine oocytes have been fertilized by ICSI. The results showed that 8% of the injected oocytes developed two pronuclei and 42% had a female pronucleus and decondensed sperm chromatin. However, no further cleavage occurred (Fulton *et al.*, 1998). This means that the low availability of mature oocytes severely limits the application of ICSI in this species at the present time.

Results: cats

Successful embryo production from feline oocytes has been achieved through ICSI with ejaculated and epididymal spermatozoa. The birth of healthy live kittens following transfer of ICSI-derived embryos to synchronous recipients was first reported in 1998. The cleavage and blastocyst rates of oocytes recovered from domestic queens treated with gonadotrophin (*in vivo* matured) were similar after ICSI and IVF, but higher compared with those obtained from oocytes matured *in vitro*. Nevertheless, the transfer of embryos obtained through ICSI of *in vitro* matured oocytes also resulted in the birth of kittens (Gómez *et al.*, 2000).

It is important to note that the injection of spermatozoa obtained from teratozoospermic feline semen (see below) into oocytes matured *in vivo* resulted in the production of embryos, and that ICSI has also been performed in the cat using testicular spermatozoa.

Clinical relevance of IVF and ICSI

The production of offspring from animals that have not been conceived *in vivo*, especially in very valuable breeding stock or endangered species, is the main goal of IVF and ICSI associated with ET. Assisted reproduction can be particularly useful in cases where migration of embryos into the uterus is not possible, in ovulation failure or permanent anoestrus, and in cases that do not respond to hormonal stimulation. In addition, IVF may be successful where poor semen quality does not allow fertilization *in vivo*.

The advantage of ICSI compared with IVF is that only one spermatozoon is needed for injection into the oocyte, and even in cases where the sperm are of poor quality ICSI may achieve fertilization. Some of the attributes and functions of spermatozoa, such as motility, sperm–zona pellucida binding, acrosome reaction, and penetration of the zona pellucida and oolemma, which are essential for fertilization *in vivo*, are not required in ICSI. Ejaculated, epididymal or even testicular spermatozoa, indeed almost any live sperm or elongated spermatid, can be used. In addition, ICSI can be used in cases of oligozoospermia, teratozoospermia and where standard IVF is not successful.

Domestic cats, as well as non-domestic felids, are often affected by teratozoospermia, a condition in which >60% of the spermatozoa show abnormalities. High proportions of such spermatozoa have a retained cytoplasmatic droplet with resultant angulations of the midpiece and flagellum. Acrosomal defects and abnormalities in the nuclear chromatin are often observed in teratozoospermic samples. These abnormalities seriously compromise the ability of the sperm to capacitate and to undergo the acrosome reaction *in vitro*, thus limiting their use in IVF, although they can be used for ICSI.

Careful clinical examination, a diagnosis of the aetiology of infertility, and conventional treatments should be attempted first. However, in cases that do not respond, where the cause of infertility is unknown, semen tests do not show gross abnormalities, and there is no other pathology to explain the infertility, *in vitro* embryo production may be an option.

IVF and ICSI with oocytes obtained from isolated ovaries and/or epididymal (or testicular) spermatozoa retrieved from excised organs play an important role in the preservation of endangered species. At present, only a few laboratories are equipped and sufficiently skilled to attempt IVF/ICSI in carnivores. Shipping of chilled and frozen gametes to distant laboratories has proven to be useful in allowing processing of them within sensible time limits. Further improvements in the success rates of these techniques will increase their application for clinical purposes.

In vitro sperm–oocyte interaction: *in vitro* sperm function tests

In vitro tests of the sperm–oocyte interaction can provide important information concerning the ability of the spermatozoa to fertilize the oocytes, which cannot be obtained by routine microscopic evaluation of semen. Factors such as the capacity to bind to the zona pellucida and penetrate the oocyte are dependent on motility, membrane integrity, expression of binding sites and ability to undergo the acrosome reaction, which can be assessed *in vitro* (Figure 21.7).

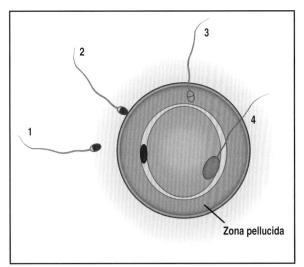

21.7 *In vitro* tests of sperm function. 1 = acrosome intact spermatozoon. 2 = zona pellucida binding. 3 = zona pellucida penetration. 4 = oocyte penetration.

Zona binding and hemizona assays

In vivo, spermatozoa reach the zona pellucida, bind to its surface and penetrate it. The binding of spermatozoa to the zona pellucida is one of the first steps of the fertilization process, and evaluating the number of spermatozoa that bind *in vitro* to an oocyte is an important marker of gamete functionality.

To perform the zona binding assay (ZBA) routinely, it is necessary to have a constant supply of oocytes; these can be recovered fresh, or from frozen (long-term) or chilled (short-term) stored ovaries. Oocytes can also be stored long term in a hypertonic salt solution, in which the biological and physiological properties of the zona pellucida are well preserved. Oocytes from frozen ovaries or salt-stored oocytes have a reduced sperm binding capacity compared with that of fresh oocytes, which should be taken into consideration when assessing the sperm.

The hemizona assay (HZA) is based on the same principles as the ZBA, the only difference being that instead of incubating spermatozoa with whole oocytes, the male gametes are incubated with zona hemispheres obtained by bisection of the zona. This avoids the variability in binding capacity between individual zona pellucida; the results obtained with the tested spermatozoa can be compared with those of a normal fertile donor (control). If a reliable source of oocytes is available, ZBA is more practical to perform than HZA, which is technically more demanding.

In both tests, spermatozoa are diluted to the desired concentration in capacitation medium and incubated with oocytes or hemizonae in microdroplets. After incubation, the oocytes are washed gently to remove loosely attached spermatozoa. The number of spermatozoa attached is counted under a phase-contrast microscope, or the sample is fixed and stained with a fluorescent stain (Figure 21.8) and the bound spermatozoa counted using an epifluorescent microscope.

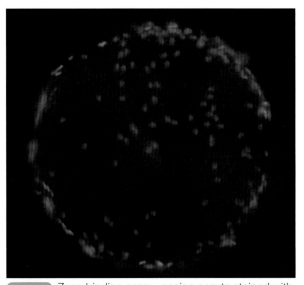

21.8 Zona binding assay: canine oocyte stained with fluorescent dye (Hoechst 33342) to visualize spermatozoa bound to the zona pellucida.

Binding to the zona pellucida is a specific feature of active live spermatozoa, but it does not necessarily mean that they are able to penetrate the oocyte.

Zona pellucida/oocyte penetration test

The ability of spermatozoa to undergo the acrosome reaction is essential for oocyte penetration. The zona pellucida/oocyte penetration test (Z/OPT) provides information about the ability of the spermatozoa to accomplish the process of vesiculation of its membranes and fusion with the oolemma.

Oocyte penetration tests can be performed with homologous (same species) mature oocytes or, given that homologous oocytes may not always be available, with heterologous gametes (hamster oocytes). Hamster oocytes are treated with trypsin solution in order to remove the zona pellucida that would otherwise prevent heterologous spermatozoa from penetrating the oocytes. After co-incubation, the sperm–oocyte complexes are fixed and stained with a fluorescent stain and examined using an epifluorescent microscope in order to count the number of spermatozoa that have penetrated into the zona pellucida or ooplasm.

Owing to oocyte variability a second test should be performed to confirm the results. Oocyte defects may also be responsible for defective sperm–zona pellucida binding, but low sperm–zona pellucida binding and failure of zona pellucida/oocyte penetration is usually a result of spermatozoon defects.

Results: dogs and cats

Sperm function tests have only been applied quite recently to the analysis of canine and feline semen. The main purpose of the ZBA or OPT has been the evaluation of cold-shock induced by chilling or freezing the spermatozoa, rather than the diagnosis of infertility. However, a significant difference between fertile and infertile dogs in the number of tightly bound spermatozoa has been demonstrated with the HZA.

Due to the low efficiency of IVM in dogs, mature oocytes would only be available in very limited numbers for OPT. However, unlike those of most other mammalian species, capacitated canine spermatozoa are able to penetrate immature oocytes and so these can be used for *in vitro* penetration tests.

In the cat, *in vitro* tests have been used to evaluate teratozoospermic semen samples and it has been shown that abnormal cat spermatozoa are capable of binding to and entering the outer zona pellucida, but their ability to penetrate the inner zona and reach the perivitelline space to fertilize the oocyte is compromised.

Clinical relevance of *in vitro* sperm function tests

Although only a few laboratories are equipped to perform these tests in carnivores, the assessment of semen samples *in vitro* could provide clinically useful information when investigating male infertility. If *in vitro* production of embryos is considered to be an option, the choice between standard IVF and ICSI should be based on the results of the sperm analyses. When defective sperm–zona pellucida interaction impairs fertilization rates ICSI should be performed.

For men, ≥100 spermatozoa bound per zona pellucida, at a sperm concentration of 2×10^6 motile sperm/ml is considered to indicate fertility. Cases with <40 bound spermatozoa are defined as abnormal. Threshold values for carnivores are currently not available owing to the lack of studies. However, comparisons between samples from dogs or cats with poor conception and proven fertile semen may give useful information when tested with Z/OPT.

Penetration of the zona pellucida and/or the ooplasm by spermatozoa is closely associated with motility, but no correlation has been found between the acrosomal status of the spermatozoa and sperm penetration. It seems that both acrosome-intact and acrosome-activated dog spermatozoa are capable of binding to the zona pellucida. Thus, further investigations are needed in order to better understand the clinical value of these *in vitro* tests in the dog.

Routine semen analysis should always be performed first, but in future the combination with *in vitro* tests may give a more accurate diagnosis of infertility and a more reliable planning of assisted reproductive techniques.

Cloning

The term cloning describes the process of producing offspring through asexual reproduction. A somatic cell or its nucleus (karyoplast) is transferred into an enucleated oocyte (cytoplast). Therefore, the embryo is not derived from the fusion of oocyte and spermatozoon, but from the nuclear genome of a differentiated somatic cell and an oocyte deprived of its nuclear genome. Several replicates of an adult can potentially be produced through this technology.

Procedure

Cloning consists of several steps from the selection and preparation of donor cells and recipient oocytes, to the transfer of reconstructed embryos into surrogate mothers. Each step is a complex procedure, which is summarized in Figure 21.9.

Somatic cells may be adult or fetal fibroblasts (from cutaneous or muscular tissue), or from cumulus cells obtained from cumulus–oocyte complexes. Fetal fibroblasts are characterized by high developmental competence and maintenance of their characteristics prior to cellular senescence, and they are therefore particularly useful for generating transgenic clones (clones obtained after genetic modification of the donor cells).

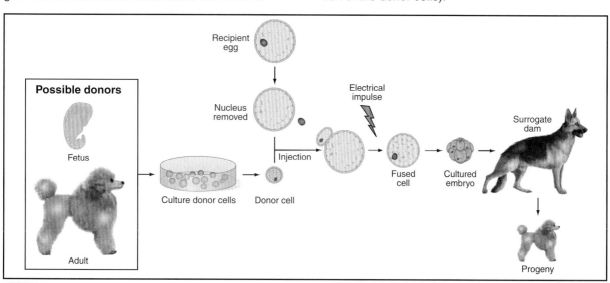

21.9 Cloning procedure.

A source of matured oocytes (cytoplasts) is also needed. The oocytes can be matured either *in vivo* or *in vitro* and can be homologous or heterologous depending on availability. Heterologous oocytes can be extremely useful when the supply of recipient matured homologous oocytes is limited. The limited number of homologous oocytes is due to a number of different factors, including:

- A low efficiency of IVM of oocytes (e.g. in the dog)
- A limited number of oocytes matured *in vivo* and collected surgically
- Application of cloning for preservation of rare and endangered species.

The result of cloning using heterologous cytoplasts is an interspecies nuclear transfer (NT). When recipients of the same species as the cell donor are not available, as in the case of endangered animals, an interspecies ET could follow the production of reconstructed embryos.

The first step in cloning is the enucleation of the mature oocyte. Oocyte DNA is usually stained with a fluorescent dye to improve visualization and to ascertain its removal. This is followed by injection of the somatic cell or its nucleus into the perivitelline space of the oocyte with the aid of microtools and a micromanipulator. Fusion of the membranes (oolemma and cell plasmalemma) is induced by electrical impulses. Activation of the oocyte, which happens *in vivo* as the sperm penetrates, is induced *in vitro* by electrical impulses or brief exposure to chemical agents.

The reconstructed embryos are cultured *in vitro* to obtain embryos at optimal developmental stages to transfer into a recipient's oviduct (early stages) or uterus (late stages). Short-term culture followed by ET into the oviduct of the recipient is often practised to limit the negative effects of *in vitro* culture on the embryo.

Results: dogs

The first cloned dog was born in Korea in 2005. The embryo was produced by transfer of somatic nuclei from fibroblasts cultured *in vitro* (originating from a biopsy of ear skin) into enucleated oocytes. More than 1000 reconstructed embryos were transferred into 123 recipient bitches. Three bitches became pregnant and two cloned puppies were born, of which only one survived. His name is 'Snuppy' (Seoul National University Puppy). Owing to the difficulties in obtaining canine oocytes matured *in vitro*, oocytes matured *in vivo* (recovered from the oviduct 72 hours after ovulation) were used as cytoplasts for the cloning.

Since then other groups have produced three female Afghan hounds, five male Beagles, four Golden Retrievers and five crossbred dogs using the same procedure (Jang *et al.*, 2007; Hossein *et al.*, 2009ab; Kim *et al.*, 2009). In most of the studies, young animals were selected as donors for the somatic cells, but cloned puppies can also be produced through transfer of fibroblasts isolated from older dogs (Jang *et al.*, 2008). The birth of a cloned

poodle confirmed that older bitches can be selected as donors for karyoplasts, but monitoring of the growth and health of the offspring is still in progress.

Interspecies NT has resulted in the birth of two and, later, three cloned wolves, using somatic cells (from ear or abdominal skin) isolated *in vivo* or post mortem from endangered grey wolves (*Canis lupus*), which were injected into canine oocytes (Kim *et al.*, 2007; Oh *et al.*, 2008) and transferred into dog recipients. The birth of two cloned puppies has been achieved recently through transfer of fetal fibroblasts (Hong *et al.*, 2009). Transgenic clones have not yet been produced in the dog.

Results: cats

The successful application of NT in domestic cats was achieved for the first time in 2002 with the birth of 'Copy cat' or 'CC'. This domestic cloned kitten was created by transferring embryos reconstructed from the nuclei of cumulus cells and enucleated oocytes (matured *in vitro*). Adult fibroblasts have also been used, and a second generation of cloned cats has already been produced (Yin *et al.*, 2008a).

The feasibility of interspecies NT for endangered cats has been confirmed by the birth of African Wildcats (*Felis silvestris lybica*) and the successful pregnancy of Black-footed Cats (*Felis nigripes*) produced by nuclear transfer of adult fibroblasts into enucleated domestic cat oocytes and transferred into recipient queens (Gomez *et al.*, 2009b).

Transgenic clones have also been produced in the cat. Genetically modified adult or fetal fibroblasts (expressing the gene of red or green fluorescent protein) have been used as karyoplasts and transferred into enucleated oocytes. After ET, three and one live transgenic kittens expressing red and green fluorescence, respectively, in some tissues have been born (Yin *et al.*, 2008b; Gomez *et al.*, 2009a).

Applications of cloning

Cloning of cats and dogs represents a milestone in biotechnology and opens many new possibilities. However, at this time the efficiency of the procedure is very low, the costs are very high and widespread use is a long way out of sight. Looking to the future, cloning of dogs and cats will satisfy different needs: the emotional requests of affluent owners to replicate a beloved pet; the availability of animal models for biomedical research; and the preservation of rare and endangered wild carnivores.

Dogs are already used as models for biomedical research in areas such as cardiovascular physiology, diabetes and cancer. Cats are phylogenetically close to humans and many feline genetic diseases are analogous to inherited disorders of humans (e.g. autosomal dominant polycystic kidney disease and retinal atrophy). The possibility of producing genetically modified animals carrying genes related to specific human disorders would help biomedical research greatly, but it also raises animal welfare concerns. Cloning also has great potential as a tool to prevent the extinction of some of the rarest species, whose numbers have declined below those required to sustain a breeding population.

References and further reading

Dunbar BS and O'Rand MG (1991) *A Comparative Overview of Mammalian Fertilization.* Plenum Press, New York

England GCW, Verstegen JP and Hewitt DA (2001) Pregnancy following *in vitro* fertilisation of canine oocytes. *Veterinary Record* **148**, 20–22

Fulton RM, Keskintepe L, Durrant BS *et al.* (1998) INtracytoplasmic sperm injection (ICSI) for the treatment of canine infertility. *Theriogenology* **48**, 366 (abstract)

Gómez MC, Pope CE and Dresser BL (2006) Nuclear transfer in cats and its application. *Theriogenology* **66**, 72–81

Gómez MC, Pope CE, Harris R *et al.* (2000) Birth of kittens by intracytoplasmic sperm injection of domestic cat oocytes matured *in vitro. Reproduction Fertility and Development* **12**, 423–433

Gómez MC, Pope CE, Kutner RH *et al.* (2009a) Generation of domestic transgenic cloned kittens using lentivirus vectors. *Cloning and Stem Cells* **11**, 167–175

Gómez MC, Pope CE, Ricks DM *et al.* (2009b) Cloning endangered felids using heterospecific donor oocytes and interspecies embryo transfer. *Reproduction, Fertility and Development* **21**, 76–82

Hong SG, Jang G, Kim MK *et al.* (2009) Dogs cloned from fetal fibroblasts by nuclear transfer. *Animal Reproduction Sciences* **115**, 334–339

Hossein MS, Jeong YW, Park SW *et al.* (2009a) Birth of Beagle dogs by somatic cell nuclear transfer. *Animal Reproduction Sciences* **114**, 404–414

Hossein MS, Jeong YW, Park SW *et al.* (2009b) Cloning Missy: obtaining multiple offspring of a specific canine genotype by somatic cell nuclear transfer. *Cloning and Stem Cells* **11**, 123–130

Jang G, Hong SG, Oh HJ *et al.* (2008) A cloned toy poodle produced from somatic cells derived from an aged female dog. *Theriogenology* **69**, 556–563

Jang G, Kim MK, Oh HJ *et al.* (2007) Birth of viable female dogs produced by somatic cell nuclear transfer. *Theriogenology* **67**, 941–947

Kim MK, Jang G, Oh HJ *et al.* (2007) Endangered wolves cloned from adult somatic cells. *Cloning and Stem Cells* **9**, 130–137

Kim S, Park SW, Hossein MS *et al.* (2009) Production of cloned dogs by decreasing the interval between fusion and activation during somatic cell nuclear transfer. *Molecular Reproduction and Development* **76**, 483–489

Luvoni GC and Chigioni S (2006) Cultural strategies for maturation of carnivore oocytes. *Theriogenology* **66**, 1471–1475

Luvoni GC, Chigioni S, Allicvi E *et al.* (2005) Factors involved *in vivo* and *in vitro* maturation of canine oocytes. *Theriogenology* **63**, 41–59

Luvoni GC, Chigioni S and Beccaglia M (2006) Embryo production in dogs: from in vitro fertilization to cloning. *Reproduction in Domestic Animals* **41**, 286–290

Oh HJ, Kim MK, Jang G *et al.* (2008) Cloning endangered gray wolves (*Canis lupus*) from somatic cells collected postmortem. *Theriogenology* **70**, 638–647

Pelican KM, Wildt DE, Pukazhenthi B *et al.* (2006) Ovarian control for assisted reproduction in the domestic cat and wild felids. *Theriogenology* **66**, 37–48

Pope CE, Crichton EG, Gómez MC *et al.* (2009) Birth of domestic cat kittens of predetermined sex after transfer of embryos produced by *in vitro* fertilization of oocytes with flow-sorted sperm. *Theriogenology* **71**, 864–871

Pope CE, Gómez MC and Dresser BL (2006) *In vitro* production and transfer of cat embryos in the 21st century. *Theriogenology* **66**, 59–71

Pukazhenthi BS, Neubauer K, Jewgenow K *et al.* (2006) The impact and potential etiology of teratospermia in the domestic cat and its wild relatives. *Theriogenology* **66**, 112–121

Rijsselaere T, Van Soom A, Tanghe S *et al.* (2005) New techniques for the assessment of canine semen quality: a review. *Theriogenology* **64**, 706–719

Ström Holst B, Larsson B, Rodriguez-Martinez H *et al.* (2001) Zona pellucida binding-assay – A method for evaluation of canine spermatozoa. *Journal of Reproduction and Fertility Supplement* **57**, 137–140

Wolf DP and Zelinski-Wooten M (2001) *Assisted Fertilization and Nuclear Transfer in Mammals.* Humana Press, Totowa, New Jersey

Yin XJ, Lee HS, Yu XF *et al.* (2008a) Production of second-generation cloned cats by somatic cell nuclear transfer. *Theriogenology* **69**, 1001–1006

Yin XJ, Lee HS, Yu XF *et al.* (2008b) Generation of cloned transgenic cats expressing red fluorescent protein. *Biology of Reproduction* **78**, 425–431

Index

Page numbers in *italics* indicate figures

Abdominal distension, queen 189–90
Abortion *see* Pregnancy failure, Pregnancy termination
Actinomyces pyogenes, neonatal disease 152
Adriamycin, in vaginal neoplasia, bitch *171*
Agalactia 157–8
Aglepristone
    in mammary gland neoplasia 165
    in mammary hyperplasia 162
    in mucometra, queen 187
    pregnancy termination
        bitch 107, 109, 110, 111, 112
        queen 107, 113
    in pyometra
        bitch 176–7
        queen 187
AI *see* Artificial insemination
Alopecia, bitch *179*
Altrenogest, in hypoluteoidism 116
Anaemia, in pregnancy, queen 100
Analgesia
    after Caesarean operation 130
    in neonates 154
Anasarca, puppy *124*
Androgen drugs
    control of oestrus 25
    side effects *26*, 27
    spermatogenesis disruption 36
    (see also specific drugs)
Anoestrus (phase)
    bitch 3
        abnormalities, and infertility 56, 58
        cytology *47*
        vaginoscopy 48
    queen 9
        abnormalities, and infertility 63–4
Anti-androgens, spermatogenesis disruption 37
    (see also specific drugs)
Anti-progestogens, pregnancy termination 107
    (see also specific drugs)
Apgar scoring, neonates 149
Aromatase inhibitors, in mammary gland neoplasia 165
Artificial insemination, dogs
    ethical considerations 87–8
    indications 80
    insemination
        techniques 85–7
        timing 85
    regulations 87–8

semen
    collection 80–1
    dilution 81–2
    freezing 82–4
    transport 84–5
Aspermia 78–9
Asthenozoospermia 75
Azoospermia 75–6

Bacillus, orchitis/epididymitis, dog 191
Balanitis
    dog 197–8
    tom cat 198
BCG, spermatogenesis disruption 38
Benign prostatic hyperplasia/hypertrophy, dog 202–3
Biotechnology *see specific procedures*
Birth *see* Parturition
Bisdiamines, spermatogenesis disruption *35*, 40
Blastomyces dermatitidis
    orchitis/epididymitis, dog 191, *192*
    prostatitis 205
Blood sampling, neonates 148
Body temperature
    bitch
        as parturition predictor 91
        postpartum 96, 130
    neonates 147
Bodyweight
    neonates 146
    queen, during pregnancy 99, *102*
Bottle feeding neonates 141
Breeding
    optimal time
        bitch 44–9
        queen 49–50
    prevention
        females
            drugs 23–5
            surgical 29–33
        males
            drugs 34–40
            surgical 40–3
    preventive healthcare 145–6
    season, queen 8–9
Bromocriptine
    pregnancy termination 108
        bitch *112*
    in pseudopregnancy 160
Brucella, prostatitis 205
Brucella abortus, orchitis/epididymitis, dog 191

Index

Index

Levonorgestrol, infertility induction, tom cat *35*
Leydig cell tumours *see* Interstitial cell tumours
Leydig cells 14, 18
    dysfunction *16, 17*
    stimulation tests 73–4
Litter size, puppies 94
Lochia, bitch 94, 130
Luteal cysts, bitch 58, 182
Luteal phase
    bitch 5–8
        abnormalities, and infertility 58
    queen 11–12
Luteinizing hormone (LH)
    bitch
        in infertility 52, *53*
        and ovulation 4, 45, 46
        in pregnancy 6
    males
        levels 18
        stimulation tests *16–17*
        testosterone regulation 15
        vaccines 38
    queen
        in oestrous cycle 9
        and ovulation 10, 66
    regulation *1*

Male behaviour, persistence after neutering 211
Mammary fibroadenomatous hyperplasia 161–2
Mammary glands
    abnormalities 160–1
    anatomy and physiology 155–6
    asymmetry 161
    development 155
    disease diagnosis 156–7
    hyperplasia 161–2
    lactation problems 157–9
    neoplasia 30, *157, 162*–5
        and pseudopregnancy, bitch 183
    palpation 52
Masculinization, queen 185, 189, 190
Mastectomy 164
Mastitis 158–9
Maternal behaviour 130, 131
    abnormalities 131
    bitch 91, 96
    queen 103, 104
Mating
    cats 20–1
        failure 66
    dogs 20, *21*
        failure 60–1
        unwanted, clinical approach 109
Medroxyprogesterone acetate
    in benign prostatic hyperplasia 203
    control of oestrus 24
    spermatogenesis disruption, dog *35,* 36
Megestrol acetate
    in benign prostatic hyperplasia 203
    control of oestrus 24–5
    spermatogenesis disruption *35, 36*
Melatonin, breeding prevention, queen 28
Melengestrol acetate, infertility induction, tom cat *35*
Metergoline
    pregnancy termination, bitch *112*
    in pseudopregnancy 160

Methyltestosterone
    spermatogenesis disruption 36
    suppression of oestrus, bitch 25
Metoclopramide, in agalactia *158*
Metoestrus, bitch, cytology *47*
Metritis
    postpartum 131–2
    queen 186
Metritis/endometritis/pyometra syndrome, queen 67
Mibolerone, oestrus prevention, bitch 25
Milk
    absence of ejection 158
    composition *141*
    excessive production 158
    replacers 141
    (*see also* Lactation)
Mitoxantrone, in mammary gland neoplasia *164*
Moraxella, pyometra, queen 186
Mucometra
    bitch 177–8
    queen 187
Mucous membrane colour, neonates 147
Mycoplasma
    orchitis/epididymitis, dog 191
    prostatitis 205
Myometrial activity and fetal loss 117

Nafarelin, infertility induction, males 36
Neonates
    acid–base balance 136–7
    adrenal gland 140
    analgesia 154
    antibiotic choice 154
    biochemical data 138–9
    blood sampling 148
    body temperature 147
    catheterization 144–5
    congenital problems 151, 153
    dehydration 143–4
    development 135–6
    diarrhoea 152
    energy requirements *141*
    'fading' 150–1
    fluid requirements 136, *141*
    fluid therapy 144–5
    gastrointestional tract 140
    glucose metabolism 138
    haematology 137–8
    health scoring system 149–50
    heart 137
        rate 135
    hepatic function 138
    history taking 147
    hypoglycaemia 143
    hypothermia 142–3
    hypothyroidism 153
    hypoxia 145
    imaging 148
    immune system 139
    infections 150–2
    isoerythrolysis 153
    monitoring 146
    mortality rates 96, 104, *151*
    neurological examination *135–6*

Index

Index

BSAVA Manual of
Practical Animal Care

Edited by

Paula Hotston Moore and Alan Hughes

The *BSAVA Manual of Practical Animal Care* has been written principally for animal nursing assistants, veterinary care assistants and veterinary nurses beginning their training in practice. This edition has been designed with the Animal Nursing Assistant syllabus in mind but is also suitable for others embarking on a career in animal care within a veterinary practice.

- Care of dogs, cats and exotic pets
- Use of medicines
- First aid
- Organization and communication skills
- Veterinary terminology

Contents:
An introduction to veterinary practice; General care and management of the dog; General care and management of the cat; General care and management of other pets and wildlife; Management of an animal ward; Introduction to veterinary care; Use of medicines; Animal first aid; Organization and communication skills; Veterinary terminology; Index

Published August 2007
208 pages
ISBN   978 0 905214 90 0

Price to non-members: £50

MEMBER PRICE: £30

ORDERING DETAILS

British Small Animal Veterinary Association
Woodrow House, 1 Telford Way, Waterwells Business Park,
Quedgeley, Gloucester GL2 2AB

Tel: 01452 726700
Fax: 01452 726701
Email: administration@bsava.com
Web: www.bsava.com/shop

BSAVA reserves the right to change these prices at any time

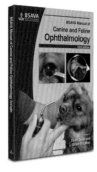

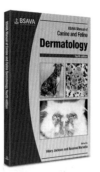

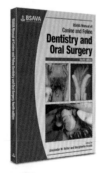

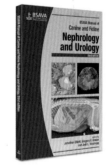
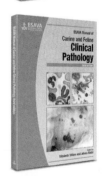

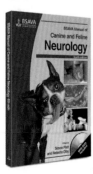

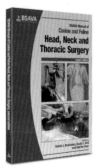

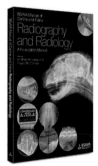